Noise in nonlinear dynamical systems

Volume 1
Theory of continuous Fokker–Planck systems

Noise in nonlinear dynamical systems

Volume 1
Theory of continuous Fokker–Planck systems

Edited by
Frank Moss, *Professor of Physics,*
University of Missouri at St Louis
and
P. V. E. McClintock, *Reader in Physics,*
University of Lancaster

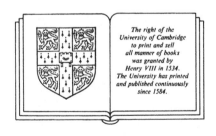

The right of the
University of Cambridge
to print and sell
all manner of books
was granted by
Henry VIII in 1534.
The University has printed
and published continuously
since 1584.

CAMBRIDGE UNIVERSITY PRESS
Cambridge   New York   New Rochelle
Melbourne   Sydney

CAMBRIDGE UNIVERSITY PRESS
Cambridge, New York, Melbourne, Madrid, Cape Town, Singapore, São Paulo, Delhi

Cambridge University Press
The Edinburgh Building, Cambridge CB2 8RU, UK

Published in the United States of America by Cambridge University Press, New York

www.cambridge.org
Information on this title: www.cambridge.org/9780521118507

First published 1989
This digitally printed version 2009

A catalogue record for this publication is available from the British Library

Library of Congress Cataloguing in Publication data

Theory of continuous Fokker–Planck systems.
(Noise in nonlinear dynamical systems; v. 1)
Includes index.
1. Fokker–Planck equation. I. Moss, Frank,
1934- . II. McClintock, P. V. E. III. Series.
QC6.4.F58N64   vol. 1   003 s   87-36802
[QC20.7.D5]   [003]

ISBN 978-0-521-35228-4 hardback
ISBN 978-0-521-11850-7 paperback

Contents

Contents

Contents

Contributors

A. Andronov (deceased, 1952)
Scientific Research Institute
Moscow
USSR

W. Ebeling
Sektion Physik
Bereich 04 Humboldt-Universität zu Berlin
Invalidenstrasse 42
DDR-1040 Berlin

Robert Graham
Fachbereich Physik
Universität-Gesamthochschule-Essen
Essen
FRG

Paolo Grigolini
Dipartmento di Fiscia e GNSM del CNR
Piazza Torricelli 2
56100 Pisa
Italy

Peter Hänggi
Lehrstuhl für Theoretische Physik
Universität Augsburg
Memminger Strasse 6
D-8900 Augsburg
FRG

Rolf Landauer
IBM Thomas J. Watson Research Center
Yorktown Heights
NY 10598
USA

Contributors

Katja Lindenberg
Department of Chemistry, B-014, and
Institute for Nonlinear Science, R-002
University of California
San Diego
La Jolla
CA 92093
USA

Jaume Masoliver
Department of Chemistry, B-014, and
Institute for Nonlinear Science, R-002
University of California
San Diego
La Jolla
CA 92093
USA

L. Pontryagin
Present address: Stecklov Institute
GSP-1 Vavilov Street 42
117966 Moscow
USSR

H. Risken
Abteilung für Theoretische Physik
Universität Ulm
D-7900 Ulm
FRG

M. San Miguel
Dept de Fiscia
Universität de les Illes Balears
07071 Palma de Mallorca
Spain

J. M. Sancho
Dept de Estructura y Constituyentes de la Materia
Universidad de Barcelona
Diagonal 647
08028 Barcelona
Spain

L. Schimansky-Geier
Sektion Physik
Bereich 04 Humboldt-Universität zu Berlin
Invalidenstrasse 42
DDR-1040 Berlin

Contributors

R. L. Stratonovich
Moscow State University
117234 Moscow
USSR

A. Vitt (deceased, 1937)
Scientific-Research Institute
Moscow
USSR

H. D. Vollmer
Abteilung für Theoretische Physik
Universität Ulm
D-7900 Ulm
FRG

Bruce J. West
Department of Chemistry, B-014, and
Institute for Nonlinear Science, R-002
University of California
San Diego
La Jolla
CA 92093
USA

Preface

All macroscopic physical systems are subject to fluctuations or noise. One of the most useful and interesting developments in modern statistical mechanics has been the realization that even complex nonequilibrium systems can often be reduced to equivalent ones of only a few degrees of freedom by the elimination of dynamically nonrelevant variables. Theoretical descriptions of such contracted systems necessarily begin with a set of either continuous or discrete dynamical equations, which can then be used to describe noise driven systems with the inclusion of random terms. Studies of these stochastic dynamical equations have expanded rapidly in the past two decades, so that today an exuberant theoretical activity, a few experiments, and a remarkably large number of applications, some with challenging technological implications, are evident.

The purpose of these volumes is twofold. First we hope that their publication will help to stimulate new experimental activity by contrasting the smallness of the number of existing experiments with the many research opportunities raised by the chapters on applications. Secondly, it has been our aim to collect together in one place a complete set of authoritative reviews with contributions representative of all the major practitioners in the field. We recognize that as an inevitable consequence of the intended comprehensiveness, there will be few readers who will wish to digest these volumes in their entirety. We trust, instead, that readers will be stimulated to choose from the many possibilities for new research represented herein, and that they will find all the specialised tools, be they experimental or theoretical, that they are likely to require.

Although there is a strong underlying theme running through all three volumes – the influence of noise on dynamical systems – each chapter should be considered as a self-contained account of the authors' most important research in the field, and hence can be read either alone or in concert with the others. In view of this, the editors have chosen not to attempt to impose a uniform style; nor have they insisted on any standard set of mathematical symbols or notation. The discerning reader will detect points of detail on which full concordance appears to be lacking, especially in Volume 1. This should certainly be regarded as the signature of an active and challenging theoretical activity in a rapidly expanding field of study. In selecting the contributors, the editors have made special efforts to include younger authors with new ideas and perspectives along with the more active seasoned veterans, in the confident belief that the field will be invigorated by their contributions.

Finally, it is our pleasure to acknowledge the many valuable suggestions made by our colleagues and contributors, from which we have greatly benefited. Special thanks are due to R. Mannella for constructive criticism and helpful comments at various stages of this enterprise.

Completion of this work owes much to the generous support provided by the North Atlantic Treaty Organisation (under grant RG85/0070) and the UK Science and Engineering Research Council (under GR/D/61925).

Frank Moss and P. V. E. McClintock
Lancaster, May 1987

Introduction to Volume 1

Over the last few years there has been a quite remarkable increase of activity in the study of nonlinear dynamical systems exposed to external noise. This is especially true in relation to *colored* noise, that is noise whose correlation time τ is nonzero. The field as it has developed stands on a firm foundation of pioneering, mostly Russian, theoretical work on white noise driven systems dating back to the early 1930s, an exemplar being the landmark 1933 paper by Pontryagin, Andronov and Vitt, of which a first-ever translation into English appears as an appendix to this volume. The early development of the field is reviewed in Chapter 1.

The recent intensification of research effort has in large part been stimulated by two distinct but closely related factors. First, it has become increasingly apparent that most (perhaps all) real physical systems need to be considered in the context of colored noise: they cannot be described adequately within the framework of white noise theory. Secondly, it is the case that, with few exceptions, physically interesting Fokker–Planck systems cannot be solved exactly when driven by colored noise. To meet this challenge, a variety of approximation schemes, carried out within several theoretical frameworks, have been developed, and much effort has been devoted to the testing of their efficacy under different conditions. These approximations fall into three broad categories: (i) the so-called 'small τ' theories wherein the corrections to an essentially Markovian (white noise) theory are obtained by series expansions, usually in terms of the noise parameters; (ii) forms of 'statistical linearization' or mean-field approximation, which effectively reduce the dimensionality of the relevant Fokker–Planck equation; and (iii) a recently proposed unifying approximation that becomes exact in the limits of both short and long correlation time (see Chapter 9). The four main theoretical frameworks within which quantities of physical interest have been calculated to date are: the cummulant summation, functional calculus, path integral, and projection operator techniques; in addition, there are the methods of matrix continued fraction expansion, and the macroscopic potential approach. All of these techniques are described in detail in the chapters that follow. A point that the reader should constantly bear in mind is that, although a particular approximate theory may yield accurate results in a specific application, it may nonetheless turn out to be inaccurate and inadequate in other applications.

The original small τ theory of Stratonovich, the spiritual predecessor of most modern non-Markovian (colored noise) theories is reviewed and developed in Chapter 2; contemporary adaptations and extensions of the Stratonovich approach are described in the three chapters that follow. Chapter 3 deals with the calculation of a variety of statistical quantities such as stationary densities, relaxation times, and decay rates, as well as nonstationary decays of unstable and metastable states; the situation of nonlinear noise, where the noise itself enters the stochastic differential equation in a nonlinear way, is also treated. Techniques for the calculation of first passage times under the influence of colored Gaussian noise, dichotomous noise and shot noise are presented and discussed in Chapter 4. Chapter 5 is devoted to the projection operator formalism, its application to the solution of a range of colored noise problems, and to a critical discussion of its particular advantages and disadvantages in comparison to other methods.

Although the small τ theories have been highly successful in particular applications, they are not of general applicability and, as already indicated, are liable to fail under particular sets of circumstances. An alternative approach to the problem of colored noise, that of matrix continued fraction (MCF) expansion, is reviewed in Chapter 6. This method has the great virtue that the solutions obtained are, in principle, exact. Because they involve the numerical truncation and inversion of an infinite matrix, however, some degree of approximation is unavoidable in practice. Even for two-dimensional stochastic systems huge amounts of computer time may be needed and, for higher dimensional systems, the required procedure represents a daunting problem in numerical analysis.

The use of the macroscopic potential is reviewed in Chapter 7. A parallel is drawn between the familiar potentials of classical equilibrium thermodynamics and a quantity that plays the same role in nonequilibrium systems: the macroscopic potential itself. The necessary formalism is developed, and then applied to fluctuations around steady states and at Hopf, pitchfork and codimension-two bifurcations. The influence of noise color on the shape of the macroscopic potentials is investigated within a small τ approximation. In Chapter 8, a similar approach is applied to treat the effect of colored noise on bistable systems: it is shown *inter alia* that a characteristic skewing effect in the two-dimensional density, already seen in analogue experiments and MCF calculations, can be derived theoretically by this means.

Finally, in this volume, the functional calculus approach to the problem of colored noise is reviewed in Chapter 9. A colored noise master equation is obtained and a number of example flows are treated. The relative advantages and disadvantages of the three approximation schemes, referred to above, are discussed.

1   Noise-activated escape from metastable states: an historical view

ROLF LANDAUER

Historical comments by a participant in the field represent a personal viewpoint. They are not a detached and scholarly contribution to the history of science. The field of this volume, and the (hopefully) closely related field of this discussion, involves a number of 'schools', of varying degrees of distinctness, and I cannot represent them all in this brief discussion. The *Brussels School* associated with Ilya Prigogine is one example. Another group, which is much more loosely knit and less clearly identifiable, can be associated with the work and influence of the late Elliott Montroll. This work comes closer to my viewpoint, represented below, but has had public visibility through recent books and conferences (Montroll and Lebowitz, 1979; Shlesinger and Weiss, 1985; Weiss, 1983, 1986). Both of the aforementioned 'schools' have a close relationship to physical chemistry, and we will, later in our discussion, return to some of the pioneering (but long neglected) insights which arose in this area. On the other hand, our field has equally deep, or perhaps deeper, roots in electronics. We will take it upon ourselves to represent that heritage. The approach we will stress describes some degrees of freedom and their dynamics explicitly; the remaining degrees of freedom are regarded as a source of noise and of damping. It is a viewpoint which has wide applicability, but is particularly directly and accurately applicable to electrical circuits. At the very beginning we have to stress the pioneering role of R. L. Stratonovich in the development of our subject. His two-volume treatise (Stratonovich, 1963, 1967) is a landmark. Undoubtedly, much of the contents of those two volumes appeared in the Soviet periodical literature long before then. (Some of these items will be cited subsequently.) In an editorial, some years ago (Landauer, 1981), I referred to a *Stratonovich test* for the integrity of citation practices in the field. Do authors acknowledge the relationship of their work to that of Stratonovich?

Unintentional repetition in science is unavoidable; we cannot spend all of our time in the library guarding against that. This author once repeated a seventeen-year-old piece of work, unknowingly. Nevertheless, the field of this volume seems to have had more than its share of rediscovery. Originally the overlap arose, presumably, because the investigators came from different backgrounds and disciplines. In an unfashionable field it is hard to become aware of earlier literature in strange journals and unfamiliar languages. Also, in a relatively inactive field, it is hard to latch on to a meaningful citation trail. The later examples of repeated discovery, after the field became well established, are less easily explained. Repeated discovery will be a recurrent theme in our subsequent discussion.

Our subject has a diversity of roots. It has become fashionable and acclaimed within the past ten or fifteen years, but has connections to work that took place many decades ago. The concern with fluctuation induced motion, at a serious analytical level, started near the turn of the century, with contributions by Pearson, Lord Rayleigh, Einstein, v. Smoluchowski, and others. These pioneering investigations are cited by Chandrasekhar (1943). A more detailed discussion of early Brownian motion studies is given by Coffey (1985), in a review paper which in many ways complements our account. Five Einstein papers on Brownian motion have been translated and supplied with supplementary comments (Fürth, 1956). Pais's (1982) biography of Einstein*, with its marvelous mixture of scholarship and insight, provides an excellent account of the history of Brownian motion studies. We learn there that Smoluchowski's 1906 publication date for his derivation of the 1905 'Einstein' relation may have been the result of a conservative attitude toward publication. Pais also points to a more unexpected fact: William Sutherland (1905), from Australia, submitted a paper with the same result, for publication, a few weeks before Einstein's submission date. Indeed, Sutherland's paper refers to his 1904 communication to the Australian Association for the Advancement of Science, in which 'The formula obtained made the velocity of diffusion of a substance through a liquid vary inversely as the radius a of its molecule and inversely as the viscosity of the liquid.' Readers and authors of this present volume may also be amused by Pais's remarkable understatement: 'Brownian motion was still a subject of active research in the 1970's.'

The concern with fluctuations reached a particularly developed form, at an early stage, in electronics. Einstein (1907) pointed out that an open circuited capacitor exhibits thermal equilibrium voltage fluctuations, which have an average energy $\frac{1}{2}kT$, and are conceivably measurable. Schottky (1918, 1922) described the fluctuations in emission limited current flow in thermionic

---

* For the benefit of future Einstein historians we allude here to a little mystery, unrelated to the real theme of this paper. Most Einstein accounts tell us he was born in Ulm, where his family actually stayed only briefly. Mohn (1970) tells us that Einstein's real origins were in Buchau, where his parents resided until the move to Ulm, and where Einstein's father's family had lived for many generations. Einstein was probably the descendant of Baruch Mosis Aynstein, who in 1665 started the thriving pre-World War II Jewish community of Buchau. Einstein's parents left Buchau toward the end of 1878; Einstein was born in Ulm on March 14, 1879. Mohn's description agrees with the view which this author has heard repeatedly from former members of the Jewish community of Buchau. In contrast, Specker (1979) has Einstein's father already residing in Ulm at the time of his marriage in 1876. I am indebted to H. Risken, whose own contribution appears in chapter 6 of this volume, for a copy of Specker's *Einstein and Ulm.*

diodes, due to the graininess in the current, carried by single and independently emitted electrons. Johnson (1927) and Nyquist (1928) at Bell Laboratories, a decade later, introduced us to thermal equilibrium noise arising in resistors, through their respective experimental and theoretical investigations. In this work, Nyquist adapted the standard black-body radiation discussion to an electrical transmission line terminated by a resistor. Despite the widely prevalent use of these results by electrical engineers, physicists did not develop an equivalent concern with fluctuation–dissipation theory until the mid-1950s. Kubo (1986), one of the principal exponents of this later viewpoint, has described this approach and its history in a recent account. Despite the great impact of this work on the physics community, it was, in a way, a step back from the understanding already achieved in circuit theory. In circuit theory a duality between current and voltage sources had long been recognized. The *linear response* theory, associated with the developments described by Kubo (1986), was more limited. There, fields were taken to be *the* cause of transport, and current flow was the response. Furthermore, the typical versions of linear response theory treat conservative Hamiltonian systems, and still require subtle cheating to yield dissipative behavior.

By 1938 the first book appeared on noise in electrical circuits (Moullin, 1938). The key points of understanding that had come out of the physical electronics literature were summarized, effectively and compactly, by Pierce (1948). Pierce emphasized the viewpoint of one concerned with vacuum tubes and electron beams, and can still be read with profit today.

Systems of serious engineering interest, which do not just sit there statically, but perform a useful function, are usually non-linear. In electronics, only straightforward amplification and signal propagation have limited ranges of linearity. Other functions, such as modulation, detection, heterodyning, and generation of a fixed amplitude oscillation, are strongly, and intrinsically, nonlinear. It is, therefore, no great surprise that the concern with noise in nonlinear circuits eventually received attention, particularly in connection with World War II radar development (Boonimovich, 1946; Lawson and Uhlenbeck, 1950; Rice, 1945).

After World War II the concern with noise in non-linear systems gradually broadened beyond the specialized communications and radar applications, regaining a more conceptual and statistical mechanics flavor. There were, for example, discussions whether rectification of spontaneously arising noise signals could generate power (Brillouin, 1950). The concern with diode kinetics was extended to vacuum diodes, which pass one electronic charge at a time (Alkemade, 1958; Van Kampen, 1960, 1961, 1981). By 1960 there was also a review by Lax (1960), emphasizing fluctuations about a state which is not necessarily one of thermal equilibrium.

The development of the laser acted as a further spur to the study of noise. We can cite only a few of the many key items in this field (Armstrong and Smith, 1967; Haken, 1964; Lax, 1966; Risken, 1965, 1966). Eventually the insights

3

gained in this field were generalized, and led to analogies with ordinary thermodynamic phase transitions (DeGiorgio and Scully, 1970; Graham and Haken, 1968, 1970; Haken, 1970). Haken's many subsequent conferences and books helped to create public awareness of stochastic phenomena in systems far from equilibrium.

The early interest in noise was a result of its role as a source of confusion and error in measurement and communication, and, with limited exceptions, that characterizes the history of the subject we have presented so far. In more recent times there has been a growing awareness that noise can be *a controlling qualitative influence on the development of systems*. In systems which are multistable, i.e. have two or more competing states of local stability, noise determines the motion of the system between these competing states. Small systems, as found in molecular biology, and in the drive toward miniaturization of computer components, can show a strong susceptibility to noise. Furthermore, such noise induced transitions may be of utility, as invoked in the simulated annealing technique for optimization in the presence of complex constraints (Kirkpatrick and Toulouse, 1985). A similar application occurs in a model for the mechanism of perception and visual recognition (Ackley, Hinton and Sejnowski, 1985; Kienker, Sejnowski, Hinton and Schumacher, 1986; Sejnowski, Kienker and Hinton, 1986), in which fluctuations are invoked to allow escape from premature fixation. First order phase transitions can occur by homogeneous nucleation, followed by expansion of the nucleus through motion of the interface. The formation of a critical nucleus is a fluctuation which takes the system from one state of local stability (i.e. the original metastable phase) towards a new state of local stability. Indeed, in recent years, such nucleated transitions have become of cosmological interest in connection with the *inflationary universe* (Barrow and Tipler, 1986). In our general discussion of competing states of local stability, and stochastically induced transitions between them, it is appropriate to point out the connection to discussions of evolution and the origin of life. This, also, can be viewed as an optimization process, in which fluctuations (e.g. mutations) take us from one metastable ecology to a new one. (For citations see ref. 40 in Landauer (1987a).) Evolution can be viewed as hill climbing (or valley search) in a multi-valley 'Fitness Landscape', a concept often associated with the names S. Wright and R. A. Fisher.

Indeed, evolution bears a strikingly detailed analogy to the nucleation process. In nucleation a fluctuation is required which takes us to the formation of the critical nucleus; not just any fluctuation will do. After that the macroscopic system laws, which determine interface motion, take over. Genetic mutations are like the critical nucleus, and represent the necessary consequence of fluctuations. After that the normal biological machinery can take over and control the further time development. In contrast to the discussions cited in Landauer (1987a), which elaborate on this theme, and represent a longstanding orientation in ecology, physical scientists have been inclined to use the word *self-organization* very loosely. To apply this expression to the Benard instability, for example, to the Zhabotinskii reaction, or to the oscillations of a laser, under conditions where these temporal and spatial patterns are the only allowed behavior, seems strained. We might, with almost equal justice, refer to the revolution of an electron around a hydrogen nucleus, or the rotation of a water wheel, as self-organization.

The modern analysis of noise induced transitions between states of local stability has a number of precursors. Many escape problems, e.g. the thermionic emission of electrons from a metal surface, or the vaporization of an atom from the surface of a condensed phase, have the character of noise-assisted escape from a state of local stability. Thus, for example, the electrons in a metal, at the absolute zero of temperature (i.e. in the absence of noise), have no chance to escape. In the case of electron emission, however, the typical analysis did not greatly concern itself with the way fluctuations gave the electrons their required escape energy; it was simply assumed that gaining the required energy was not the bottleneck in the process. A pioneering paper by Becker and Döring (1935) went well beyond that and treated the formation of a critical nucleus, in the case of droplet formation, by treating the detailed statistics of atoms arriving at the droplet and evaporating from it.

The key step, however, in the treatment of escape from a locally stable state and in the treatment of transition kinetics between competing states, came through the paper by H. A. Kramers (1940). Kramers treated the escape from a potential well as a problem of Brownian motion in a non-uniform force field. Most of the papers in this volume can be considered to be descendants of Kramers' work. Table 1.1, adapted from Landauer (1987b), illustrates the diversity this field has achieved. No review covering most of these topics exists; it can be hoped that the present volume will help to correct this gap. Kramers' paper was a model scientific paper. It was compact by modern standards, but clear and very physical. It touched upon several of the most basic aspects of the area, some of which took decades to be appreciated. Yet, it left room for many later Ph.D. theses. The real question: Why did it take several decades for the flood of papers based on Kramers' work to start?

In connection with our emphasis on Kramers' paper, we need to add a qualification. R. L. Stratonovich, in correspondence, drew my attention to a 1933 paper by Pontryagin, Andronov and Vitt (1933). An English translation of this paper is provided in the Appendix to this volume. Quoting Stratonovich's correspondence:

In this work mean exit time was exactly found for the one-dimensional case, initial point being given. If we take the point where double-well potential has local maximum (the point lies between the wells) as a boundary point and the centre of a well as an initial point, then we immediately find the mean lifetime of metastable state.

It is left up to the reader to judge the exact extent to which this work pioneered. These authors, however, understood first passage time evaluation, understood

Table 1.1 *Escape from the metastable state*

Dimensions and degrees of freedom
Particle moving in one dimension
Many degrees of freedom
  All on similar time scale
  Born–Oppenheimer surfaces
Infinite number of particles
  Sine–Gordon chain, ϕ^4 chain, solitons
  One-dimensional Ising

Law of motion
Completely Hamiltonian (molecular dynamics, MQT theory)
Hamiltonian with noise and friction
More general dynamics, e.g. multistable circuit, laser, ecology

Number of competing locally stable states
Single well, escape into unlimited range of motion
Bistable well
  Symmetrical
  Unsymmetrical
Many competing states
  No symmetries (e.g. spin glass, competing states of ecology)
  All alike: particle in sinusoidal potential
  Biased sinusoidal potential

Quantization
Classical continuous case
Classical discrete case
QM case
  Quantization in well
  Escape by tunneling with or without friction

Noise
Thermal equilibrium
More general, e.g. $1/f$ noise
Big occasional jumps (chemical collision, quantized radiation)
Markovian v. correlated noise
Noise independent of state of system v. state-dependent noise

Time dependence
Time independent law of motion
Modulated barrier and/or sinusoidal excitation in well

Damping in well, noise
Strong, weak, or very weak

Purpose; answer desired
Escape probability
Equilibration within a well
Average escape energy
Dwell time near top of barrier, or tunneling through barrier
Approach to optimum well (spin glass annealing)

bistable potentials, and understood that noise could depend on the state of the system. They also addressed motion in a general dynamical system, not just motion of a particle in a potential. The paper, however, does not seem to be widely known. *Science Citation Index* for 1986 provides only three citations of this paper. None of these were from the Soviet Union (but 1986 is atypical in that respect). None of the three listed citations relate to the full scope of the paper, e.g. to the bistable case.

A few words about the authors of this pioneering 1933 paper are in order. Pontryagin is a well known blind mathematician, particularly noted for his work on topological groups. He received the Stalin prize in 1941, the first year for that prize*. When a new prize is issued there is always a backlog of candidates, and inclusion in this first year must signify an unusual distinction. In 1962 Pontryagin received the Lenin prize, which in the intervening years had replaced the Stalin prize. Pontryagin also was decorated with the order of Lenin; an honor he shares with Pravda, among other Soviet institutions.

Let me also add a few remarks related to Pontryagin's collaborators. A book entitled *Theory of Oscillations*, with Andronow and Chaikin listed as authors, was published in the USSR in 1937, and in English translation in 1949 (Andronow and Chaikin, 1949). The second edition, published in Moscow in 1959, bears the names of Andronov, Vitt and Khaikin. It also appeared in an English translation (Andronov, Vitt and Khaikin, 1966). The translated preface to this second Russian edition, signed by Khaikin, tells us:

> The writer of this Preface is the only one of the three authors of this book who is still alive. Aleksandr Adol'fovich Vitt, who took part in the writing of the first edition of this book equally with the other two authors, but who by an unfortunate mistake was not included on the title page as one of the authors, died in 1937.

This belated correction, after twenty-two years, without real explanation, is remarkable. The fact that the English editor, in his own notes, provides no further elucidation, is even more remarkable. The year of the original appearance of this book, 1937, was at, or near, the peak of Stalin's purges; 1956 was the beginning of Khrushchev's reaction against Stalin, and 1962 its high point. We cannot but help compare Vitt's misfortune with Pontryagin's ability to earn high honors under a diversity of regimes.

Remarkably enough, Kramers' paper gained attention, at first, primarily for its development of equations for Brownian motion in a non-uniform force field, and not for the paper's real target, the escape rate problem. Actually, the equations for Brownian motion in a non-uniform force field had already been discussed by others, e.g. Klein (1922). Quite likely, the first important paper which carried Kramers (1940) further was that of Brinkman (1956), who generalized Kramers' work for a one-dimensional potential to many dimensions. A few years later Landauer and Swanson (1961), unaware of Brinkman's

---

* I am indebted to T. Walsh of Dublin, Ireland, expert on Soviet history, for the interpretation in terms of broader Soviet history, given in this paragraph, and in the next one.

work, repeated it but also went beyond Brinkman and discussed escape from an extremely underdamped multi-dimensional well. The next rediscovery was by Langer (1968). We can hardly afford to list the multitude of subsequent rediscoveries, and mention only one recent paper to illustrate this tendency (Barma and Ramaswamy, 1986). The authors of that paper state: 'The single particle problem has been generalized relatively recently to allow for higher dimensional potentials.' The authors append a citation to a 1980 paper, in connection with that statement. Another example of rediscovery: Kramers' basic notions were found again in 1978, under the label *Stochastic Catastrophe Models* (Cobb, 1978).

Kramers' discussion, in the very heavily damped case, pointed out that escape over a potential barrier was controlled by diffusive motion over the top of the barrier. The motion of injected minority carriers, through the base of a bipolar transistor, is a similar diffusion controlled escape over a barrier. The transistor literature found its own way to an understanding of that problem (Shockley, 1950), without reference to Kramers. Eventually the transistor related literature became extensive, with many variations on the basic problem, including time-dependent cases, non-uniform mobilities, and trapping phenomena. The transistor literature and the statistical mechanics community concerned with escape from the metastable state have continued on their independent courses with few cross-citations. The simplicity and power of the viewpoint used in the transistor literature has been stressed elsewhere (Büttiker and Landauer, 1982, Appendix, p. 138; Landauer, 1983).

Active kinetic systems, following their own law for time evolution and influenced by noise, can show great similarity to the motion of a particle in a noisy force field. In many cases an exact equivalence can be found. This was understood in connection with the treatment of electronic oscillator synchronization, in the presence of noise, in the mid 1950s (Kuznetsov, Stratonovich and Tikhonov 1954, 1955; Raevskii and Khokhlov, 1958; Stratonovich, 1958). These ideas were subsequently explored in greater detail by Stratonovich and other Soviet colleagues (Kuznetsov, Stratonovich and Tikhonov, 1965; Stratonovich, 1967). Stratonovich (1958) treated a problem equivalent to Brownian motion in a sinusoidal potential, with an additional spatially uniform force, and this discussion was expanded by Tikhonov (1959). There have been numerous independent rediscoveries of this analysis since that time. In 1978, for example, papers by two separate groups published results virtually identical to the Figure 3 of Tikhonov (1959).

I asked Stratonovich to explain the relationship between the three names that occur on a number of our citations: Kuznetsov, Stratonovich and Tikhonov. These papers are identified in the journals as coming from Moscow State University. Stratonovich writes:

At that time P. I. Kuznetsov was the senior professor and the head of the Chair of Physical Department of the Moscow University. I was the junior member of his Chair.

8

V. I. Tikhonov was the friend of Kuznetsov. He never was the member of the Moscow University. Afterwards he became my friend despite the age gap: he was nine years older than I. Tikhonov was interested in the theory of fluctuations. Because of acquaintance and contacts with him I got interested in the fluctuation theory and began to work on this subject. Tikhonov stated problems. Following his advice, I wrote my first book *Topics in the Theory of Random Noise* (Russian version of the book was published in 1961).

The series of papers (which we have not cited in its entirety) started only after Stratonovich came on the scene. With awareness of Stratonovich's later work, I cannot but help guess Stratonovich was, in fact, the key participant. Clearly, however, such a speculation by someone far from the scene, unfamiliar with the language, and who has never met the participants, must remain a guess.

R. Landauer (1962) returned to the subject of active multi-stable systems (without awareness of the Soviet work), in an analysis of the stability of information in very small, active, memory circuits. The particular vehicle used in that analysis was a bistable Esaki diode circuit. That circuit, once, was a serious candidate for memory and logic applications. Landauer (1962) was anticipated in an earlier qualitative discussion (Landauer, 1961) of 'structures which are in a steady (time invariant) state, but in a dissipative one, while holding on to information'. (The occurrence of the words *structure* and *dissipative* in this quotation is not in the order which subsequently became popular.) In a narrow technical sense Landauer (1962) can be considered to be an amalgamation of Kramers' work and that of Van Kampen (1960, 1961). Nevertheless, it took a dozen years after Landauer (1962) appeared before these notions were again taken up in the basic science literature. Indeed, as late as 1984, a set of papers (Knessl, *et al.*, 1984; Knessl, Matkowsky, Schuss and Tier, 1986; Matkowsky *et al.*, 1984), rediscovered one of the key points of Landauer (1962), that Kramers' view of escape from the metastable state could be extended to systems in which there is a discrete ladder of states, rather than a continuum. (As in a number of other rediscovery cases discussed in this paper, the rediscovery included a good deal not present in the earlier version.) The engineering literature turned out to be more receptive than the more basic areas. See, for example, Lindsey (1969), which demonstrated the influence of Stratonovich's (1963, 1967) work. The Esaki diode paper (Landauer, 1962) analyzed a system in which the fluctuations are not simply determined by a temperature, but depend on the state of the system. The paper explained that in such cases questions about relative stability cannot be answered by a simple appeal to the noiseless, deterministic, equations of motion, e.g. by a version of Maxwell's equal area rules. This same point is also contained in equation (4.49) of Stratonovich (1963) and in much of the earlier Soviet literature we have cited. The point is totally explicit in some of the cases discussed in Stratonovich (1967). The general understanding of this point, elsewhere, came more than a decade later, typically associated with the labels *multiplicative noise, external noise* and *noise induced phase transitions*.

We will not take our history past 1970. After that the field became very active, and historical perspective takes time to develop. But we will point out the connection of some early work to a few of the subjects in Table 1.1, which became fashionable later on. One of these is the *statistical mechanics* of sine–Gordon solitons. We have in mind a set of particles, strung out along a line, with each particle in a biased sinusoidal potential, i.e. with a spatially uniform force added to the sinusoidal potential. The particles are subject to damping and to thermal fluctuations. The particles are coupled to each other, so that adjacent particles tend to be in alignment in their motion along the sinusoidal potential.

Consider such a chain of particles, in which all the particles are aligned, lying along the bottom of a particular trough in the tilted sinusoidal potential, analogous to a chain lying along a trough in a tilted washboard. This is a metastable state; after all, one of the potential valleys next to the occupied one is at a lower energy. To make a transition to chain configurations of lower energy we must first cross a saddle point, i.e. form a critical nucleus of the 'new' phase. This nucleus consists of a portion of the chain which has been transferred to the adjacent lower valley. The formation of the critical nucleus is, of course, noise-activated escape from a many-dimensional potential well (Brinkman, 1956; Landauer and Swanson, 1961; Langer, 1968). This problem was faced first by the dislocation theorists, concerned with the behavior of kinks in edge dislocations (Hirth and Lothe, 1968; Lothe and Hirth, 1959; Seeger and Schiller, 1966). (Strictly speaking, the dislocation theory problem differs slightly from the sine–Gordon soliton problem. The difference arises from the fact that the dislocation theorists invoke an attractive force between kink and anti-kink which represents elastic interactions in a *three-dimensional* medium.) Some of the dislocation theorists knew about Kramers, but not about Brinkman (1956), and Landauer and Swanson (1961). The dislocation theorists developed their own picture of this multi-dimensional process. Finally, McCumber and Halperin (1970) solved a problem of this type (in the treatment of resistive fluctuations in thin superconducting wires) by combining the correct approach to the saddle point configuration, with the kinetics of Brinkman (1956), and of Landauer and Swanson (1961). The correct saddle point configuration had, of course, already been discussed by the dislocation theorists and also by Langer (1967), in a paper which preceded Langer (1968). The synthesis provided by McCumber and Halperin (1970) was also provided in the dislocation theory literature by Petukhov and Pokrovskii (1973). In the mid-1970s, the statistical mechanics of sine–Gordon solitons became widely recognized and fashionable. For review articles written after the initial burst of this activity see Bishop, Krumhansl and Trullinger (1980), and Büttiker and Landauer (1982). The extent to which even conference organizers, as well as authors of books and review articles, have managed to ignore the early dislocation theory work is truly amazing. In fact, the dislocation theorists had pioneered not only in the statistical mechanics of sine–Gordon solitons, which is our concern here, but also in the earlier deterministic theory. One of the key participants in all this, Alfred Seeger, has provided his own historical observations (Seeger, 1980a, b, 1983, 1986). Seeger's historical accounts, however, emphasize the deterministic case, which is not our principal interest.

The escape from a very underdamped potential well has already been mentioned in our discussion of the multi-dimensional well, and has become a subject of considerable attention recently. We refer the reader to Büttiker (Chapter 3, Volume 2) for a description of the modern work. A particle in a well without damping and fluctuations will just stay at its initial energy, and cannot escape. As the coupling to the reservoir, and therefore the noise and damping, is increased, the chances for a change in energy, and therefore for escape, will improve. At sufficiently high damping the particle will execute diffusive motion over the barrier, and in that range further increases in damping will decrease the escape probability. All this was clear already to Kramers (1940). It seems, however, to have escaped the attention of many theoretical physicists, starting with Chandrasekhar's (1943) otherwise definitive review paper. The continuing failure to understand the point may, in part, have been an unfortunate and unintended consequence of the elegant viewpoint introduced by Langer in the late 1960s, and reviewed in Langer (1980), which invoked analytic continuation, and the imaginary part of the free energy, to calculate escape rates. That, however, introduces no real kinetics. The physical chemists were more perceptive. After all, molecules in a gas, going through some internal rearrangement which involves barrier crossing, need to be jostled by other molecules to get the necessary energy. This has been understood, at least, since the early 1920s, as reflected in the work of Lindemann (1922) and of J. A. Christiansen (1921, Ph.D. Thesis, University of Copenhagen. Note: this author has not inspected Christiansen's thesis and has copied this citation from other sources). It is, perhaps, no great surprise that Kramers, a collaborator of Christiansen (Christiansen and Kramers, 1923), later returned to a more analytically definitive discussion of the subject (Kramers, 1940). In recent years theoretical physics has finally caught up with the physical chemists, though there are also a few older papers in the physics literature which appreciated the behavior of very lightly damped systems (Bak and Lebowitz, 1963; Iche and Nozières, 1976; Landauer and Swanson, 1961; Landauer and Woo, 1971; Lee, 1971).

Christiansen (1936) published a perceptive paper which treated the successive steps in an activated chemical reaction very much like steady state current transport through a set of series resistors, with the size of the local conductivity representing the level of response to a gradient in chemical potential. This is the approach which the transistor community rediscovered (Shockley, 1950) and which this author has repeatedly advocated (Büttiker and Landauer, 1982, Appendix, p. 138; Landauer, 1983). Unfortunately, Christiansen's paper had little impact on the literature that followed it. Lifson and Jackson (1962),

11

for example, derived an effective diffusion constant for particle motion in a periodic field. This is a problem which becomes very simple when handled by the series resistance methods of Christiansen (1936). Furthermore, via the Einstein relation, it becomes equivalent to the problem of drift in a weakly biased sinusoidal potential, a special case of the problem already treated in Stratonovich (1958) and Tikhonov (1959). Nevertheless, Lifson and Jackson (1962) was still ahead of many others in its correct and clear treatment of this problem. The subject was revisited on many later occasions, e.g. by Keck and Carrier (1965). At a much later stage, Festa and d'Agliano (1978) rediscovered the result of Lifson and Jackson by a most elegant and complex method. Gunther, Revzen and Ron (1979) responded to this by correctly and succinctly pointing out the simplicity of the problem, but presumed they were the first authors to understand that. A still later return to the question, without awareness of this long history, is exemplified by Derrida (1983). Again we stress that we emphasize the continued rediscovery episodes not in an attempt to blame particular authors, but only to characterize the overall evolution of the subject.

I do not want to overdo the case for the perception of the physical chemistry community; they, in turn, largely ignored Kramers (1940). For example: three otherwise very authoritative texts on unimolecular reactions (Forst, 1973; Robinson and Holbrook, 1972; Slater, 1959) do not cite Kramers (1940) in their comprehensive bibliographies.

Our account has emphasized the prevalence of unintentional rediscovery in this field, without even remotely exhausting my list of relevant case histories. Rediscovery is not limited to this field. Elsewhere (Landauer, 1978), I have described the history of the internal field in dielectric theory; over many decades, different scientists gave us the same answer to that problem. It does seem, however, that the huge volume of work done in modern science has made the communications process ineffective and has increased the probability of unintentional reinvention. In fact, we have positive feedback in our publication system: the flood of publications makes any one item invisible. As a result, the author has to repeat the message in closely related manuscripts, thus helping to cause the flood.

Is unintentional rediscovery a serious problem, aside from the obvious credit questions? I believe that it is; it is a symptom of an inefficient method of doing science. In addition to the costs of repeating the work, even the repeated publication, in itself, is a costly process. It ties up referees, editors and authors, and permits the existence of journals which cost of the order of one-thousand dollars for an annual subscription. As scientists, we all tend to turn to our sources of financial support and plead: Give us more money to do a better job. But an even larger number of scientists is likely to give us an even more inefficient process. It may be in order for the scientific community to put its house in order, and first learn to be more effective at our current level of support.

References

Ackley, D. H., Hinton, G. E. and Sejnowski, T. J. 1985. *Cognitive Science* **9**, 147–68.

Alkemade, C. T. J. 1958. *Physica* **24**, 1029–34.

Andronov, A. A., Vitt, A. A. and Khaikin, S. E. 1966. *Theory of Oscillators*. Oxford: Pergamon.

Andronow, A. A. and Chaikin, C. E. 1949. In *Theory of Oscillations* (S. Lefschetz, ed.). Princeton University Press.

Armstrong, J. A. and Smith, A. W. 1967. In *Progress in Optics*, (E. Wolf, ed.), vol. VI, pp. 212–57. Amsterdam: North Holland.

Bak, T. A. and Lebowitz, J. L. 1963. *Phys. Rev.* **131**, 1138–48.

Barma, M. and Ramaswamy, R. 1986. *J. Stat. Phys.* **43**, 561–70.

Barrow, J. D. and Tipler, F. J. 1986. *The Anthropic Cosmological Principle*, sect. 6.12, p. 430. Oxford: Clarendon.

Becker, R. and Döring, W. 1935. *Ann. der Phys.* **24**, 719–52.

Bishop, A. R., Krumhansl, J. A. and Trullinger, S. E. 1980. *Physica* **1D**, 1–44.

Boonimovich, V. 1946. *J. Phys.* (*Moscow*) **10**, 35–48.

Brillouin, L. 1950. *Phys. Rev.* **78**, 627.

Brinkman, H. C. 1956. *Physica* **22**, 149–55.

Büttiker, M. and Landauer, R. 1982. In *Nonlinear Phenomena at Phase Transitions and Instabilities* (T. Riste, ed.), pp. 111–43. NATO ASI Series. New York: Plenum Press.

Chandrasekhar, S. 1943. *Rev. Mod. Phys.* **15**, 1–89.

Christiansen, J. A. 1936. *Z. Phys. Chemie B* **33**, 145–55.

Christiansen, J. A. and Kramers, H. A. 1923. *Z. Phys. Chem.* **104**, 451–71.

Cobb, L. 1978. *Behavioral Science* **23**, 360–74.

Coffey, W. 1985. In *Development and Application of the Theory of Brownian Motion* (M. Evans, ed.), pp. 69–252. New York: Wiley.

DeGiorgio, V. and Scully, M. O. 1970. *Phys. Rev. A* **2**, 1170–7.

Derrida, B. 1983. *J. Stat. Phys.* **31**, 433–50.

Einstein, A. 1907. *Ann. der Phys.* (*Leipzig*) **22**, 569–72.

Festa, R. and d'Agliano, E. G. 1978. *Physica* **90A**, 229–44.

Forst, W. 1973. *Theory of Unimolecular Reactions*. New York: Academic Press.

Fürth, R., ed. 1956. In *A. Einstein, Investigations on the Theory of the Brownian Movement* (A. D. Cowper trans.). New York: Dover.

Graham, R. and Haken H. 1968. *Z. Phys.* **213**, 420–50.

Graham, R. and Haken H. 1970. *Z. Phys.* **237**, 31–46.

Gunther, L., Revzen, M. and Ron, A. 1979. *Physica* **95A**, 367–9.

Haken, H. 1964. *Z. Phys.* **181**, 96–124.

Haken, H. 1970. In *Festkörper Probleme X* (O. Madelung, ed.), pp. 351–65. Braunschweig: Pergamon.

Hirth, J. P. and Lothe, J. 1968. *Theory of Dislocations*. New York: McGraw-Hill.

Iche, G. and Noziéres, P. 1976. *J. Phys.* (*Paris*) **37**, 1313–23.

Johnson, J. B. 1927. *Phys. Rev.* **29**, 367–8.

Keck, J. and Carrier, G. 1965. *J. Chem. Phys.* **43**, 2284–98.

Kienker, P. K., Sejnowski, T. J., Hinton, G. E. and Schumacher, L. E. 1986. *Perception* **15**, 197–216.

Kirkpatrick, S. and Toulouse, G. 1985. *J. Phys.* (*Paris*) **46**, 1277–92.

Klein, O. 1922. *Arkiv Mat. Astr. Fys.* **16**, (5), 1–51.

Knessl, C., Mangel, M., Matkowsky, B. J., Schuss, Z. and Tier, C. 1984. *J. Chem. Phys.* **81**, 1285–95.

Knessl, C., Matkowsky, B. J., Schuss, Z. and Tier, C. 1986. *J. Stat. Phys.* **45**, 245–66.

Kramers, H. A. 1940. *Physica* **7**, 284–304.

Kubo, R. 1986. *Science* **233**, 330–4.

Kuznetsov, P. I., Stratonovich, R. L. and. Tikhonov, V. I. 1954. *Dokl. Akad. Nauk SSSR* **97**, 639–42.

Kuznetsov, P. I., Stratonovich, R. L. and Tikhonov, V. I. 1955. *Sov. Phys. JETP* **1**, 510–19.

Kuznetsov, P. I., Stratonovich, R. L. and Tikhonov, V. I., eds. 1965. *Non-Linear Transformations of Stochastic Processes*. Oxford: Pergamon. (Contains English translations of several of the Soviet papers cited in this chapter.)

Landauer, R. 1961, *IBM J. Res. Dev.* **5**, 183–91.

Landauer, R. 1962. *J. Appl. Phys.* **33**, 2209–16.

Landauer, R. 1978. In *Electrical Transport and Optical Properties of Inhomogeneous Media* (J. C. Garland and D. B. Tanner, eds.), pp. 2–45. New York: American Institute of Physics.

Landauer, R. 1981. *Am. J. Physiology* **10**, R107–13.

Landauer, R. 1983. *Helv. Phys. Acta* **56**, 847–861. (See table, p. 852–5.)

Landauer, R. 1987a. *Physica Scripta* **35**, 88–95.

Landauer, R. 1987b. In *Nonlinearity in Condensed Matter* (A. R. Bishop, D. K. Campbell, S. E. Trullinger and P. Kumar, eds.), pp. 2–22. Heidelberg: Springer.

Landauer, R. and Swanson, J. A. 1961. *Phys. Rev.* **121**, 1668–74.

Landauer, R. and Woo, J. W. F. 1971. *J. Appl. Phys.* **42**, 2301–8.

Langer, J. S. 1967. *Ann. Phys.* **41**, 108–57. (Langer actually treated a closely related bistable potential, not the sine–Gordon case.)

Langer, J. S. 1968. *Phys. Rev. Lett.* **21**, 973–6.

Langer, J. S. 1980. In *Systems Far from Equilibrium* (L. Garrido, ed.), pp. 12–47. Heidelberg: Springer.

Lawson, J. L. and Uhlenbeck, G. E., eds. 1950. *Threshold Signals*. New York: McGraw-Hill.

Lax, M. 1960. *Rev. Mod. Phys.* **32**, 25–64.

Lax, M. 1966. *Phys. Rev.* **145**, 110–29.

Lee, P. A. 1971. *J. Appl. Phys.* **42**, 325–34.

Lifson, S. and Jackson, J. L. 1962. *J. Chem. Phys.* **36**, 2410–14.

Lindemann, F. A. 1922. *Trans. Faraday Soc.* **17**, 598–606.

Lindsey, W. C. 1969. *Proc. IEEE* **57**, 1705–22.

Lothe, J. and Hirth, J. P. 1959. *Phys. Rev.* **115**, 543–50.

McCumber, D. E. and Halperin B. I. 1970. *Phys. Rev.* **B 1**, 1054–70.

Matkowsky, B. J., Schuss, Z., Knessl, C., Tier, C. and Mangel, M. 1984. *Phys. Rev.* **A 29**, 3359–69.

Mohn, J. 1970. *Der Leidensweg unter dem Hakenkreuz*. Bad Buchau: Sandmaier & Sohn.

Montroll, E. W. and Lebowitz, J. L., eds. 1979. *Fluctuation Phenomena*. Amsterdam: North-Holland.

Moullin, E. B. 1938. *Spontaneous Fluctuations of Voltage*. New York: Oxford.

Nyquist, H. 1928. *Phys. Rev.* **32**, 110–13.

Pais, A. 1982. *Subtle is the Lord.* New York: Oxford. See Chap. 5, p. 79.

Petukhov, B. V. and Pokrovskii, V. L. 1973. *Sov. Phys. JETP* **36**, 336–42.

Pierce, J. R. 1948. *Bell Syst. Tech. J.* **27**, 158–74.

Pontryagin, L., Andronov, A. and Vitt, A. 1933. *Zh. Eksp. Teor. Fiz.* **3**, 165–80.

Raevskii, S. Ya. and Khokhlov, R. V. 1958. *Radiotekh. Electron.* **3**, 507–11.

Rice, S. O. 1945. *Bell Syst. Tech. J.* **24**, 109–57.

Risken, H. 1965. *Z. Phys.* **186**, 85–98.

Risken, H. 1966. *Z. Phys.* **191**, 302–12.

Robinson, P. J. and Holbrook, K. A. 1972. *Unimolecular Reactions.* New York: Wiley.

Schottky, W. 1918. *Ann. der Phys.* **57**, 541–67.

Schottky, W. 1922. *Ann. der Phys.* **68**, 157–76.

Seeger, A. 1980a. *Proc. Roy. Soc. Lond.* **A371**, 173–7.

Seeger, A. 1980b. In *Continuum Models of Discrete Systems* (E. Kröner and K.-H. Anthony, eds.), pp. 253–327. Ontario: University of Waterloo Press.

Seeger, A. 1983. In *Forschung in der Bundesrepublik Deutschland* (C. Schneider, ed.), pp. 587–609. Weinheim: Verlag Chemie.

Seeger, A. 1986. In *Trends in Applications of Pure Mathematics to Mechanics.* Lecture Notes in Physics 249. (E. Kröner and K. Kirchgassner, eds.), pp. 114-55. Heidelberg: Springer.

Seeger, A., and Schiller, P. 1966. In *Physical Acoustics* (W. P. Mason, ed.), Vol. III, p. 361. New York: Academic Press.

Sejnowski, T. J., Kienker, P. K. and Hinton, G. E. 1986. *Physica* **22D**, 260–75.

Shlesinger, M. F. and Weiss, G. H., eds. 1985. *Wonderful World of Stochastics: A Tribute to Elliott W. Montroll.* Amsterdam: North-Holland.

Shockley, W. 1950. *Electrons and Holes in Semiconductors.* New York: Van Nostrand.

Slater, N. B. 1959. *Theory of Unimolecular Reactions.* New York: Cornell University Press.

Specker, H. E., ed. 1979. *Einstein and Ulm.* Stuttgart: Kohlhammer.

Stratonovich, R. L. 1958. *Radiotekh. Elektron.* **3**, 497–506.

Stratonovich, R. L. 1963. *Topics in the Theory of Random Noise*, Vol. I. New York: Gordon and Breach.

Stratonovich, R. L. 1967. *Topics in the Theory of Random Noise*, Vol. II. New York: Gordon and Breach.

Sutherland, W. 1905. *Phil. Mag.* **9**, 781–5.

Tikhonov, V. I. 1959. *Automat. i Telemekh.* **20**, 1188–96.

Van Kampen, N. G. 1960. *Physica* **26**, 585–604.

Van Kampen, N. G. 1961. *J. Math. Phys.* **2**, 592–601.

Van Kampen, N. G. 1981. *Stochastic Processes in Physics and Chemistry.* Amsterdam: North-Holland. (This gives a modern account of this subject.)

Weiss, G. 1983. *J. Stat. Phys.* **30**, 249. (See papers that follow, reporting the *Symposium on Random Walks*, US National Bureau of Standards, Gaithersburg, Maryland, June 28–July 1, 1982.)

Weiss, G. 1986. *J. Stat. Phys.* **42**, 3–37. (See papers that follow, reporting the *Proceedings of the Symposium on Rate Processes and First Passage Times*, National Institutes of Health, Bethesda, Maryland, May 6–8, 1985.)

2 Some Markov methods in the theory of stochastic processes in nonlinear dynamical systems

R. L. STRATONOVICH

2.1 Introduction

During the long development time of the theory of stochastic processes in nonlinear dynamical systems many approximate methods of calculation of various stochastic characteristics have been worked out for the Markov and non-Markov cases. There are two different equivalent approaches to the problem of the Markov case. Stochastic differential equations are used in the first approach, and the Fokker–Planck equation for time-dependent probability density is applied in the second. One must know how to pass from the stochastic differential equation to the Fokker–Planck equation and vice versa, so choosing the most appropriate method of solving the problem. Therefore at the beginning of this chapter we consider the rules governing the choice of method mentioned above and the different interpretations of stochastic differential equations are also considered.

It is a well-known fact that there are two basic interpretations, each of them having its own advantages. The first is the Ito interpretation. Using this interpretation one can easily write the drifts if the stochastic differential equation is known (generally speaking, the rules of averaging are simpler in this interpretation). This is the advantage of the Ito interpretation. The other interpretation is the S-interpretation. In it the stochastic differential equation is more simply transformed when the nonlinear change of variables is made, and the form of the stochastic S-equation coincides with the form of the pre-limit stochastic differential equation. This is the advantage of the S-interpretation.

There are numerous methods of analysing stochastic processes in nonlinear dynamical systems; here we choose, and consider, only a small number of them. The method based on using the functional probability density will be described later at greater length. At present, the path integral methods attract much attention, and much work is devoted to them. This, we hope, justifies the detailed account of the problem here.

The concept of the functional probability density will be defined in detail below, and we shall point out in which sense one must understand its 'functionality'. This question is nontrivial for the following reasons:

(a) If we divide the time axis into the points $\{t_n\}, t_{n+1} - t_n = \varepsilon$, then the joint probability density, for all $x(t_n)$, will assume the 'many-storeyed' form

$$
\begin{aligned}
p[x(t)] = \text{const} \cdot \exp\{ &- \varepsilon^{-1} F_0[x(t)] \\
&- F_1[x(t)] - \varepsilon F_2[x(t)] \}.
\end{aligned} \tag{2.1.1}
$$

Here $F_0[x]$, $F_1[x]$ and $F_2[x]$ are the functions of $\{x(t_n)\}$; each of them tends to a definite functional for $\varepsilon \to 0$ (for a smooth sample path $x(t)$). In particular cases $F_0[x]$ and/or $F_2[x]$ may be absent. In spite of the 'many-storeyed' form (2.1.1), the results of the calculations obtained by using (2.1.1) must be independent of the parameter ε provided that ε is sufficiently small.

(b) In the Markov case the limit

$$
\lim_{\varepsilon \to 0} F_1[x(t)]
$$

is equal to infinity for a typical non-smooth sample path $x(t)$ (the set of such paths has the probability 1). This also makes the progression to the limit $\varepsilon \to 0$ in (2.1.1) impossible.

Early results on functional probability densities will also be considered in this chapter. In addition, the question of how the functional probability density is transformed under the curvilinear coordinate transformation will be discussed. We shall show that the functional probability density has correct transformational properties.

The problem of an approximate calculation of the conditional probability density $p_{T0}(x(T)|x(0))$ and also of stationary probability density $p_0(x)$ (the last in the non-potential case) will be considered as an example of applying the functional probability density. A multicomponent saddle-point method will be used for calculating the abovementioned functions. The method of calculating the stationary probability density by using the path integral method naturally supplements the methods based on solving the stationary Fokker–Planck equation. In solving the above problem we shall see the advantages of the path integral method. The fact is that it is difficult to calculate the behavior of the stationary probability density $p_0(x)$ in the non-potential case and the probability density $p_{T0}(x(T)|x(0))$ in the regions where these probability densities are very small by using other methods.

We shall show, relatively simply, that the results of the path integration are independent of the discretization parameter λ which specifies the interpretation of the stochastic differential equation.

In conclusion we shall consider the problem of calculating the mean first-passage time including the mean exit time from a metastable state, i.e. an exit from the potential well. In the approximate solution of the problem, for the case of small random forces, the mean exit time is determined by the relative height of the potential barrier or by its equivalent, the ratio of stationary probability densities. Thus the stationary probability density, which is calculated by one of the methods considered above, helps to solve the problem of an approximate calculation of the mean exit time.

2.2   Stochastic differential equations and their relation to the Fokker–Planck equation

2.2.1   Various interpretations of the stochastic differential equation

Let a multicomponent stochastic process $x(t) = \{x_\alpha(t), \; \alpha = 1,\ldots,r\}$ corresponding to the time evolution of a dynamic system be described by a set of differential equations

$$\dot{x}_\alpha(t) = g_\alpha(x(t)) + h_{\alpha i}(x(t))\xi_i(t), \tag{2.2.1}$$

where $g_\alpha(x)$ and $h_{\alpha i}(x)$ are differentiable functions, and $\xi_i(t)$, $i = 1, 2, \ldots,$ are Gaussian random processes describing external random forces acting on the dynamic system or internal noises in it. In (2.2.1) and in all following equations summation over repeated subscripts is to be understood.

Without restriction of generality we can suppose that $\langle \xi_i(t) \rangle = 0$ (otherwise $h_{\alpha i}\langle \xi_i \rangle$ can be included in g_α). In the Markov case $\xi_i(t)$ must be delta-correlated in time:

$$\langle \xi_i(t)\xi_j(t') \rangle = N_{ij}\delta(t - t'). \tag{2.2.2}$$

In other words, $\xi_i(t)$ must be the time derivatives of the Wiener processes, i.e.

$$\xi_i(t) = \dot{W}_i(t). \tag{2.2.3}$$

As a consequence of (2.2.2), the values of $\xi_i(t)$ and $\dot{x}_\alpha$ must almost always be equal to $\pm \infty$, since they have infinite variance. This means that functions $x_\alpha(t)$ alter so quickly that one must define the expression on the right-hand side of (2.2.1) more exactly. Let us choose a certain value of λ, $0 \leqslant \lambda \leqslant 1$, and specify (2.2.1) as a limit of the equation

$$[x_\alpha(t) - x_\alpha(t - \varepsilon)]/\varepsilon = g_\alpha(\lambda x(t) + (1 - \lambda)x(t - \varepsilon))$$
$$+ h_{\alpha i}(\lambda x(t) + (1 - \lambda)x(t - \varepsilon))[W_i(t) - W_i(t - \varepsilon)]/\varepsilon \tag{2.2.4}$$

for $\varepsilon \to 0$. Denoting $x_\alpha(t) - x_\alpha(t - \varepsilon) = \Delta x_\alpha$, and $W_i(t) - W_i(t - \varepsilon) = \Delta W_i$, we have

$$\Delta x_\alpha = g_\alpha(x(t - \varepsilon) + \lambda \Delta x)\varepsilon + h_{\alpha i}(x(t - \varepsilon) + \lambda \Delta x)\Delta W_i. \tag{2.2.5}$$

Expanding the functions $h_{\alpha i}$, g_α into a series in $\lambda \Delta x$, we get

$$\Delta x_\alpha = g_\alpha(x(t - \varepsilon))\varepsilon + h_{\alpha i}(x(t - \varepsilon))\Delta W_i + \lambda \frac{\partial h_{\alpha i}}{\partial x_\beta}(x(t - \varepsilon))\Delta x_\beta \Delta W_i$$
$$+ O(\varepsilon \Delta x) + O((\Delta x)^2 \Delta W). \tag{2.2.6}$$

If by iteration we substitute the whole equation (2.2.6) for Δx_β into the right-hand side of (2.2.6), we obtain

$$\Delta x_\alpha = g_\alpha(x(t-\varepsilon))\varepsilon + h_{\alpha i}(x(t-\varepsilon))\Delta W_i + \lambda\frac{\partial h_{\alpha i}}{\partial x_\beta}(x(t-\varepsilon))$$

$$\times h_{\beta j}(x(t-\varepsilon))\Delta W_i\Delta W_j + O(\varepsilon\Delta W) + O((\Delta W)^3). \qquad (2.2.7)$$

One can easily understand why we write down the term of the order of $(\Delta W)^2$ and omit the terms of the orders of $\varepsilon\Delta W$ and $(\Delta W)^3$. The fact is that, on integrating (2.2.2) and (2.2.3), we readily find

$$\langle \Delta W_i\Delta W_j\rangle = N_{ij}\varepsilon. \qquad (2.2.8)$$

Hence $\Delta W_i \sim \varepsilon^{1/2}$, so that $\Delta W_i\Delta W_j$ has the order ε, which must be taken into account. The terms $\varepsilon\Delta W$, $(\Delta W)^3,\ldots$ have higher orders in ε and may be neglected.

Equation (2.2.7) can be written in integral form. Performing ε-division of the time axis by the points t_m, writing (2.2.7) for each elementary interval and summing up over m from 1 to n, gives

$$x_\alpha(t_n) - x_\alpha(0) = \sum_{m=1}^{n}\left\{g_\alpha(x(t_{m-1})) + h_{\alpha i}(x(t_{m-1}))\Delta W_i(m)\right.$$

$$\left. + \lambda\frac{\partial h_{\alpha i}}{\partial x_\beta}h_{\beta j}\Delta W_i(m)\Delta W_j(m)\right\}, \qquad (2.2.9)$$

where $\Delta W_i(m) = W_i(t_m) - W_i(t_{m-1})$. The following fact from the theory of probability is well known: in the case (2.2.2) and (2.2.3), for any function satisfying the simple conditions (continuity and so on) the equation

$$\lim_{\varepsilon\to 0}\sum_m f(t_m)\Delta W_i(m)\Delta W_j(m) = \int f(t)N_{ij}\,dt \qquad (2.2.10)$$

holds with probability 1. We shall not prove this theorem; however, it is easy and useful to give some substantiation of the analogous one-component equation

$$\lim_{\varepsilon\to 0}\sum_m f(t_m)[\Delta W(m)]^2 = \int f(t)N_{11}\,dt. \qquad (2.2.11)$$

We first find the average of the right-hand side of (2.2.11). Using (2.2.8) in the one-component case yields

$$\lim_{\varepsilon\to 0}\sum_m\langle f(t_m)[\Delta W(m)]^2\rangle = \lim_{\varepsilon\to 0}\sum f(t_m)N_{11}\varepsilon. \qquad (2.2.12)$$

The limit on the right-hand side is equal to $\int f(t)N_{11}\,dt$. Thus, there is a convergence of the type (2.2.11) for the mean value. Evaluating the mean square of the difference

$$D = \sum_m f(t_m)[\Delta W(m)]^2 - \sum_m f(t_m)N_{11}\varepsilon, \qquad (2.2.13)$$

we easily find

$$\langle D^2 \rangle = \sum_{m_1 l} f(t_m) f(t_l) \{ [\Delta W(m)]^2 [\Delta W(l)]^2 - N_{11}^2 \varepsilon^2 \}. \qquad (2.2.14)$$

But

$$\langle \Delta W(n) \Delta W(m) \Delta W(l) \Delta W(k) \rangle$$

$$= \varepsilon^2 N_{11}^2 (\delta_{nm} \delta_{lk} + \delta_{nl} \delta_{mk} + \delta_{nk} \delta_{ml}). \qquad (2.2.15)$$

Therefore

$$[\Delta W(m)]^2 [\Delta W(l)]^2 = \varepsilon^2 N_{11}^2 (1 + 2\delta_{ml}), \qquad (2.2.16)$$

and (2.2.14) takes the form

$$\langle D^2 \rangle = 2 \sum_m f^2(t_m) N_{11}^2 \varepsilon^2. \qquad (2.2.17)$$

The expression on the right-hand side of (2.2.17) tends to zero for $\varepsilon \to 0$. The convergence of the limit (2.2.12) to $\int f(t) N_{11} \, dt$ and the convergence of $\langle D^2 \rangle$ to zero prove the limit (2.2.11) in the mean. In reality the stronger convergence takes place with probability 1.

Because of the convergence (2.2.11), in (2.2.7) and (2.2.9) we can transform the term with $\Delta W_i \Delta W_j$. Instead of (2.2.7) we obtain

$$\frac{\Delta x_\alpha}{\varepsilon} = g_\alpha(x(t-\varepsilon)) + \lambda \frac{\partial h_{\alpha i}}{\partial x_\beta}(x(t-\varepsilon)) h_{\beta j}(x(t-\varepsilon)) N_{ij}$$

$$+ h_{\alpha i} \frac{\Delta W_i}{\varepsilon}. \qquad (2.2.18)$$

Thus we see that the choice of λ in (2.2.4) essentially changes the form of (2.2.18) and, therefore, it changes the statistical properties of the random process in the dynamic system if $g_\alpha(x)$ does not change.

The stochastic differential equation (2.2.1) is called the Ito equation if we take $\lambda = 0$ in the interpretational formula (2.2.4). If we assume $\lambda = \frac{1}{2}$, (2.2.1) is called the S-equation. In this case (2.2.1) may be written as

$$\text{(S)} \qquad \dot{x}_\alpha = g_\alpha(x) + h_{\alpha i}(x) \xi_i(t). \qquad (2.2.19)$$

If we want the stochastic properties of the process to remain unchanged, we must take the Ito equation with another function

$$\text{(I)} \qquad \dot{x}_\alpha = a_\alpha + h_{\alpha i} \xi_i(t). \qquad (2.2.20)$$

Using (2.2.18), where λ must be equal to $\frac{1}{2}$, we easily find the connection between g_α and a_α:

$$a_\alpha = g_\alpha + \frac{1}{2} \frac{\partial h_{\alpha i}}{\partial x_\beta} h_{\beta j} N_{ij}. \qquad (2.2.21)$$

If we take $h_{\alpha j}(N^{1/2})_{ji}$ as $h_{\alpha i}$, then (2.2.21) is somewhat simplified:

$$a_\alpha = g_\alpha + \frac{1}{2} \frac{\partial h_{\alpha i}}{\partial x_\beta} h_{\beta i}. \qquad (2.2.22)$$

For the arbitrary interpretation (2.2.4) of equation (2.2.1) we must set the functions a_α entering into (2.2.1) to be equal to

$$a_\alpha = g_\alpha + \lambda \frac{\partial h_{\alpha i}}{\partial x_\beta} h_{\beta j} N_{ij}, \qquad (2.2.23)$$

as seen from (2.2.18). Only then do the statistical properties of $x(t)$ remain unchanged.

The advantage of the S-equation (2.2.19) over (2.2.20) is that it is simply transformed under curvilinear transformations of variables. Let $y(x)$ be new variables, then in the S-equation

(S) $\qquad \dot{y}_\sigma = G_\sigma(y) + H_{\sigma i}(y) \xi_i(t) \qquad (2.2.24)$

the new functions G_σ, $H_{\sigma i}$ will be connected with the old ones by simple formulas

$$G_\sigma = \frac{\partial y_\sigma}{\partial x_\alpha} g_\alpha, \qquad (2.2.25a)$$

$$H_{\sigma i} = \frac{\partial y_\sigma}{\partial x_\alpha} h_{\alpha i}. \qquad (2.2.25b)$$

This means that $g_\alpha, h_{\alpha 1}, \ldots, h_{\alpha r}$ have vector transformational properties (such as properties of dx_α). For other values of the functions $g_\alpha, h_{\alpha i}$ do not have these properties.

2.2.2   *Connection between the stochastic differential equation and the Fokker–Planck equation*

As is well known the coefficient functions of the Fokker–Planck equation

$$\dot{p}(x) = -\frac{\partial}{\partial x_\alpha}[K_\alpha(x)p(x)] + \frac{1}{2}\frac{\partial^2}{\partial x_\alpha \partial x_\beta}[b_{\alpha\beta}(x)p(x)] \qquad (2.2.26)$$

are defined by

$$K_\alpha(x) = \lim_{\varepsilon \to 0} \{\varepsilon^{-1}\langle x_\alpha(t) - x_\alpha(t-\varepsilon)\rangle_{x(t-\varepsilon)=x}\}, \qquad (2.2.27a)$$

$$b_{\alpha\beta}(x) = \lim_{\varepsilon \to 0} \{\varepsilon^{-1}\langle [x_\alpha(t) - x_\alpha(t-\varepsilon)] \\ \times [x_\beta(t) - x_\beta(t-\varepsilon)]\rangle_{x(t-\varepsilon)=x}\}. \qquad (2.2.27b)$$

The vector K_α is called the drift vector, and $b_{\alpha\beta}$ is called the diffusion matrix. The conditional mean values, i.e. the mean values taken under the condition $x(t-\varepsilon) = x$, stand on the right-hand sides of (2.2.27a, b). We easily see that owing to this condition it is convenient to take the Ito stochastic differential equation (2.2.20) or the corresponding difference equation

$$x_\alpha(t) - x_\alpha(t-\varepsilon) = a_\alpha(x(t-\varepsilon))\varepsilon + h_{\alpha i}(x(t-\varepsilon))\Delta W_i \qquad (2.2.28)$$

21

for calculating the functions (2.2.27a, b). After the substitution of (2.2.28) into (2.2.27a), the term with ΔW_i falls out when the averaging is carried out, and we obtain

$$K_\alpha(x) = a_\alpha(x) \qquad (2.2.29)$$

After the substitution of (2.2.28) into (2.2.27b), the term with a_α will be of no importance, and we get

$$b_{\alpha\beta}(x) = h_{\alpha i}(x) h_{\beta j}(x) N_{ij}. \qquad (2.2.30)$$

If we include $\hat{N}^{1/2}$ in $\hat{h}$, in addition to (2.2.22) we will have

$$b_{\alpha\beta}(x) = h_{\alpha i}(x) h_{\beta j}(x). \qquad (2.2.31)$$

Thus, when the initial equation is the S-equation (2.2.19), we must use (2.2.22), (2.2.29) and (2.2.31) for writing the Fokker–Planck equation. When the initial equation is the Ito equation, it is sufficient to use the simpler formula (2.2.29) and also to consider (2.2.31). For an arbitrary value of λ in the initial equation (2.2.4), we must also use (2.2.23) and (2.2.29).

Now we consider the reverse problem: in what form do we have to write down the stochastic differential equation when the Fokker–Planck equation is known? Of course, in so doing one should use the same transition equations. A problem of determining the matrix $h_{\alpha i}$ by using (2.2.31) requires special comment. Naturally, equation (2.2.31) determines $h_{\alpha i}$ non-uniquely. In the stochastic differential equation, the matrix $h_{\alpha i}$ may be non-square, i.e. the number of the functions $\xi_i(t)$ may be different from the number of the components x_α. However, the case of the non-square matrix $h_{\alpha i}$ is inconvenient since in this case there is no inverse (and also unique) matrix $(h^{-1})_{i\alpha}$ for the non-degenerate matrix $b_{\alpha\beta}$. We shall see later that the inverse matrix $\hat{h}^{-1}$ is very useful. One can get rid of non-squareness of the matrix $h_{\alpha i}$, firstly, by calculating $b_{\alpha\beta}$ using (2.2.31). Then one can find a new matrix $h_{\alpha i}$, which is now square, with the help of the same formula. But the square matrix $h_{\alpha i}$ is also determined by (2.2.31) non-uniquely. We now prove this. Let $h_{i\alpha}^0$ be a certain solution of (2.2.31). We can verify whether the matrix $h_{\beta j} = h_{\alpha i}^0 S_{ij}$ or

$$\hat{h} = \hat{h}^0 \hat{S} \qquad (2.2.32)$$

is the solution of (2.2.31). Substituting (2.2.32) into (2.2.31), which may be written in the matrix form

$$\hat{h}\hat{h}^{\mathrm{T}} = \hat{b}, \qquad (2.2.33)$$

gives

$$\hat{h}^0 \hat{S} \hat{S}^{\mathrm{T}} \hat{h}^0 = \hat{h}^0 \hat{h}^{0\mathrm{T}}. \qquad (2.2.34)$$

In the case of a non-degenerate matrix $\hat{h}^0$, we have

$$\hat{S}\hat{S}^{\mathrm{T}} = \hat{I}. \qquad (2.2.35)$$

This means that the transposed matrix $\hat{S}^{\mathrm{T}}$ is equal to the inverse matrix S^{-1}.

Consequently, $\hat{S}$ is the matrix of orthogonal transformation. In all cases, besides the one-component case, there are various real orthogonal transformations, which proves the above non-uniqueness.

There is one especially simple method of solving equation (2.2.31). Postulating the symmetry of $\hat{h}$ ($\hat{h}^T = \hat{h}$), from (2.2.31) we obtain $\hat{h}^2 = \hat{b}$, i.e. $\hat{h}$ is the symmetrical square root of $\hat{b}$. Because of the non-negative definiteness of matrix $b_{\alpha\beta}$, and of its symmetry, such a root exists. The solution

$$h_{\alpha i} = (\hat{b}^{1/2})_{\alpha i} \tag{2.2.36}$$

has one disadvantage connected with transformational properties. This is that the square root of the tensor of rank two is not a vector. Therefore (2.2.36) can be applied only in one coordinate system. When the variables are changed, the matrix $\hat{h}$ should be transformed according to (2.2.25b).

Instead of (2.2.26) one can use the S-form of the Fokker–Planck equation:

$$\dot{p} = -\frac{\partial}{\partial x_\alpha}(g_\alpha p) + \frac{1}{2}\frac{\partial}{\partial x_\alpha}\left[h_{\alpha i}\frac{\partial(h_{\beta i}p)}{\partial x_\beta} \right]. \tag{2.2.37}$$

We can easily verify that (2.2.37) is equivalent to (2.2.36) when (2.2.22) is taken into account.

2.2.3 Generalization to the case of the non-Markov process that differs little from the Markov one and to the case of non-Gaussian random forces

If the random functions $\xi_i(t)$ in (2.2.1) are non-Gaussian, the Fokker–Planck equation is replaced by the more general Kramers–Moyal equation

$$\dot{p}(x) = \sum_{n=1}^{\infty}\frac{(-1)^n}{n!}\frac{\partial^n}{\partial x_{\alpha_1}\dots\partial x_{\alpha_n}}[K_{\alpha_1\dots\alpha_n}(x)p], \tag{2.2.38}$$

where, by analogy with (2.2.27),

$$K_{\alpha_1\dots\alpha_n}(x) = \lim_{\varepsilon\to 0}\left\{\varepsilon^{-1}\left\langle\prod_{i=1}^{n}[x_{\alpha_i}(t)-x_{\alpha_i}(t-\varepsilon)]\right\rangle_{x(t-\varepsilon)=x}\right\}. \tag{2.2.39}$$

We note that instead of (2.2.39) we can use

$$
\begin{aligned}
K_{\alpha_1\dots\alpha_n}(x) &= \lim_{\varepsilon\to 0}\frac{d}{d\varepsilon}\left\langle\prod_{i=1}^{n}[x_{\alpha_i}(t)-x_{\alpha_i}]\right\rangle_{x(t-\varepsilon)=x} \\
&= n\lim_{\varepsilon\to 0}\left\{\left\langle \dot{x}_{\alpha_1}(t)\prod_{i=2}^{n}[x_{\alpha_i}(t)-x_{\alpha_i}]\right\rangle_{x(t-\varepsilon)=x}\right\}_{\text{sym}},
\end{aligned} \tag{2.2.40}
$$

where $\{\cdots\}_{\text{sym}}$ denotes the operation of symmetrization in subscripts $\alpha_1,\dots,\alpha_n$. In order to verify the equivalence of (2.2.39) to (2.2.40), we represent the function entering into the right-hand sides of these equations in the form of the expansion

$$\left\langle \prod_{i=1}^{n} [x_{\alpha_i}(t) - x_{\alpha_i}] \right\rangle_{x(t-\varepsilon)=x} = \varphi_{\alpha_1...\alpha_n}(x)\varepsilon$$
$$+ \tfrac{1}{2}\psi_{\alpha_1...\alpha_n}\varepsilon^2 + \cdots, \qquad (2.2.41)$$

where $\varepsilon > 0$. Substituting (2.2.41) into (2.2.39) and (2.2.40), we make sure that both formulas lead to the same result $K_{\alpha_1...\alpha_n} = \varphi_{\alpha_1...\alpha_n}$.

When $\xi_i(t)$ is non-Gaussian, the various cumulants

$$\langle \xi_{i_1}(t_1), \dots, \xi_{i_s}(t_s) \rangle = M_{i_1...i_s} f(t_1, \dots, t_s), \quad s \geqslant 2, \qquad (2.2.42)$$

where $M_{ij} = N_{ij}$, are not equal to zero. In the general case we require that the functions $f(t_1, \dots, t_s)$ in (2.2.42) be symmetrical in all of their arguments and the following normalization conditions hold:

$$\int f(t_1, \dots, t_s)\,dt_2 \dots dt_s = 1. \qquad (2.2.43)$$

In the Markov case all $f(t_1, \dots, t_s)$ are equal to the multidimensional delta-function

$$f(t_1, \dots, t_s) = \delta(t_1, \dots, t_s)$$
$$\equiv \delta(t_{12})\delta(t_{13})\dots\delta(t_{1s}) \quad (t_{1i} = t_1 - t_i). \qquad (2.2.44)$$

In the non-Markov case close to the Markov one, the functions $f(t_1, \dots, t_s)$ markedly differ from zero only for $|t_i - t_j| \lesssim \mu (i,j = 1, \dots, r)$ and are practically equal to zero if $|t_i - t_j| \gg \mu$, at least for one pair (i,j). Here μ is a very small quantity. If τ_1 is the characteristic time of the relaxation process described by the equations $\dot{x}_\alpha = g_\alpha(x)$, then the inequality

$$\mu \ll \tau_1 \qquad (2.2.45)$$

must be valid. If this condition is not met, the process is not close to the Markov one. It should be noted that in (2.2.39) and (2.2.40) one should not take the limit by explicitly letting $\varepsilon \to 0$, but it is sufficient to put $\varepsilon = (\mu\tau_1)^{1/2}$. In the case of the limit $\mu \to 0$ corresponding to passing to the Markov process, the limit $\varepsilon \to 0$ in (2.2.39) and (2.2.40) will be ensured automatically.

For calculating the function (2.2.40) it is necessary to express $x(t)$ in terms of $x(t-\varepsilon) = x^0$ with some desired accuracy using (2.2.1). Let me explain how to do this. Introducing the formal small parameter κ, we denote the right-hand side of (2.2.1) by $\kappa F_\alpha(x(t), t)$, i.e.

$$\kappa F_\alpha(x,t) = g_\alpha(x) + h_{\alpha i}(x)\xi_i(t). \qquad (2.2.46)$$

Time $t_0 = t - \varepsilon$ plays the role of the initial time, and the condition $x(t_0) = x^0$ plays the role of the initial condition. We introduce the designation

$$\Delta_\alpha(t) = x_\alpha(t) - x_\alpha^0. \qquad (2.2.47)$$

Integrating the equation $\dot{x} = \kappa F(x, t)$ from t_0 to t, for the difference (2.2.47), we obtain

$$\Delta_\alpha(t) = \kappa \int_{t_0}^t F_\alpha(x(t'), t')\,dt' \qquad (2.2.48)$$

or, if we expand $F_\alpha(x(t'), t') = F_\alpha(x^0 + \Delta(t'), t')$ into a series in $\Delta(t')$,

$$\Delta_\alpha(t) = \kappa \int_{t_0}^t dt' \sum_{m=0}^\infty \frac{1}{m!} \frac{\partial^m F_\alpha}{\partial x_{\beta_1} \ldots \partial x_{\beta_m}} (x^0, t') \Delta_{\beta_1}(t') \ldots \Delta_{\beta_m}(t'). \quad (2.2.49)$$

The solution of this equation should be sought in the form of an expansion in the formal small parameter

$$\Delta_\alpha(t) = \sum_{i=1}^\infty \kappa^i \Delta_\alpha^{(i)}(t). \qquad (2.2.50)$$

Substituting (2.2.50) into (2.2.49) and equating the terms of the various orders in κ, we get the various approximations

$$\Delta_\alpha^{(1)}(t) = \int_{t_0}^t dt'\, F_\alpha(x^0, t'),$$

$$\Delta_\alpha^{(2)}(t) = \int_{t_0}^t dt'\, \frac{\partial F_\alpha}{\partial x_\beta}(x^0, t') \Delta_\beta^{(1)}(t'), \qquad (2.2.51)$$

$$\Delta_\alpha^{(3)}(t) = \int_{t_0}^t dt'\, \frac{\partial F_\alpha}{\partial x_\beta}(x^0, t') \Delta_\beta^{(2)}(t')$$

$$+ \frac{1}{2} \int_{t_0}^t dt'\, \frac{\partial^2 F_\alpha}{\partial x_\beta \partial x_\gamma}(x^0, t') \Delta_\beta^{(1)}(t') \Delta_\gamma^{(1)}(t'),$$

etc. Using (2.2.51) and considering (2.2.46), (2.2.47) and (2.2.50) gives

$$x_\alpha(t) - x_\alpha^0 = g_\alpha(x^0)(t - t_0) + h_{\alpha i}(x^0) \int_{t_0}^t dt'\, \xi_i(t')$$

$$+ \frac{\partial h_{\alpha i}}{\partial x_\beta}(x^0) h_{\beta i}(x^0) \int_{t_0}^t dt'\, \xi_i(t') \int_{t_0}^{t'} dt''\, \xi_j(t'') + \cdots. \qquad (2.2.52)$$

Here a number of unessential terms are omitted. If in (2.2.1) we use the expansion

$$h_{\alpha i}(x) = h_{\alpha i}(x^0) + \frac{\partial h_{\alpha i}}{\partial x_\beta}(x^0)\Delta_\beta + \frac{1}{2}\frac{\partial^2 h_{\alpha i}}{\partial x_\beta \partial x_\gamma}(x^0)\Delta_\beta \Delta_\gamma + \cdots \qquad (2.2.53)$$

(and analogously for $g_\alpha(x)$), where (2.2.50) and (2.2.51) should be inserted, then (2.2.1) is transformed to

$$\dot{x}_\alpha(t) = g_\alpha(x^0) + h_{\alpha i}(x^0)\xi_i(t) + \left(\frac{\partial h_{\alpha i}}{\partial x_\beta} h_{\beta j}\right)_{x^0} \xi(t) \int_{t_0}^t dt'\, \xi_j(t')$$

$$+\left(\frac{\partial h_{\alpha i}}{\partial x_\beta}\frac{\partial h_{\beta j}}{\partial x_\gamma}h_{\gamma k}\right)_{x^0}\xi_i(t)\int_{t_0}^{t}dt'\,\xi_j(t')\int_{t_0}^{t'}dt''\,\xi_k(t'')$$

$$+\frac{1}{2}\left(\frac{\partial^2 h_{\alpha i}}{\partial x_\beta\,\partial x_\gamma}h_{\beta j}h_{\gamma k}\right)_{x^0}\xi_i(t)\int_{t_0}^{t}dt'\,\xi_j(t')\int_{t_0}^{t'}dt''\,\xi_k(t'').\quad(2.2.54)$$

Unessential terms with derivatives of g_α are not included. Substituting (2.2.52) and (2.2.54) into the pre-limit version,

$$K_{\alpha_1\ldots\alpha_n}(x^0)=n\left\{\left\langle\dot{x}_{\alpha_1}(t)\prod_{i=2}^{n}(x_{\alpha_i}(t)-x_{\alpha_i}^0)\right\rangle_{x(t-\varepsilon)=x^0}\right\}_{\mathrm{sym}}$$
$$(\varepsilon=(\mu\tau_1)^{1/2})\qquad\qquad\qquad(2.2.55)$$

of (2.2.40) and using (2.2.42), we will discard fourth and higher moments. For $n=1$ we get

$$K_\alpha(x^0)=g_\alpha(x^0)+\left(\frac{\partial h_{\alpha i}}{\partial x_\beta}h_{\beta j}\right)_{x^0}N_{ij}\int_{t-\varepsilon}^{t}f(t,t')\,dt'$$

$$+\left(\frac{\partial h_{\alpha i}}{\partial x_\beta}\frac{\partial h_{\beta j}}{\partial x_\gamma}h_{\gamma k}\right)_{x^0}M_{ijk}\int_{t-\varepsilon}^{t}dt'\int_{t-\varepsilon}^{t'}dt''\,f(t,t',t'')$$

$$+\frac{1}{2}\left(\frac{\partial^2 h_{\alpha i}}{\partial x_\beta\,\partial x_\gamma}h_{\beta j}h_{\gamma k}\right)_{x^0}M_{ijk}\int_{t-\varepsilon}^{t}dt'$$

$$\times\int_{t-\varepsilon}^{t'}dt''\,f(t,t',t'').\qquad(2.2.56)$$

In consequence of the condition $\varepsilon\gg\mu$ we can here take $-\infty$ as a lower limit. From (2.2.43) and the condition of symmetry of the functions $f(t_1,\ldots,t_s)$, we have

$$\int_{t_1>t_2>\cdots>t_s}f(t_1,\ldots,t_s)\,dt_2\ldots dt_s=1/s!.\qquad(2.2.57)$$

Therefore, in the chosen approximation, from (2.2.56) we finally obtain

$$K_\alpha(x)\equiv a_\alpha(x)=g_\alpha(x)+\frac{1}{2}\frac{\partial h_{\alpha i}}{\partial x_\beta}(x)h_{\beta j}(x)N_{ij}$$

$$+\frac{1}{6}\left[\frac{\partial h_{\alpha i}(x)}{\partial x_\beta}\frac{\partial h_{\beta j}(x)}{\partial x_\gamma}h_{\gamma k}(x)+\frac{\partial^2 h_{\alpha i}(x)}{\partial x_\beta\,\partial x_\gamma}h_{\beta j}h_{\gamma k}(x)\right]M_{ijk}.\qquad(2.2.58)$$

Analogously, it is easy to find

$$K_{\alpha\beta}(x)\equiv b_{\alpha\beta}(x)$$

$$=h_{\alpha i}h_{\beta i}N_{ij}+\frac{1}{2}\left(\frac{\partial h_{\alpha i}}{\partial x_\gamma}h_{\gamma j}h_{\beta k}+h_{\alpha i}\frac{\partial h_{\beta k}}{\partial x_\gamma}h_{\gamma j}\right)M_{ijk}\qquad(2.2.59a)$$

$$K_{\alpha\beta\gamma}(x)=h_{\alpha i}(x)h_{\beta j}(x)h_{\gamma k}(x)M_{ijk}.\qquad(2.2.59b)$$

Of course, these formulas just obtained also hold in the precise Markov case. The term with M_{ijk} in (2.2.58) and (2.2.59a) makes the expressions in (2.2.21) and (2.2.30) more nearly exact. The non-zero value of $K_{\alpha\beta\gamma}(x)$ means that the approximation of the Fokker–Planck equation is now insufficient. If we make a still more precise calculation, i.e. if we take into account a greater number of terms of the expansion (2.2.50), new terms specifying the functions $K_\alpha(x)$, $K_{\alpha\beta}(x)$ and $K_{\alpha\beta\gamma}(x)$ appear in (2.2.58) and (2.2.59) and new non-zero terms also appear in (2.2.38).

If, however, we restrict ourselves to a less precise approximation and discard the terms with M_{ijk} in (2.2.58) and (2.2.59), then we obtain the former formulas

$$K_\alpha = g_\alpha + \frac{1}{2}\frac{\partial h_{\alpha i}}{\partial x_\beta} h_{\beta j} N_{ij}, \quad K_{\alpha\beta} = h_{\alpha i} h_{\beta j} N_{ij}, \tag{2.2.60}$$

which are valid (according to the above considerations) both for non-zero (but small) values of μ and in the Markov limit $\mu \to 0$. This implies that the pre-limit, non-Markov equation (2.2.1) coincides with the S-form of the Markov stochastic differential equation. This is the advantage of the S-form over the Ito form. The Ito equation was first suggested, and then developed, in Ito (1944, 1946, 1951). The S-equation was first published in Stratonovich (1964). However, in the one-component case the formulas of the type (2.2.60) were obtained in 1961 in the Russian original of a two-volume treatise (Stratonovich, 1963, 1967). (Pre-limit stochastic differential equations were also considered.) Since then this problem has been considered in many works.

2.3   Concept of functional probability density. Simple particular cases

A number of effective approximate methods of analysis of stochastic dynamical systems are based on using the functional probability density to describe the processes going on in these systems. Some of these methods will be described below. We begin by elucidating the concept of functional probability density.

For simple diffusion processes the functional probability densities and path integrals were first introduced by Wiener in the 1920s (Wiener, 1923, 1924). In the case of the simple linear Markov process the path integral was considered in Onsager and Machlup (1953). In the non-linear Markov case the functional probability density was obtained in Stratonovich (1962) and elsewhere (see below).

2.3.1   *Functional probability densities for the Wiener process and white noise*

Historically, the Markov processes were the first for which functional probability densities were written down. The simplest Markov process is the

Wiener process, because the functional probability density is defined for this process most simply. The Gaussian process $W(t)$ is called the Wiener process if all of its increments $(\Delta W)_1, \ldots, (\Delta W)_m$ per non-overlapping intervals $\Delta t_1, \ldots, \Delta t_m$ are statistically independent of one another, for each increment $(\Delta W)_i = W(t_i') - W(t_i)$ $(\tau = t_i' - t_i > 0)$ the probability density having the form

$$p_\tau((\Delta W)_i, \tau) = (2\pi K\tau)^{-1/2} \exp\left[\frac{(\Delta W)_i^2}{2K\tau}\right], \tag{2.3.1}$$

where K is a constant.

Suppose that the Wiener process is defined over the interval $[0, T]$. Let us divide this interval by points

$$t_n = n\varepsilon, \quad n = 0, 1, \ldots, N, \tag{2.3.2}$$

where $\varepsilon = T/N$.

Using the independence of increments and also (2.3.1), we can readily write the joint probability density of random variables $W(t_1), \ldots, W(t_N)$ provided that $W(0) = 0$. We have

$$p_W(W(t_1), \ldots, W(t_N))$$

$$= (2\pi K\varepsilon)^{-N/2} \exp\left\{-\frac{1}{2K} \sum_{n=1}^{N} \left[\frac{W(t_n) - W(t_{n-1})}{\varepsilon}\right]^2 \varepsilon\right\}. \tag{2.3.3}$$

For non-uniform division of interval $[0, T]$ the following sum

$$S_N = \sum_{n=1}^{N} \left[\frac{W(t_n) - W(t_{n-1})}{t_n - t_{n-1}}\right]^2 (t_n - t_{n-1}) \tag{2.3.4}$$

stands under the exponent sign. It is obvious that for $\varepsilon = \max_n (t_n - t_{n-1}) \to 0$ the sum (2.3.4) tends to the integral

$$\int_0^T [\dot{W}(t)]^2 \, dt \tag{2.3.5}$$

if the limit

$$\lim_{\varepsilon \to 0} \sum_{n=1}^{N} \left[\frac{\Delta W}{\Delta t}\right]^2 \Delta t \tag{2.3.6}$$

and also the derivative

$$\lim_{t_n - t_{n-1} \to 0} \frac{W(t_n) - W(t_{n-1})}{t_n - t_{n-1}} = \dot{W} \tag{2.3.7}$$

exists. If the condition of existence of (2.3.6) and (2.3.7) is fulfilled, we have

$$\lim_{\varepsilon \to 0} [(2\pi K\varepsilon)^{N/2} p_W(W(t_1), \ldots, W(t_N))]$$

$$= \exp\left[-\frac{1}{2K} \int_0^T (\dot{W})^2 \, dt\right]. \tag{2.3.8}$$

28

This limit can be interpreted as a functional space probability density $P_W[W(t)]$ multiplied by a certain constant. Therefore (2.3.8) can be written as

$$P_W[W(t)] = \text{const} \cdot \exp\left\{ -\frac{1}{2K} \int_0^T [\dot{W}(t)]^2 \, dt \right\}, \qquad (2.3.9)$$

where the normalization constant has no precise sense.

We now return to the abovementioned condition of existence of limits (2.3.6) and (2.3.7). These limits exist only for the very small subset A of the whole set of sample paths $W(t)$. As is well known, typical sample paths $W(t)$ are non-differentiable. Therefore (2.3.8) is invalid for typical paths $W(t)$. Averaging the sum (2.3.4) gives

$$\langle S_N \rangle = \sum_{n=1}^N \frac{\langle [W(t_n) - W(t_{n-1})]^2 \rangle}{t_n - t_{n-1}} = \sum_{n=1}^N K = KN, \qquad (2.3.10)$$

so that the limit $\lim_{N \to \infty} \langle S_N \rangle$ does not exist. Consequently, $\lim_{N \to \infty} S_N$, i.e. the limit (2.3.9), also does not exist. One can state that the measure of subset A is equal to zero. The fact is that, as is well known in the theory of probability, the sum

$$\sum_{n=1}^N [W(t_n) - W(t_{n-1})]^2$$

converges to KT with probability 1 (see (2.2.11)). Therefore the sum in (2.3.3) diverges with probability 1. Consequently, (2.3.8) and (2.3.9), in which the integral has precise sense, are of no importance. We shall not use these formulas. Instead of that we write (2.3.3) in symbolic form

$$P_W[W(t)] = \text{const} \cdot \exp\left\{ -\frac{1}{2K} \int_0^T {}^*[\dot{W}(t)]^2 \, dt \right\}. \qquad (2.3.11)$$

Here $P_W[W(t)]$ denotes the probability density $p_W(W(t_1), \ldots, W(t_N))$ for large N. The asterisk in (2.3.11) indicates that the integral is symbolic, i.e. it should be interpreted as the prelimit sum (2.3.4) for very small

$$\varepsilon = \max_n (t_n - t_{n-1}).$$

Although $W(t)$ is undifferentiable in the sense of ordinary analysis, the derivative

$$\xi(t) = \dot{W}(t) \qquad (2.3.12)$$

exists if it is understood in the sense of the generalized function, just as we understand, say, the Dirac delta-function $\delta(t)$. The random function (2.3.12) is called the Gaussian white noise. It is easy to obtain its correlation function

$$\langle \xi(t)\xi(t') \rangle = K\delta(t - t'). \qquad (2.3.13)$$

Using (2.3.12), from (2.3.11) we readily get the 'functional' probability density for white noise

$$P[\xi(t)] = \text{const} \cdot \exp\left[-\frac{1}{2K} \int_0^T *\xi^2(t)\,dt \right].$$ (2.3.14)

This equation should also be understood in the symbolic sense. We can easily write the multidimensional generalization of formulas (2.3.11) and (2.3.14). In the multicomponent case, (2.3.13) is replaced by (2.2.2), and instead of (2.3.14) we will have

$$P[\xi(t)] = \text{const} \cdot \exp\left[-\frac{1}{2} \int_0^T *N_{ij}^{-1}\xi_i(t)\xi_j(t)\,dt \right].$$ (2.3.15)

Here $N_{ij}^{-1} = (\hat{N}^{-1})_{ij}$, and summation over repeated subscripts is to be assumed. According to (2.3.15), for the multidimensional Wiener process $\{W_i(t)\}$, which is related to $\{\xi_i(t)\}$ by (2.2.3), we get

$$P_W[W(t)] = \text{const} \cdot \exp\left[-\frac{1}{2} \int_0^T *N_{ij}^{-1} \dot{W}_i\dot{W}_j\,dt \right].$$ (2.3.16)

2.3.2 The more complex Markov process. On one particular result of the theory of probability

Now we move on to more complex processes. At first we will consider one rather old probability theory result related to functional probability density.

Let $x(t)$ be the one-component Markov process determined by

$$\dot{x} = g(x) + \xi(t) = g(x) + \dot{W},$$ (2.3.17)

where $\xi(t)$ is the white noise (2.3.12) with correlation function (2.3.13). The initial value $x(0)$ is assumed to be non-random.

In the theory of probability the concept of the derivative of one probability measure with respect to another one is considered. We will denote the Wiener measure, i.e. the measure corresponding to the Wiener process, by P_W, and the probability measure corresponding to the process $x(t)$ will be denoted by P. We take, for example, the division (2.3.2), for which the derivative dP/dP_W is the function of arguments $x_1 = x(t_1), \ldots, x_N = x(t_N)$ and is equal to

$$\frac{dP}{dP_W}(x_1, \ldots, x_N) = \frac{p(x_1, \ldots, x_N)}{p_W(x_1, \ldots, x_N)}.$$ (2.3.18)

Here p_W is the function (2.3.3) taken with other arguments.

It is essential that we can extend (2.3.18) to the continuum limit $\varepsilon \to 0$, $N \to \infty$. The limit derivative of one measure with respect to another one will be the functional of $x(t)$:

$$\frac{dP}{dP_W}[x(t)] = \lim_{\varepsilon \to 0} \frac{dP}{dP_W}(x(t_1), \ldots, x(t_N)).$$ (2.3.19)

In the continuum case the derivative dP/dP_W is the functional of $x(t)$, and can be mathematically introduced without resorting to the limit (2.3.19). This is a consequence of a Radon–Nicodim theorem in the theory of probability. This theorem states that for any σ-algebra, i.e. for any probability space, the derivative of one measure with respect to another exists if these measures are mutually absolutely continuous. The functional derivative dP/dP_W cannot be written in the form of the ratio of functional probability densities $p[x(t)]$ and $p_W[x(t)]$ since these functional probability densities are not (in contrast to dP/dP_W) precisely definable concepts for $\varepsilon = 0$. In other words, in the rigorous continuum case the derivative dP/dP_W is defined in the theory of probability without the definition of functional probability densities.

For (2.3.17), when $g(x)$ is independent of time, the continuum derivative is of the form

$$\frac{dP}{dP_W}[x(t)] = \exp\left\{\frac{1}{K}[U(x(0)) - U(x(T))]\right.$$
$$\left. - \frac{1}{2K}\int_0^T [g^2(x(t)) + K\frac{dg}{dx}(x(t))]dt\right\}, \qquad (2.3.20)$$

where the function $U(x)$ is defined by

$$dU/dx = -g(x). \qquad (2.3.21)$$

Equation (2.3.20) is valid when the integral on the right-hand side exists. It is essential that the integral in (2.3.20) exists for a much broader class of sample paths than does the integral in (2.3.9). It is known that in the case (2.3.17) the function $x(t)$ is continuous with probability 1; therefore the integral in (2.3.20) exists with probability 1 (but not with probability 0, as in (2.3.9)).

For a much more narrow set of sample paths we can use (2.3.21) to perform the transformation

$$U(x(T)) - U(x(0)) = \int_0^T \frac{dU}{dx}dx = -\int_0^T g(x)\dot{x}\,dt. \qquad (2.3.22)$$

Substituting (2.3.22) into (2.3.20) gives

$$\frac{dP}{dP_W}[x(t)] = \left\{-\frac{1}{2K}\int_0^T [g^2(x) - 2g(x)\dot{x} + \frac{dg}{dx}(x)]dt\right\}. \qquad (2.3.23)$$

If we try to apply (2.3.23) to typical non-differentiable paths whose set has the measure 1, then the question arises in what sense the 'strange' term $-2g(x)\dot{x}$ should be understood. To answer this question we should again use the ε-division of interval $[0, T]$. We have

$$U(x(T)) - U(x(0)) = \sum_{n=1}^N [U(x_n) - U(x_{n-1})], \quad (x_n = x(t_n)), \qquad (2.3.24)$$

where

$$U(x_n) - U(x_{n-1}) = -\int_{x_{n-1}}^{x_n} g(x)\,dx$$

$$= -\int_{-\Delta x/2}^{\Delta x/2} g(\tilde{x}_n + y)\,dy, \qquad (2.3.25)$$

and $\Delta x = x_n - x_{n-1}$, $\tilde{x}_n = (x_n + x_{n-1})/2$. Before taking the last integral we expand $g(\tilde{x}_n + y)$ into a series in y:

$$U(x_n) - U(x_{n-1}) = -\int_{-\Delta x/2}^{\Delta x/2} [g(\tilde{x}_n) + g'(x_n)y$$

$$+ \tfrac{1}{2}g''(\tilde{x}_n)y^2 + \tfrac{1}{6}g'''(\tilde{x}_n)y^3 + \cdots]\,dy$$

$$= -g(\tilde{x}_n)\Delta x - \tfrac{1}{24}g''(\tilde{x}_n)(\Delta x)^3 + O((\Delta x)^5). \quad (2.3.26)$$

Substituting (2.3.26) into (2.3.24) yields

$$U(x(T)) - U(x(0)) = -\sum_{n=1}^{N}\left[g\left(\frac{x_n + x_{n-1}}{2}\right)\Delta x + O((\Delta x)^3) \right].$$

$$(2.3.27)$$

If we pass to the limit $\varepsilon \to 0$, then the terms $O((\Delta x)^3)$, etc., become unsignificant, and we obtain

$$U(x(T)) - U(x(0)) = -S\int_0^T g(x)\dot{x}\,dt, \qquad (2.3.28)$$

where S designates the integral in symmetrized sense

$$S\int_0^T g(x)\dot{x}\,dt = \lim_{\varepsilon \to 0}\sum_{n=1}^{N} g\left(\frac{x_n + x_{n-1}}{2}\right)\left(\frac{x_n - x_{n-1}}{\varepsilon}\right)\varepsilon. \qquad (2.3.29)$$

We see that we must use the concept of the stochastic integral if we want to apply (2.3.23) to a broad set of sample paths with measure 1. Thus

$$\frac{dP}{dP_W}[x(t)] = \exp\left\{ -\frac{1}{2K}S\int_0^T\left[g^2(x) - 2g(x)\dot{x} + K\frac{dg(x)}{dx} \right]dt \right\}.$$

$$(2.3.30)$$

Formula (2.3.30) is the particular case of a much more general result contained in Ghirsanov (1960), where the Ito stochastic integral is used in contrast to (2.3.30). Let us now find the pre-limit probability density $p[x(t)]$ corresponding to small ε. Applying (2.3.18) gives

$$p[x(t)] = p_W[x(t)]\exp\left\{ -\frac{1}{2K}\int_0^T *\left[g^2 - 2g\dot{x} + K\frac{dg}{dx} \right]dt \right\}.$$

$$(2.3.31)$$

or, if (2.3.11) is substituted,

$$p[x(t)] = \text{const} \cdot \exp\left\{ -\frac{1}{2} \int_0^T *\left[\frac{1}{K}(\dot{x} - g)^2 + \frac{dg}{dx} \right] dt \right\}. \tag{2.3.32}$$

Here we don't write the symmetrization symbol S in front of symbolic integral $\int *$ since the symmetrization is automatically implied by this notation if there are no other designations.

It should be noted that, unlike (2.3.3), (2.3.32), where the integral is interpreted as a sum, is inexact for any $\varepsilon > 0$. However, the less the value of ε (though we cannot let $\varepsilon = 0$), the higher is the accuracy of this formula. Hence we see in what sense we should understand the adjective 'functional' applied to the term 'probability density': it is necessary that ε should be non-zero, but it is desirable that it should be very small.

The derivation of (2.3.20) can be found in Stratonovich (1962), where the functional probability density for the process determined by a stochastic differential equation

$$\text{(S)} \qquad \dot{y} = g_1(y) + h(y)\xi(t) \tag{2.3.33}$$

(the equation in the S-sense) was also found. However, one should take into account that a more simple equation may be considered if a new variable x is introduced by

$$dx = \frac{dy}{h(y)}. \tag{2.3.34}$$

Then (2.3.33) is reduced to the form $\dot{x} = g_1(y(x))/h(y(x)) + \xi(t)$, i.e. to (2.3.17). Therefore the functional probability density $P[y(t)]$ can be obtained from (2.3.32) by a change of variables $x = x(y)$.

2.4 The non-Markov case of functional probability density when the random forces acting on the system are Gaussian

2.4.1 *Calculation of functional probability density in the general non-Markov case with Gaussian random forces acting on the system*

For the general, i.e. not only Markov, case the functional probability densities were obtained (Phythian, 1977). Now we show how to do this. The calculation is carried out in three parts.

First let the process $x(t) = \{x_\alpha(t), \alpha = 1, \dots, r\}$ be characterized by the set of differential equations (2.2.1), where $\xi_i(t)$ are Gaussian processes with zero mean and the correlation function

$$\langle \xi_i(t)\xi_j(t') \rangle = r_{ij}(t, t'), \tag{2.4.1}$$

$r_{ij}(t, t')$ is independent of x. The functions r_{ij} may also be of the δ-form. Let us assume that for every x matrix $h_{\alpha i}(x)$ is non-degenerate, i.e. it has the inverse matrix

$$\| h_{i\alpha}^{-1}(x) \| = \| h_{\alpha i}(x) \|^{-1}. \tag{2.4.2}$$

Let us now find the functional probability density $p[x(t)]$, i.e., as is pointed out above, the multidimensional probability density $p(x(t_1), \ldots, x(t_N))$ corresponding to ε-division (see (2.3.2)) for very small (as small as desired) ε. For ε-division, (2.2.1) should be written as a difference equation

$$\frac{x_\alpha(t_n) - x_\alpha(t_{n-1})}{\varepsilon} = g_\alpha(\lambda x(t_n) + (1-\lambda)x(t_{n-1}))$$

$$+ h_{\alpha i}(\lambda x(t_n) + (1-\lambda)x(t_{n-1}))\xi_{in}, \tag{2.4.3}$$

where

$$\xi_{in} = \frac{1}{\varepsilon} \int_{t_{n-1}}^{t_n} \xi_i(t) \, dt \tag{2.4.4}$$

($x_\alpha(0)$ is supposed to be non-random). In consequence of (2.4.3), all $x_\alpha(t_n)$ are, in essence, the functionals of $\xi_{\beta n}$, i.e.

$$x_\alpha(t_n) = X_{\alpha n}[\xi]. \tag{2.4.5}$$

The inverse dependence is written as

$$\xi_{in} = \Xi_{in}[x]. \tag{2.4.6}$$

Rearranging (2.4.3) to solve for ξ_{in}, we readily obtain this dependence:

$$\Xi_{in}[x] = h_{i\alpha}^{-1}(\tilde{x}_n) \left[\frac{x_\alpha(t_n) - x_\alpha(t_{n-1})}{\varepsilon} - g_\alpha(\tilde{x}_n) \right], \tag{2.4.7}$$

where

$$\tilde{x}_n = \lambda x(t_n) + (1-\lambda)x(t_{n-1}). \tag{2.4.8}$$

We easily find $p[x(t)]$ if the probability density for forces $p_\xi[\xi(t)]$ is known. Taking into account the routine rules of transformation of the probability density under a non-degenerate transformation of random variables, we have

$$p[x(t)] = p_\xi[\Xi[x(t)]] \frac{D\xi}{Dx}, \tag{2.4.9}$$

where $D\xi/Dx$ is the Jacobian of (2.4.7). This Jacobian is the inverse of the Jacobian of (2.4.5).

Secondly, we now find the explicit form of the functional $p_\xi[\xi]$. Since all ξ_{in} are Gaussian, this functional is of the form

$$p_\xi[\xi] = \text{const} \cdot \exp[-\tfrac{1}{2} R_{im,jn}^{-1} \xi_{im} \xi_{jn}], \tag{2.4.10}$$

where R^{-1} is the matrix inverse of the correlation matrix:

$$\| R_{im,jn}^{-1} \| = \| R_{im,jn} \|^{-1}. \tag{2.4.11}$$

Here

$$R_{im,jn} = \langle \xi_{im} \xi_{jn} \rangle.$$ (2.4.12)

In (2.4.10) all subscripts i, m, j, n, run through the appropriate values. Using (2.4.1), (2.4.4) and (2.4.12) gives

$$R_{im,jn} = \varepsilon^{-2} \int_{t_{m-1}}^{t_m} dt \int_{t_{n-1}}^{t_n} dt' \, r_{ij}(t, t').$$ (2.4.13)

If the functions $r_{ij}(t, t')$ are smooth, then for small ε (2.4.13) differs from $r_{ij}(t_m, t_n)$ very little.

It is convenient to introduce the function $s_{ji}(t', t)$ defined by

$$\int_0^T s_{ji}(t', t) v_i(t) \, dt = u_j(t')$$ (2.4.14)

provided that

$$\int_0^T r_{ij}(t, t') u_j(t') \, dt' = v_i(t).$$ (2.4.15)

Here $u_j(t')$ are arbitrary functions. From (2.4.14) and (2.4.15) it is seen that the kernel $r_{ij}(t, t')$ is the inverse of the kernel $s_{ji}(t', t)$ (and vice versa) in the continuum sense, while the matrices (2.4.11) and (2.4.12) are the inverse ones in the discrete sense. We assume that the inverse kernel $s_{ji}(t', t)$ exists. From (2.4.15) we have

$$\sum_n \int_{t_{m-1}}^{t_m} \int_{t_{n-1}}^{t_n} r_{ij}(t, t') u_j(t') \, dt \, dt' = \int_{t_{m-1}}^{t_m} v_i(t) \, dt.$$ (2.4.16)

If $u_j(t')$ and $v_i(t)$ do not change too fast, then for small ε they may be regarded as constant on every elementary interval in (2.4.16). This yields

$$\varepsilon^2 R_{im,jn} u_j(t_n) = \varepsilon v_i(t_m).$$ (2.4.17)

The analogous equation can be written for (2.4.14):

$$\varepsilon S_{jn,im} v_i(t_m) = u_j(t_n),$$ (2.4.18)

where

$$S_{jn,im} = \varepsilon^{-2} \int_{t_{n-1}}^{t_n} dt' \int_{t_{m-1}}^{t_m} dt \, s_{ji}(t', t).$$ (2.4.19)

Comparing (2.4.17) with (2.4.18) gives

$$\| R_{im,jn} \|^{-1} = \varepsilon^2 \| S_{jn,im} \|.$$ (2.4.20)

By virtue of (2.4.20) the probability density (2.4.10) assumes the form

$$p_\xi[\xi] = \text{const} \cdot \exp\left[-\tfrac{1}{2} \varepsilon^2 S_{jn,im} \xi_{jn} \xi_{im} \right].$$ (2.4.21)

For small ε the integral in (2.4.19) differs little from $s_{ji}(t_n, t_m)$, and the sum in

(2.4.21) coincides with the pre-limit sum corresponding to the integral

$$\int_0^T \int_0^T dt' \, dt \, s_{ji}(t', t)\xi_j(t')\xi_i(t). \tag{2.4.22}$$

Therefore, designating the pre-limit sum at small ε by $\int *$, we write (2.4.21) in the form

$$p_\xi[\xi(t)] = \text{const} \cdot \exp\left[-\frac{1}{2} \int_0^T \int_0^T *dt' \, dt \, s_{ji}(t', t)\xi_j(t')\xi_i(t) \right]. \tag{2.4.23}$$

Thereby the explicit form of the functional entering into (2.4.9) is obtained.

In the final part of our calculations, we busy ourselves with calculating the Jacobian $D\xi/Dx$. At first we find the inverse Jacobian

$$\frac{Dx}{D\xi} = \text{Det} \left\| \frac{\partial X_{\alpha m}}{\partial \xi_{jn}} \right\| \tag{2.4.24}$$

(see (2.4.5)). Using (2.4.3), we easily verify that the causality law holds: $x_\alpha(t_m)$ are independent of future values of ξ_{in}, $n > m$. This makes the calculations of the determinant (2.4.24) easy. It is equal to the product of 'small' determinants, each of which corresponds to fixed m:

$$\frac{Dx}{D\xi} = \prod_{m=1}^N \det \left\| \frac{\partial X_{\alpha m}}{\partial \xi_{jm}} \right\|. \tag{2.4.25}$$

Under the det operation the elements of the matrix differ from one another in that they have different values of α and j. To explain (2.4.25), we consider the situation when the matrix has the block form

$$A = \begin{pmatrix} B_{11} & 0 & 0 & \cdots & 0 \\ B_{21} & B_{22} & 0 & \cdots & 0 \\ \cdots & \cdots & \cdots & \cdots & \cdots \\ B_{N1} & B_{N2} & B_{N3} & \cdots & B_{NN} \end{pmatrix},$$

where B_{mn} are the 'small' matrices (in our case the elements of each of them differ in that they have different values of α and j), and 0 are the zero 'small' matrices. In this case

$$\text{Det } A = \prod_{m=1}^N \det B_{mm}. \tag{2.4.26}$$

Let us find the 'small' determinants. Multiplying (2.4.3) by ε and summing over n from one to m gives

$$X_{\alpha m} = x_{0\alpha} + \sum_{n=1}^m \varepsilon[g_\alpha(\lambda X_n + (1-\lambda)X_{n-1})$$
$$+ h_{\alpha i}(\lambda X_n + (1-\lambda)X_{n-1})\xi_{in}], \tag{2.4.27}$$

the initial condition $x_\alpha(0) = x_{a0} = x_{0\alpha}$ being used. Differentiating (2.4.27) with respect to ξ_{jm} and using the law of causality yields

$$\frac{\partial X_{\alpha m}}{\partial \xi_{jm}} = \varepsilon[g_{\alpha,\beta}(\tilde{X}_m)$$

$$+ h_{\alpha i,\beta}(\tilde{X}_m)\xi_{im}]\lambda \frac{\partial X_{\beta m}}{\partial \xi_{jm}} + \varepsilon h_{\alpha j}(\tilde{X}_m), \qquad (2.4.28)$$

where

$$g_{\alpha,\beta}(x) = \frac{\partial g_\alpha(x)}{\partial x_\beta}, \quad h_{\alpha i,\beta}(x) = \frac{\partial h_{\alpha i}(x)}{\partial x_\beta}, \qquad (2.4.29)$$

$$\tilde{X}_m = \lambda X_m + (1-\lambda) X_{m-1}.$$

Hence, we obtain

$$\{\delta_{\alpha\beta} - \varepsilon\lambda[g_{\alpha,\beta}(\tilde{X}_m) + h_{\alpha i,\beta}(\tilde{X}_m)\xi_{im}]\}\frac{\partial X_{\beta m}}{\partial \xi_{jm}} = \varepsilon h_{\alpha j}(\tilde{X}_m). \qquad (2.4.30)$$

Here, expressing ξ in terms of x, we can change the argument. In so doing, $X_{\alpha m}$ and $\xi_{\beta m}$ should be replaced by x and $\Xi_{\beta m}[x]$, respectively. Taking the 'small' determinant of both sides of (2.4.30), we get

$$\det\left\|\frac{\partial x_\alpha(t_m)}{\partial \xi_{jm}}\right\| = \varepsilon^r \det \|h_{\alpha j}(\tilde{x}_m)\|$$

$$\times \det^{-1}\|\delta_{\alpha\beta} - \varepsilon\lambda[g_{\alpha,\beta}(\tilde{x}_m) + h_{\alpha i,\beta}(\tilde{x}_m)\Xi_{im}\|. \qquad (2.4.31)$$

In (2.4.31) we will take into account that

$$\prod_m \det \|\delta_{\alpha\beta} - \varepsilon a_{\alpha\beta}\| = \exp\left[\sum_m \operatorname{tr}\{\ln\|\delta_{\alpha\beta} - \varepsilon a_{\alpha\beta}(m)\|\}\right], \qquad (2.4.32)$$

where tr is 'small' trace and $a_{\alpha\beta}(m)$ is a sequence of matrices independent of ε. Expanding $\ln\|\delta_{\alpha\beta} - \varepsilon a_{\alpha\beta}\|$ into a series and retaining the lowest terms gives

$$\ln\|\delta_{\alpha\beta} - \varepsilon a_{\alpha\beta}\| = -\varepsilon a_{\alpha\beta} - \tfrac{1}{2}\varepsilon^2 a_{\alpha\gamma} a_{\gamma\beta} - \cdots. \qquad (2.4.33)$$

Consequently

$$\operatorname{tr}\{\ln\|\delta_{\alpha\beta} - \varepsilon a_{\alpha\beta}\|\} = -\varepsilon a_{\alpha\alpha} - \tfrac{1}{2}\varepsilon^2 a_{\alpha\gamma} a_{\gamma\alpha} - \cdots, \qquad (2.4.34)$$

and also

$$(2.4.32) = \exp\left[-\sum_m \varepsilon a_{\alpha\alpha}(m) - \frac{\varepsilon^2}{2}\sum_m a_{\alpha\gamma}(m) a_{\gamma\beta}(m) - \cdots\right]. \qquad (2.4.35)$$

Therefore after substituting (2.4.31) into (2.4.25) we arrive at

$$\frac{Dx}{D\xi}[x] = \text{const}\cdot \prod_{m=1}^N \det\|h_{\alpha\beta}(\tilde{x}_m)\|$$

$$\times \exp\left\{\lambda \sum_{m=1}^N \varepsilon[g_{\alpha,\alpha}(\tilde{x}_m) + h_{\alpha i,\alpha}(\tilde{x}_m)\Xi_{im}]\right\}. \qquad (2.4.36)$$

Terms of higher orders in ε are omitted. The sum appearing in (2.4.36) can also be written by using the symbolic integral. Substituting (2.4.23) and (2.4.36) into (2.4.9) and considering (2.4.7) and (2.4.8) yields

$$p[x(t)] = \text{const} \cdot \prod_{m=1}^{N} \det{}^{-1} \| [h_{\alpha j}(\tilde{x}_m)]_\lambda \|$$

$$\times \exp\left\{ -\frac{1}{2} \int_0^T \int_0^T *dt\, dt'\, s_{ij}(t,t') \right.$$

$$\times [h_{i\alpha}^{-1}(x)(\dot{x}_\alpha - g_\alpha(x))h_{j\beta}^{-1}(x')(\dot{x}_\beta - g_\beta(x'))]_\lambda$$

$$\left. - \lambda \int_0^T *dt\, [g_{\alpha,\alpha}(x) + h_{\alpha i,\alpha}(x)h_{i\gamma}^{-1}(x)(\dot{x}_\gamma - g_\gamma(x))] \right\}.$$

$$(2.4.37)$$

Here $x' = x(t')$, and subscript λ indicates that we take the expression $\lambda x(t_n) + (1 - \lambda)x(t_{n-1})$ as the argument of functions $h_{i\alpha}^{-1}$, g_α entering into the pre-limit sum. The exact value of λ is non-essential in the second integral; therefore λ is not included.

2.4.2 The functional probability density in the Markov case

Formula (2.4.37) can also be applied in the Markov case when the functions $r_{ij}(t, t')$ in (2.4.1) have the δ-forms:

$$r_{ij}(t,t') = N_{ij}\delta(t - t').$$

$$(2.4.38)$$

If the stochastic equation (2.2.1) is interpreted in the S-sense, then one should set $\lambda = \frac{1}{2}$ in (2.4.3) and (2.4.37). We will not write λ if its value is equal to $\frac{1}{2}$.

For (2.4.38) the inverse kernel $s_{ij}(t, t')$ is found trivially as

$$s_{ij}(t,t') = N_{ij}^{-1}\delta(t - t')$$

$$(2.4.39)$$

The matrix

$$s_{ij}(t,t')h^{-1}(x)h_{j\beta}^{-1}(x) = N_{ij}^{-1}h_{i\alpha}^{-1}(x)h_{j\beta}^{-1}(x)\delta(t - t')$$

$$(2.4.40)$$

$$= b_{\alpha\beta}^{-1}(x)\delta(t - t'),$$

$$(2.4.41)$$

where $b_{\alpha\beta}^{-1}(x)$ is the matrix inverse of (2.2.30). Therefore in the Markov case, for $\lambda = \frac{1}{2}$, (2.4.37) can be written as

$$p[x(t)] = \text{const} \cdot \prod_{n=1}^{N} \det{}^{-1} \| h_{\alpha i}(\tilde{x}_n) \| \cdot \exp\left\{ -\frac{1}{2} \int_0^T *dt \right.$$

$$\left. \times [b_{\alpha\beta}^{-1}(\dot{x}_\alpha - g_\alpha)(\dot{x}_\beta - g_\beta) + g_{\alpha,\alpha} + h_{\alpha i,\alpha}h_{i\gamma}^{-1}(\dot{x}_\gamma - g_\gamma)] \right\}.$$

$$(2.4.42)$$

The vector g_α is related to drifts a_α appearing in the Fokker–Planck equation

$$\dot{p}(x) = -\frac{\partial}{\partial x_\alpha}[a_\alpha(x)p(x)] + \frac{1}{2}\frac{\partial^2}{\partial x_\alpha \partial x_\beta}[b_{\alpha\beta}(x)p(x)] \qquad (2.4.43)$$

by (2.2.21) or by

$$g_\alpha = a_\alpha - \tfrac{1}{2}h_{\alpha i,\beta}h_{\beta i} \qquad (2.4.44)$$

(see (2.2.22)).

For the one-component case the functionals obtained by substituting (2.4.44) into (2.4.42) were found (Stratonovich, 1962).

It should be noted that, in the Markov case, (2.4.36) and (2.4.37) are not quite correct. In fact, the term $-(\varepsilon^2/2)a_{\alpha\gamma}a_{\gamma\alpha}$ should also be considered in (2.4.34). The expression

$$\varepsilon^2 \Xi_{im}\Xi_{jm} = h_{i\gamma}^{-1}[x_\gamma(t_m) - x_\gamma(t_{m-1}) - \varepsilon g_\gamma]h_{j\delta}^{-1}$$
$$\times [x_\delta(t_m) - x_\delta(t_{m-1}) - \varepsilon g \delta] \qquad (2.4.45)$$

contains the term with the product $[x_\gamma(t_m) - x_\gamma(t_{m-1})][x_\delta(t_m) - x_\delta(t_{m-1})]$, which is of the order of ε (not ε^2) for typical sample path $x(t)$ in the Markov case. Therefore this term entering into $-(\varepsilon^2/2)a_{\alpha\gamma}a_{\gamma\alpha}$ must be considered, and we will have

$$\varepsilon^2 \sum_m a_{\alpha\gamma}(m)a_{\gamma\alpha}(m) = \lambda^2 \sum_m h_{\alpha i,\beta}(\tilde{x}_m)h_{\beta j,\alpha}(\tilde{x}_m)h_{i\gamma}^{-1}h_{j\delta}^{-1}\Delta x_\gamma \Delta x_\delta. \quad (2.4.46)$$

Substituting (2.4.46) into (2.4.35) and writing the new term in (2.4.36), instead of (2.4.42), for $\lambda = \tfrac{1}{2}$, we obtain the following functional probability density:

$$p[x(t)] = \text{const} \cdot \prod_{n=1}^{N} \det^{-1}\|h_{\alpha i}(\tilde{x}_n)\| \cdot \exp\left\{ -\frac{1}{2}\int_0^T *dt \right.$$

$$\times [b_{\alpha\beta}^{-1}(\dot{x}_\alpha - g_\alpha)(\dot{x}_\beta - g_\beta) + g_{\alpha,\alpha} + h_{\alpha i,\alpha}h_{i\gamma}^{-1}(\dot{x}_\gamma - g_\gamma)]$$

$$\left. -\frac{1}{8}\int_0^T *h_{\alpha i,\beta}h_{\beta j,\alpha}h_{i\gamma}^{-1}h_{j\delta}^{-1}\,dx_\gamma\,dx_\delta \right\}. \qquad (2.4.47)$$

Each cofactor determinant in (2.4.47) is easily expressed in terms of $\det\|b_{\alpha\beta}\|$. Taking a determinant of both sides of (2.2.30), which, in matrix form, can be written as

$$\hat{b} = \hat{h}\hat{N}\hat{h}^{\mathrm{T}}, \qquad (2.4.48)$$

we get

$$\det \hat{h} = \text{const} \cdot \det^{1/2} \hat{b}. \qquad (2.4.49)$$

Substituting (2.4.44) and (2.4.49) into (2.4.42) and simplifying, we have

$$p[x(t)] = \text{const} \cdot \prod_{n=1}^{N} \det^{-1/2}\|b_{\alpha\beta}(\tilde{x}_n)\| \exp\left\{ -\frac{1}{2}\int_0^T *dt \right.$$

$$\times [b_{\alpha\beta}^{-1}(\dot{x}_\alpha - a_\alpha + \tfrac{1}{2}b_{\alpha\gamma,\gamma})(\dot{x}_\beta - a_\beta + \tfrac{1}{2}b_{\beta\delta,\delta}) + a_{\alpha,\alpha} - \tfrac{1}{4}b_{\alpha\beta,\alpha\beta}]$$

$$-\frac{1}{8}\int_0^T {}^*h_{\alpha i,\beta}h_{\beta j,\alpha}[h_{i\gamma}^{-1}h_{j\delta}^{-1}\,\mathrm{d}x_\gamma\,\mathrm{d}x_\delta - N_{ij}\,\mathrm{d}t]\bigg\}. \qquad (2.4.50)$$

The fluctuational process considered is completely described by the coefficient functions $a_\alpha(x)$, $b_{\alpha\beta}(x)$ which appear in the Fokker–Planck equation. It is natural to expect that the functional probability density $p[x(t)]$ must be expressed only in terms of them, but not in terms of functions $h_{\alpha i}(x)$, which have no independent significance. In (2.4.50) $h_{\alpha i}$ appear in the last integral under the exponent, and they cannot be excluded from this integral. We can assume that the last integral does not influence the results of the calculations in which the probability density (2.4.50) is used, i.e. this integral can be omitted. This will be corroborated below. Now we note that for any typical sample path $x(t)$ belonging to the set whose probability is 1 the following equation is valid:

$$\lim_{\varepsilon \to 0}\sum f_{ij}(x)h_{i\gamma}^{-1}h_{j\delta}^{-1}\,\Delta x_\gamma\,\Delta x_\delta = \int f_{ij}(x)N_{ij}\,\mathrm{d}t \qquad (2.4.51)$$

(f_{ij} are arbitrary functions), so that the last term in the exponent in (2.4.50) falls out.

For the case of (2.3.17) we obtain the functional (2.3.32) from (2.4.50). This corroborates (2.3.20), which was not proved in Section 2.3.2.

In the general multicomponent Markov case, for the matrix $b_{\alpha\beta}$ depending on x, the functional probability density is given in Graham (1973, 1977, 1978) and Langouche, Roekaerts and Tirapegui (1979a–c) and elsewhere.

2.4.3 Transformational properties of functional probability density

Let us prove that the functional probability density (2.4.37) taken at $\lambda = \tfrac{1}{2}$ has the correct transformational properties when we make the inertialess curvilinear change of variables $x \to y = y(x)$. The probability density

$$p[x(t)] = p(x(t_1),\dots,x(t_N)) \qquad (2.4.52)$$

must be, naturally, transformed as

$$\bar{p}[y(t)] = \prod_{n=1}^{N}\left[\frac{\mathrm{D}x(y)}{\mathrm{D}y}\right]_{y=y(t_n)} \cdot p[x(y(t))], \qquad (2.4.53)$$

where $\mathrm{D}x/\mathrm{D}y$ is the Jacobian of the inertialess transformation.

In order to prove the correct transformational properties of the functional probability density, we consider how the quantities appearing in (2.2.1), i.e. in

$$\mathrm{d}x_\alpha = g_\alpha(x)\,\mathrm{d}t + h_{\alpha i}\xi_i(t)\,\mathrm{d}t, \qquad (2.4.54)$$

are transformed. The transformation does not concern the random forces $\xi_i(t)$.

We have

$$dy_\rho = c_{\rho\alpha} dx_\alpha, \tag{2.4.55}$$

where

$$c_{\rho\alpha} = \frac{\partial y_\rho}{\partial x_\alpha}. \tag{2.4.56}$$

In order that (2.4.53) should have the analogous form:

$$dy_\rho = \bar{g}_\rho(y) dt + \bar{h}_{\rho i}(y)\xi_i dt \tag{2.4.57}$$

in new coordinates, g_α and $h_{\alpha i}$ must have the same (i.e. vector) transformational properties as dx_α:

$$\bar{g}_\rho = c_{\rho\alpha} g_\alpha \tag{2.4.58a}$$

$$\bar{h}_{\rho i} = c_{\rho\alpha} h_{\alpha i}, \quad i = 1, \ldots, r. \tag{2.4.58b}$$

Inverting the matrix (2.4.58b) gives

$$\bar{h}_{i\rho}^{-1} = h_{i\alpha}^{-1} c_{\alpha\rho}^{-1}. \tag{2.4.59}$$

Using (2.4.55), (2.4.58a) and (2.4.59), we readily verify that the expression $h_{i\alpha}^{-1}(\dot{x}_\alpha - g_\alpha)$ is invariant since

$$\bar{h}_{i\rho}^{-1}(\dot{y}_\rho - \bar{g}_\rho) = h_{i\alpha}^{-1} c_{\alpha\rho}^{-1}(c_{\rho\beta}\dot{x}_\beta - c_{\rho\beta}g_\beta) = h_{i\alpha}^{-1}(\dot{x}_\alpha - g_\alpha). \tag{2.4.60}$$

Therefore the first integral in the exponent in (2.4.37) is also invariant (strictly speaking, it is scalar). It is convenient to rewrite the second integral in another form. Then from (2.4.37) taken for $\lambda = \frac{1}{2}$ we have

$$p[x(t)] = \text{const} \cdot \prod_{n=1}^{N} \det^{-1} \left\| h_{\alpha i}\left(\frac{x_n + x_{n-1}}{2}\right) \right\|$$

$$\times \exp\left\{ -\frac{1}{2} \int_0^T \int_0^T *dt\, dt'\, s_{ij}(t, t') h_{i\alpha}^{-1}(\dot{x}_\alpha - g_\alpha) h_{j\beta}^{-1} \right.$$

$$\times (\dot{x}_\beta - g_\beta) - \frac{1}{2} \int_0^T *dt\left[h_{\alpha i}\frac{\partial m_i}{\partial x_\alpha} + h_{\alpha i,\alpha} h_{i\gamma}^{-1}\dot{x}_\gamma \right] \right\},$$

$$(x_n = x(t_n)), \tag{2.4.61}$$

where

$$m_i(x) = h_{i\alpha}^{-1}(x) g_\alpha(x) \tag{2.4.62}$$

is scalar. We easily check the invariance of the expression $(\partial m_i/\partial x_\alpha) h_{\alpha i}$. In fact,

$$\frac{\partial f}{\partial y_\rho} = \frac{\partial f}{\partial x_\alpha} \cdot \frac{\partial x_\alpha}{\partial y_\rho} = \frac{\partial f}{\partial x_\alpha} c_{\alpha\rho}^{-1} \tag{2.4.63}$$

(f is arbitrary). Therefore, from (2.4.58b) and (2.4.63) we obtain

$$\frac{\partial m_i(y)}{\partial y_\rho} \bar{h}_{\rho i} = \frac{\partial m_i}{\partial x_\alpha} c_{\alpha\rho}^{-1} c_{\rho\beta} h_{\beta i} = \frac{\partial m_i}{\partial x_\alpha} h_{\alpha i}, \tag{2.4.64}$$

and consequently the first term in the second integral in (2.4.61) is scalar.

It remains to consider the second term and the determinant. Expanding $h(\tilde{x}_m)$ into a series in $\Delta x = x_n - x_{n-1}$ gives

$$h_{\alpha i}\left(\frac{x_n + x_{n-1}}{2}\right) = h_{\alpha i}\left(x_n - \frac{\Delta x}{2}\right)$$

$$= h_{\alpha i}(x_n) - h_{\alpha i,\gamma}\frac{\Delta x_\gamma}{2} + \cdots. \qquad (2.4.65)$$

Consequently,

$$\det \|h_{\alpha i}(\tilde{x}_n)\| = \det \|h_{\alpha i}(x_n)\|$$

$$\times \det \|\delta_{ji} - \tfrac{1}{2}h_{j\alpha}^{-1}h_{\alpha i,\gamma}\Delta x_\gamma + \cdots\|. \qquad (2.4.66)$$

Multiplying the expressions (2.4.66) taken for different n and applying (2.4.32)–(2.4.35) yields

$$\prod_{n=1}^N \det^{-1}\left\|h_{\alpha i}\left(\frac{x_n + x_{n-1}}{2}\right)\right\|$$

$$= \prod_{n=1}^N \det^{-1}\|h_{\alpha i}(x_n)\| \cdot \exp\left\{\frac{1}{2}\int_0^T *h_{i\alpha}^{-1}h_{\alpha i,\gamma}\,dx_\gamma\right\}. \qquad (2.4.67)$$

Substituting (2.4.67) into (2.4.61), we obtain

$$p[x(t)] = \text{const}\cdot \prod_{n=1}^N \det^{-1}\|h_{\alpha i}(x_n)\|$$

$$\times \exp\left\{-\frac{1}{2}\int_0^T\int_0^T *dt\,dt'\,s_{ij}(t,t')h_{i\alpha}^{-1}(x)\right.$$

$$\times (\dot{x}_\alpha - g_\alpha)h_{j\beta}^{-1}(x')(\dot{x}'_\beta - g'_\beta) - \frac{1}{2}\int_0^T *dt[h_{\alpha i}m_{i,\alpha}$$

$$\left. + (h_{\alpha i,\alpha}h_{i\gamma}^{-1} - h_{\alpha i,\gamma}h_{i\alpha}^{-1})\dot{x}_\gamma]\right\} \qquad (g'_\beta = g_\beta(x')). \qquad (2.4.68)$$

Now, the transformational properties (2.4.58b) of the matrix $h_{\alpha i}(x_n)$ ensure the appearance of the correct product of Jacobians in (2.4.53). In fact,

$$\prod_{n=1}^N \det^{-1}\|\bar{h}_{\rho i}(y_n)\| = \prod_{n=1}^N \det^{-1}\|c_{\rho\alpha}(y_n)\| \det^{-1}\|h_{\alpha i}(x_n)\|$$

$$= \prod_{n=1}^N \frac{Dx_n}{Dy_n}\det^{-1}\|h_{\alpha i}(x_n)\| \qquad (2.4.69)$$

((2.4.56) is also used). Now it remains to verify how the term with $\dot{x}_\gamma$ entering into the second integral in (2.4.68) is transformed. Substituting (2.4.55), (2.4.58b), (2.4.59) and (2.4.63) into the expression

$$\left(\frac{\partial \bar{h}_{\rho i}}{\partial y_\rho}\bar{h}_{i\sigma}^{-1} - \frac{\partial \bar{h}_{\rho i}}{\partial y_\sigma}\bar{h}_{i\rho}^{-1}\right)dy_\sigma, \qquad (2.4.70)$$

and after differentiation and cancellations, we have

$$(2.4.70) = \left(\frac{\partial h_{\alpha i}}{\partial x_\alpha} h_{i\gamma}^{-1} - \frac{\partial h_{\alpha i}}{\partial x_\gamma} h_{i\alpha}^{-1} \right) dx_\gamma$$

$$+ \left(\frac{\partial c_{\rho\gamma}}{\partial x_\beta} - \frac{\partial c_{\rho\beta}}{\partial x_\gamma} \right) c_{\beta\rho}^{-1} \, dx_\gamma. \tag{2.4.71}$$

But by virtue of (2.4.56)

$$\frac{\partial c_{\rho\gamma}}{\partial x_\beta} = \frac{\partial c_{\rho\beta}}{\partial x_\gamma} = \frac{\partial^2 y_\rho}{\partial x_\beta \, \partial x_\gamma}. \tag{2.4.72}$$

Therefore the expression considered is invariant. Thus, the necessary transformational properties of functional probability density (2.4.61) have been proved.

In some cases, for simplification of the expression, one can introduce new variables $\{y_i\}$ by equations

$$dy_i = h_{i\alpha}^{-1} \, dx_\alpha \tag{2.4.73}$$

(it is not always possible). In the new coordinate system the matrix $\widehat{\overline{h}}$ will be equal to the identity matrix, and instead of (2.4.37) we arrive at

$$p[y(t)] = \text{const} \cdot \exp \left\{ -\frac{1}{2} \int_0^T \int_0^T * dt \, dt' \, s_{ij}(t, t') [(\dot{y}_i - g_i(y)) \right.$$

$$\left. \times (\dot{y}_j - g_j(y))]_\lambda - \lambda \int_0^T * dt \, g_{i,i}(y) \right\}. \tag{2.4.74}$$

2.4.4   The dependence of the functional probability density on the discretization parameter λ

We discuss first the simple case of the Markov process. In the Markov case (2.4.3), for different λ, defines the process $x(t)$ having different statistical properties since drifts $a_\alpha(x)$ depend on λ. However, if $h_{\alpha i}$ are independent of x, then the statistical properties of the process are independent of λ. Therefore in this case the functional probability $p[x(t)]$, or, at least, the results of calculations with the use of $p[x(t)]$, must be independent of λ. We will verify this. For $h_{\alpha i}$ independent of x (2.4.47) becomes simpler, and we have

$$p[x(t)] = \text{const} \cdot \exp \left\{ -\frac{1}{2} \int_0^t * dt \, [b_{\alpha\beta}^{-1} (\dot{x}_\alpha - g_\alpha(x))(\dot{x}_\beta - g_\beta(x)) \right.$$

$$\left. + 2\lambda g_{\alpha,\alpha}(x)]_\lambda \right\}. \tag{2.4.75}$$

According to convention, $[\cdots]_\lambda$ here means that instead of $g_\alpha(x)$ we must take $g_\alpha(\lambda x_n + (1 - \lambda)x_{n-1})$ in the pre-limit sum (and analogously for $g_{\alpha,\alpha}$).

Expanding the right-hand side of

$$[g_\alpha(x)]_\lambda = g_\alpha\left(\frac{x_n + x_{n-1}}{2} + (\lambda - \tfrac{1}{2})\Delta x\right) \qquad (2.4.76)$$

into its Taylor series in $(\lambda - \tfrac{1}{2})\Delta x$ gives

$$[g_\alpha(x)]_\lambda = g_\alpha\left(\frac{x_n + x_{n-1}}{2}\right)$$

$$+ g_{\alpha,\beta}\left(\frac{x_n + x_{n-1}}{2}\right)(\lambda - \tfrac{1}{2})\Delta x_\beta + \cdots. \qquad (2.4.77)$$

Substituting (2.4.77) into (2.4.75), we omit the terms of the orders of $(\Delta x)^3$, $\varepsilon \Delta x$ and of higher orders. This gives

$$p[x(t)] = \text{const} \cdot \exp\left\{ -\frac{1}{2}\int_0^T *dt\, [b_{\alpha\beta}^{-1}(\dot{x}_\alpha - g_\alpha)(\dot{x}_\beta - g_\beta) + g_{\alpha,\alpha}]\right.$$

$$\left. + (\lambda - \tfrac{1}{2})\int_0^T *b_{\alpha\beta}^{-1} g_{\alpha,\gamma}(dx_\beta\, dx_\gamma - b_{\beta\gamma}\, dt)\right\}, \qquad (2.4.78)$$

where all $g_\alpha(x)$, $g_{\alpha,\gamma}(x)$ are to be understood in the symmetrized sense. It may be shown that the second integral in (2.4.78) containing $dx_\beta\, dx_\gamma - b_{\beta\gamma}\, dt$ does not influence the results of calculations made by using the functional probability density (2.4.78). Besides, for typical sample paths belonging to the set having measure 1, this integral disappears by virtue of the following equation:

$$\lim_{\varepsilon\to 0}\sum_n f(t_n)\Delta x_\alpha(t_n)\Delta x_\beta = \int f(t)b_{\alpha\beta}\, dt$$

($f(t)$ is arbitrary), which is equivalent to (2.4.51) (see also (2.2.10)).

Secondly, we discuss the relatively more complicated case of the non-Markov process. In the non-Markov case, when $\xi_i(t)$ are not delta-correlated, sample paths $x_\alpha(t)$ are smooth, and it becomes inessential which value of λ is taken in (2.4.3). This means that the statistics of the process $x(t)$ must be independent of λ, and the results of calculations made by using the functional probability density must be independent of λ. This independence will be demonstrated below. Now we reduce (2.4.37) to another form. We use (2.4.77) and also the analogous formula for $h_{\alpha i}$ and m_i:

$$[h_{\alpha i}(x)]_\lambda = h_{\alpha i}\left(\frac{x_n + x_{n-1}}{2}\right)$$

$$+ h_{\alpha i,\beta}\left(\frac{x_n + x_{n-1}}{2}\right)(\lambda - \tfrac{1}{2})\Delta x_\beta, \qquad (2.4.79a)$$

$$[m_i]_\lambda = m_i + (\lambda - \tfrac{1}{2})m_{i,\beta}\Delta x_\beta. \qquad (2.4.79b)$$

We substitute (2.4.77) and (2.4.79a, b) into (2.4.37), which, by use of (2.4.62), can be transformed to the form

$$p[x(t)] = \text{const} \cdot \prod_{n=1}^{N} \det^{-1} \| [h_{\alpha i}(\tilde{x}_n)]_{\lambda} \|$$

$$\times \exp\left\{ -\frac{1}{2} \int_0^T \int_0^T *dt\, dt'\, s_{ij}(t,t') \right.$$

$$\times [h_{i\alpha}^{-1}(x)h_{j\beta}^{-1}(x')\dot{x}_\alpha \dot{x}'_\beta - 2m_i(x)h_{j\beta}^{-1}(x')\dot{x}'_\beta + m_i(x)m_j(x')]_\lambda$$

$$\left. -\lambda \int_0^T *dt\, [h_{\alpha i}m_{i,\alpha} + h_{\alpha i,\alpha}h_{i\gamma}^{-1}\dot{x}_\gamma] \right\}. \tag{2.4.80}$$

The determinant in (2.4.80) should not be transformed according to (2.4.79a), but by using the expansion

$$[h_{\alpha i}(x)]_\lambda = h_{\alpha i}(x_n + (\lambda - 1)\Delta x) = h_{\alpha i}(x_n) + (\lambda - 1)h_{\alpha i,\beta}(x_n)\Delta x_\beta.$$

By analogy with (2.4.66) and (2.4.67) we obtain

$$\prod_{n=1}^{N} \det^{-1} \| [h_{\alpha i}(\tilde{x}_n)]_\lambda \| = \prod_{n=1}^{N} \det^{-1} \| h_{\alpha i}(x_n) \|$$

$$\times \exp\left[-(\lambda - 1) \int_0^T *dt\, h_{i\alpha}^{-1} h_{\alpha i,\gamma}\, dx_\gamma \right]. \tag{2.4.81}$$

The terms with $\Delta x_\gamma \Delta x_\delta$ are not considered here since they give zero contribution in the non-Markov case. A number of additional terms appear after the substitution of (2.4.77), (2.4.79a, b) and (2.4.81) into (2.4.80). We write down only the terms that will give non-zero contribution. The resultant expression, for which terms in the exponent are to be understood in the symmetrized sense, may be written as

$$p[x(t)] = \text{const} \cdot \prod_{n=1}^{N} \det^{-1} \| h_{\alpha i}(x_n) \|$$

$$\times \exp\left\{ -\frac{1}{2} \int_0^T \int_0^T *dt\, dt'\, s_{ij}(t,t') \right.$$

$$\times h_{i\alpha}^{-1}(x)h_{j\beta}^{-1}(x')(\dot{x}_\alpha - g_\alpha(x))(\dot{x}'_\beta - g_\beta(x'))$$

$$-\frac{1}{2} \int_0^T *dt\, [h_{\alpha i}m_{i,\alpha} + (h_{\alpha i,\alpha}h_{i\gamma}^{-1} - h_{\alpha i,\gamma}h_{i\alpha}^{-1})\dot{x}_\gamma]$$

$$\left. -(\lambda - \tfrac{1}{2})(Z_1 + \varepsilon Z_2) \right\}, \tag{2.4.82}$$

where

$$Z_1 = \int_0^T *dt\, [h_{\alpha i}m_{i,\alpha} + (h_{\alpha i,\alpha}h_{i\gamma}^{-1} + h_{i\alpha}^{-1}h_{\alpha i,\gamma})\dot{x}_\gamma], \tag{2.4.83a}$$

45

$$\varepsilon Z_2 = -\varepsilon \int_0^T \int_0^T *\mathrm{d}t\,\mathrm{d}t'\,s_{ij}(t,t')[m_{i,\alpha}(x)h_{j\beta}^{-1}(x')\dot{x}_\alpha\dot{x}'_\beta$$

$$+ h_{i\delta}^{-1}(x)h_{\delta k,\gamma}(x)h_{k\alpha}^{-1}(x)h_{j\beta}^{-1}(x')\dot{x}_\alpha\dot{x}_\gamma\dot{x}'_\beta]. \tag{2.4.83b}$$

Here we do take into account the term with $\Delta x_\alpha(t)\,\Delta x_\gamma(t)\,\Delta x_\beta(t')$, unlike in the simple Markov case. In obtaining this term we take into account that

$$\frac{\partial h_{i\alpha}^{-1}}{\partial x_\gamma} = -h_{i\delta}^{-1}h_{\delta k,\gamma}h_{k\alpha}^{-1}. \tag{2.4.84}$$

Subsequently we shall show that the terms entering into Z_1 and εZ_2 give the contributions that cancel each other out.

2.5   Using functional probability density for approximate calculation of conditional probability density

2.5.1   The general technique of calculating the conditional probability density by using the functional probability density

If we want to calculate the conditional probability density

$$p_{T0}(x(T)|x^0) = p(x(T)|x(0)), \tag{2.5.1}$$

i.e. the probability density for $x(T)$ under the condition that the values $x_\alpha(0) = x_\alpha^0$ are given, then, from the principal point of view, the recept of its calculation is simple: we must integrate the functional probability density

$$p[x(t)] = p(x(t_1), \ldots, x(t_N)|x^0) \tag{2.5.2}$$

with respect to the other variables:

$$p_{T0} = (x(T)|x^0) = \int \ldots \int \prod_{n=1}^{N-1} \prod_{\alpha=1}^r \mathrm{d}x_\alpha(t_n)p[x(t)]. \tag{2.5.3}$$

Of course, the exact evaluation of this integral of rather high multiplicity is possible only in the trivial case of the Gaussian process $x(t)$. In the other cases one should use approximate methods of the evaluation, which are based on the presence of some small or large parameter in $p[x(t)]$. We shall take the intensity of random forces $\xi_\alpha(t)$ appearing in (2.2.1) as a small parameter. We shall assume that $g_\alpha(x)$ and $h_{\alpha i}(x)$ are independent of κ and that

$$\langle \xi_i(t)\xi_j(t') \rangle = \kappa r_{ij}^0(t,t'), \tag{2.5.4}$$

where κ is a small parameter, and $r_{ij}^0(t,t')$ is independent of κ. By virtue of (2.5.4), for inverse kernel $s_{ji}(t',t)$ determining by using (2.4.14) and (2.4.15) we obtain

$$s_{ji}(t',t) = \kappa^{-1}s_{ji}^0(t',t), \tag{2.5.5}$$

where $s_{ji}^0(t',t)$ is independent of κ.

According to (2.2.30) and (2.4.38), in the Markov case we get

$$N_{ij} = \kappa N_{ij}^0, \quad b_{\alpha\beta}(x) = \kappa B_{\alpha\beta}(x), \quad b_{\alpha\beta}^{-1} = \kappa^{-1} B_{\alpha\beta}^{-1}, \tag{2.5.6}$$

where N_{ij}^0 and $B_{\alpha\beta}(x)$ are independent of κ. Substituting (2.5.5) into (2.4.68) gives

$$p[x(t)] = \text{const} \cdot \prod_{n=1}^{N} \det{}^{-1} \|h_{\alpha i}(x_n)\| \cdot \exp\{-\kappa^{-1} f_1[x(t)]$$

$$- f_2[x(t)]\}, \tag{2.5.7}$$

where

$$f_1[x(t)] = \frac{1}{2} \int_0^T \int_0^T *dt\, dt'\, s_{ij}^0(t, t') h_{i\alpha}^{-1}(x) h_{j\beta}^{-1}(x')$$

$$\times (\dot{x}_\alpha - g_\alpha)(\dot{x}_\beta' - g_\beta'), \tag{2.5.8a}$$

$$f_2[x(t)] = \frac{1}{2} \int_0^T *dt\, [m_{i,\alpha} h_{\alpha i} + (h_{\alpha i,\alpha} h_{i\gamma}^{-1} - h_{\alpha i,\gamma} h_{i\alpha}^{-1}) \dot{x}_\gamma]. \tag{2.5.8b}$$

We see that there are two terms of different orders in κ in the exponent on the right-hand side of (2.5.7), the term $-\kappa^{-1} f_1[x(t)]$ being more important. The second term with $f_2[x(t)]$ is of a lower order than the first one.

In the Markov case the functional probability density has qualitatively the same form as (2.5.7). Take, for instance, (2.4.78) corresponding to matrix $b_{\alpha\beta}$ independent of x. By use of (2.5.6) this formula can be shown to assume the form

$$p[x(t)] = \text{const} \cdot \exp\{-\kappa^{-1} f_1[x(t)] - f_2[x(t)]\}, \tag{2.5.9}$$

where

$$f_1[x] = f_1'[x] + \varepsilon f_1''[x], \tag{2.5.10a}$$

$$f_1'[x(t)] = \frac{1}{2} \int_0^T *dt\, B_{\alpha\beta}^{-1}(\dot{x}_\alpha - g_\alpha)(\dot{x}_\beta - g_\beta), \tag{2.5.10b}$$

$$\varepsilon f_1''[x(t)] = -\varepsilon(\lambda - \tfrac{1}{2}) \int_0^T *dt\, B_{\beta\alpha}^{-1} g_{\alpha,\gamma} \dot{x}_\beta \dot{x}_\gamma, \tag{2.5.10c}$$

$$f_2[x(t)] = \frac{1}{2} \int_0^T *dt\, g_{\alpha,\alpha} + (\lambda - \tfrac{1}{2}) \int_0^T *dt\, g_{\alpha,\alpha}. \tag{2.5.10d}$$

As in the exponent in (2.5.7) the term $\kappa^{-1} f_1[x(t)]$ is the main one.

Because of the presence of the term $\kappa^{-1} f_1[x(t)]$ in the exponent in (2.5.7) and (2.5.9), the integral (2.5.3) may be evaluated by using the multidimensional saddle-point method described in the appendix to this chapter. The extremum path $x_\alpha^1(t)$ minimizing the functional $f_1[x(t)]$ will be regarded as a point x^1 of multidimensional space where $f_1(x)$ has a minimum. In so doing, (2.7.15) should be written in the form

$$\frac{\partial f_1[x]}{\partial x_\alpha(t_n)} = 0 \quad \text{or} \quad \frac{\delta f_1[x(\cdot)]}{\delta x_\alpha(t)} = 0. \tag{2.5.11}$$

47

Substituting (2.5.7) into (2.5.3) and using formula (2.7.17) of the appendix for

$$G_{aJ} = h_{\alpha i}(x(t))\delta(t - t') \quad (a = (\alpha, t), J = (i, t')), \tag{2.5.12}$$

we get

$$p_{T0}(x(T)|x_0) = \text{const} \cdot \det^{-1} \|h_{\alpha i}(x(T))\|$$

$$\times \text{Det}^{-1/2} \left\| h_{i\alpha}^T(x_m^1) \left[\frac{\partial^2 f_1[x]}{\partial x_{\alpha m} \partial x_{\beta n}} \right]_{x^1(t)} h_{\beta j}(x_n^1) \right\|$$

$$\times \exp\{-\kappa^{-1} f_1[x^1(t)] - f_2[x^1(t)]\}$$

$$\times [1 + O(\kappa)], \tag{2.5.13}$$

where $x_{\alpha m} = x_\alpha(t_m)$. Summation over m and n is not made under the sign 'Det' in (2.5.13). It is essential that the terms that are implied by designation $O(\kappa)$ and that are not considered here preserve this order for any ε, i.e. each term of definite order in κ tends to a definite limit of the same order for $\varepsilon \to 0$. This can be proved by the analysis of typical terms which we have not written here. The multiplicative constant in (2.5.13) is determined from the normalization condition after other calculations.

Now we consider what form (2.5.11) takes in the case of the functional (2.5.8a). In order to find the functional derivative appearing in (2.5.11), it is advisable to substitute $x(t) = x^1(t) + y(t)$ into (2.5.8a) and select terms linear in $y(t)$. This gives

$$f_1[x^1 + y] = \int_0^T \int_0^T *dt\, dt'\, D_{\beta\gamma}(t, t')[\dot{y}_\beta(t) - g_{\beta,\delta}(x^1)y_\delta]$$

$$\times [\dot{x}_\gamma^1(t') - g_\gamma(x^1(t'))] - \int_0^T \int_0^T *dt\, dt'\, D_{\beta\gamma}(t, t')$$

$$\times h_{\beta k,\delta}(x^1(t))h_{k\alpha}^{-1}(x^1(t))[\dot{x}_\alpha^1(t) - g_\alpha(x^1(t))]$$

$$\times [\dot{x}_\gamma^1(t') - g_\gamma(x^1(t'))]y_\delta(t) + \cdots. \tag{2.5.14}$$

((2.4.84) is used). Here $\cdots$ denotes the terms of other orders in y, and

$$D_{\beta\gamma}(t, t') \equiv D_{\beta\gamma}(t, t', x^1(t), x^1(t'))$$

$$= s_{ij}^0(t, t')h_{i\beta}^{-1}(x^1(t))h_{j\gamma}^{-1}(x^1(t')) \tag{2.5.15}$$

The integral in (2.5.14) must be taken by parts:

$$\int_0^T *dt\, F(t)\dot{y}_\beta(t) = -\int_0^T *dt\, y_\beta(t)\frac{dF(t)}{dt}$$

$$+ F(T)y_\beta(T) - F(0)y_\beta(0)$$

$$\left\{ F(t) = \int_0^T *dt'\, D_{\beta\gamma}(t, t')[\dot{x}_\gamma^1(t') - g_\gamma(x^1(t'))] \right\}. \tag{2.5.16}$$

The last terms outside the integral have no influence on the desired functional derivative

$$\frac{\delta f_1[x(s)]}{\delta x_\alpha(t)} = \frac{\delta f_1[x^1 + y]}{\delta y_\alpha(t)}. \tag{2.5.17}$$

Using (2.5.14) and (2.5.16), we easily find this derivative. As a result, (2.5.11) will be of the form

$$\frac{d}{dt}\int_0^T dt'\, D_{\alpha\gamma}(t,t')[\dot{x}^1_\gamma(t') - g_\gamma(x^1(t'))]$$

$$+ \{g_{\beta,\alpha}(x^1(t)) + h_{\beta k,\alpha}(x^1(t))h_{kp}^{-1}(x^1(t))[\dot{x}^1_\rho(t) - g_\rho(x^1(t))]\}$$

$$\times \int_0^T dt'\, D_{\beta\gamma}(t,t')[\dot{x}^1_\gamma(t') - g_\gamma(x^1(t'))] = 0. \tag{2.5.18}$$

The integral in (2.5.16) and (2.5.18) can be assumed to be an ordinary one since for very small ε the pre-limit sum differs very little from the integral. Equation (2.5.18) is an integro-differential equation. In the Markov case the integrals are trivial, and (2.5.18) becomes a differential equation. In fact, in this case, as seen from (2.2.30), (2.4.39), (2.5.5), (2.5.6) and (2.5.15),

$$D_{\alpha\gamma}(t,t') = D_{\alpha\gamma}(x(t))\delta(t-t') \quad (D_{\alpha\gamma}(x) = B_{\gamma\alpha}^{-1}(x)). \tag{2.5.19}$$

Substituting (2.5.19) into (2.5.18) gives

$$\frac{d}{dt}\{D_{\alpha\gamma}(x)[\dot{x}_\gamma - g_\gamma(x)]\} - [\tfrac{1}{2}D_{\beta\gamma,\alpha}(x)(\dot{x}_\beta - g_\beta(x)$$

$$- D_{\beta\gamma}(x)g_{\beta,\alpha}(x)](\dot{x}_\gamma - g_\gamma(x)) = 0 \tag{2.5.20}$$

at $x(t) = x^1(t)$. In the Markov case the functional

$$f_1[x(t)] = \frac{1}{2}\int_0^T dt\, B_{\beta\gamma}^{-1}(x)(\dot{x}_\beta - g_\beta(x))(\dot{x}_\gamma - g_\gamma) \tag{2.5.21}$$

can be interpreted as the action

$$S = \int_0^T dt\, L(x, \dot{x}) \tag{2.5.22}$$

with the Lagrange function

$$L(x, \dot{x}) = \tfrac{1}{2}B_{\beta\gamma}^{-1}(x)(\dot{x}_\beta - g_\beta(x))(\dot{x}_\gamma - g_\gamma(x)). \tag{2.5.23}$$

In this case the equation for the extremal (2.5.20) will be just the Lagrange equation

$$\frac{d}{dt}\frac{\partial L}{\partial \dot{x}_\alpha} - \frac{\partial L}{\partial x_\alpha} = 0. \tag{2.5.24}$$

Equation (2.5.24) is the second-order differential equation, and therefore it requires two boundary conditions. In our case these conditions are quite obvious. Since in (2.5.3) the integration over $x(0)$ and $x(T)$ is not performed, these values should be regarded as fixed:

$$x_\alpha(0) = x_{0\alpha}, \quad x_\alpha(T) = x_{T\alpha}. \tag{2.5.25}$$

These conditions are the necessary boundary conditions.

2.5.2   Second derivative of $f_1[x]$ and corresponding determinant. The Markov case

We begin with the simple case when $f_1[x]$ is determined by (2.5.10a–c) taken for $\lambda = \frac{1}{2}$ and $B_{\alpha\beta} = D_{\alpha\beta}^{-1}$ independent of x. Selecting the quadratic terms in y gives

$$f_1[x^1(t) + y(t)] = \cdots + \frac{1}{2} \int_0^T *dt\, D_{\alpha\beta}[(\dot{y}_\alpha - g_{\alpha,\gamma}(x^1)y_\gamma)$$

$$\times (\dot{y}_\beta - g_{\beta,\delta}(x^1)y_\delta) - g_{\alpha,\gamma\delta}(x^1)(\dot{x}_\beta^1 - g_\beta(x^1))y_\gamma y_\delta], \tag{2.5.26}$$

where $\cdots$ denotes the terms of other orders in y. Writing the integral in (2.5.26) in the form of the pre-limit sum and considering only the quadratic terms in y, we arrive at

$$\frac{1}{2} \sum_{m,n} \frac{\partial^2 f_1[x]}{\partial x_{\alpha m} \partial x_{\beta n}} y_{\alpha m} y_{\beta n} = \frac{\varepsilon}{2} \sum_n D_{\alpha\beta} \left\{ \frac{y_{\alpha n} - y_{\alpha,n-1}}{\varepsilon} \frac{y_{\beta n} - y_{\beta,n-1}}{\varepsilon} \right.$$

$$- 2g_{\alpha,\gamma}(\tilde{x}_n^1) \frac{y_{\gamma n} + y_{\gamma,n-1}}{2} \frac{y_{\beta n} - y_{\beta,n-1}}{\varepsilon} + [g_{\alpha,\gamma}(\tilde{x}_n^1)g_{\beta,\delta}(\tilde{x}_n^1)$$

$$- g_{\alpha,\gamma\delta}(\tilde{x}_n^1)(\dot{x}_\beta^1 - g_\beta(\tilde{x}_n^1))]y_{\gamma n} y_{\delta n} \bigg\}. \tag{2.5.27}$$

From (2.5.27) we see that the second derivative of f_1 consists of three parts:

$$\frac{1}{\varepsilon} \left[\frac{\partial^2 f_1[x]}{\partial x_{\alpha m} \partial x_{\beta n}} \right]_{x=x^1} = Q_{\alpha m,\beta n} + P_{\alpha m,\beta n} + S_{\alpha m,\beta n}, \tag{2.5.28}$$

$Q_{\alpha m,\beta n}$ arising from the quadratic terms in $\dot{y}$, $R_{\alpha m,\beta n}$ arising from the linear terms in $\dot{y}$, and $S_{\alpha m,\beta n}$ arising from terms that do not contain $\dot{y}$. From (2.5.27) we have

$$\sum Q_{\alpha m,\beta n} y_{\alpha m} y_{\beta n} = 2\varepsilon^{-2} \left[\sum_{n=1}^{N-1} D_{\alpha\beta} y_{\alpha n} y_{\beta n} \right.$$

$$\left. - \sum_{n=2}^{N-1} D_{\alpha\beta} y_{\alpha n} y_{\beta,n-1} \right]. \tag{2.5.29}$$

The terms with x_0 and x_T do not appear in (2.5.29). If we introduce the designation

$$\gamma_{mn} = \begin{cases} 2 & \text{for} \quad m = n \\ -1 & \text{for} \quad |m - n| = 1 \\ 0 & \text{for} \quad |m - n| \geqslant 2, \end{cases} \tag{2.5.30}$$

then the matrix Q in (2.5.29) can be written as

$$Q_{\alpha m, \beta n} = \varepsilon^{-2} D_{\alpha\beta} \gamma_{mn}. \tag{2.5.31}$$

In (2.5.27) we now consider the terms linear in $\dot{y}$. We have

$$\sum P_{\alpha m, \beta n} y_{\alpha m} y_{\beta n} = -\varepsilon^{-1} \sum D_{\beta\alpha}[g_{\alpha,\gamma}(\tilde{x}_n^1) - g_{\alpha,\gamma}(\tilde{x}_{n+1}^1)] y_{\beta n} y_{\gamma n}$$
$$- \varepsilon^{-1} \sum [D_{\beta\alpha} g_{\alpha,\gamma}(\tilde{x}_n^1) - D_{\gamma\alpha} g_{\alpha,\beta}(\tilde{x}_n^1)] y_{\beta n} y_{\gamma, n-1}. \tag{2.5.32}$$

In consequence of the smoothness of functions $x^1(t)$ the terms with $y_{\beta n} y_{\gamma n}$ in (2.5.32) can be written as

$$D_{\beta\alpha} \left[\frac{\mathrm{d} g_{\alpha,\gamma}(x^1)}{\mathrm{d} t} \right]_{t = t_n} y_{\beta n} y_{\gamma n}. \tag{2.5.33}$$

Further, we denote

$$2 A_{\beta\gamma}(t_n) = D_{\beta\alpha} g_{\alpha,\gamma}(\tilde{x}^1(t_n)) - D_{\gamma\alpha} g_{\alpha,\beta}(\tilde{x}^1(t_n)), \tag{2.5.34a}$$

$$\varphi_{mn} = \begin{cases} -1 & \text{for} \quad m - n = 1, \\ 1 & \text{for} \quad m - n = -1, \\ 0 & \text{for} \quad |m - n| \neq 1. \end{cases} \tag{2.5.34b}$$

Then, as we see from (2.5.32) and (2.5.33), we obtain

$$P_{\alpha m, \beta n} = \varepsilon^{-1} A_{\alpha\beta}(t_n) \varphi_{mn} + \left\{ D_{\alpha\gamma} \frac{\mathrm{d} g_{\gamma,\beta}}{\mathrm{d} t}(x_n^1) \right\}_{\text{sym}} \cdot \delta_{mn}. \tag{2.5.35}$$

It should be noted that the matrix $P_{\alpha m, \beta n}$ is symmetrical since both $A_{\alpha\beta}$ and β_{mn} are antisymmetrical. The third matrix is found trivially:

$$S_{\alpha m, \beta n} = D_{\gamma\delta}[g_{\gamma,\alpha}(x_n^1) g_{\delta,\beta}(x_n^1) - g_{\gamma,\alpha\beta}(x_n^1)(\dot{x}_\delta^1 - g_\delta(x_n^1))] \delta_{mn}. \tag{2.5.36}$$

Because of the smoothness of the extremal function $x^1(t)$, we here write x_n^1 instead of $\tilde{x}_n^1 = (x_n^1 + x_{n-1}^1)/2$ and $\dot{x}^1$ instead of $(x_n^1 - x_{n-1}^1)/\varepsilon$. Thereby the matrix in (2.5.28) has been found.

In our case (2.7.14) gives

$$p_{T0}(x(T)|x_0) = \text{const} \cdot \text{Det}^{-1/2} \left\| \left[\frac{\partial^2 f_1[x]}{\partial x_{\alpha m} \partial x_{\beta n}} \right]_{x^1} \right\|$$
$$\times \exp\{ -\kappa^{-1} f_1[x^1(t)] - f_2[x^1(t)] \}. \tag{2.5.37}$$

Unfortunately there are no simple analytical methods of evaluating the determinants of the matrices in (2.5.28), (2.5.31), (2.5.35) and (2.5.36) when the function $x^1(t)$ is found beforehand. One can, however, resort to the numerical calculation of them for some fixed ε or the method based on finding eigenvalues of this matrix. The last method can be also applied in the continuum limit. The fact is that by use of symmetry of the matrix in (2.5.28) the real eigenvalues λ_i depending on x_0 and x_T exist. The determinant is expressed by them:

$$\text{Det}\,\|Q_{\alpha m,\beta n} + P_{\alpha m,\beta n} + S_{\alpha m,\beta n}\| = \prod_{i=0}^{N-1} \lambda_i(x_0, x_T). \qquad (2.5.38)$$

The eigenvalues λ_i are determined from the characteristic equation:

$$\sum_{n=1}^{N-1} (Q_{\alpha m,\beta n} + P_{\alpha m,\beta n} + S_{\alpha m,\beta n})u_{\beta n} = \lambda u_{\alpha m}. \qquad (2.5.39)$$

Substituting (2.5.31), (2.5.35) and (2.5.36) into (2.5.39), in the limit $\varepsilon \to 0$ we have the linear differential equation

$$\begin{aligned}
&-D_{\alpha\beta}\ddot{u}_\beta(t) + 2\mathrm{d}[A_{\alpha\beta}(t)u(t)]/\mathrm{d}t + \{\tfrac{1}{2}[D_{\alpha\gamma}\dot{g}_{\gamma,\beta}(x^1) \\
&+ D_{\beta\gamma}\dot{g}_{\gamma,\alpha}(x^1)] + D_{\gamma\delta}[g_{\gamma,\alpha}(x^1)g_{\delta,\beta}(x^1) \\
&- g_{\gamma,\alpha\beta}(\dot{x}_\delta^1 - g_\delta(x^1))]\}u_\beta(t) = \lambda u_\alpha(t). \qquad (2.5.40)
\end{aligned}$$

We assume that the eigenvalues are discrete and that a sequence of constants C_i independent of x_0 and x_T exists such that the limit

$$\lim_{l \to \infty} \prod_{i=1}^{l} [C_i \lambda_i(x_0, x_T)] \qquad (2.5.41)$$

exists. Denoting this limit by $F(x_0, x_T)$, we will have

$$\begin{aligned}
p_{T0}(x_T|x_0) = \text{const}\cdot[F(x_0, x_T)]^{-1/2} \\
\times \exp\{-\kappa^{-1}f_1[x^1(t)] - f_2[x^1(t)]\}. \qquad (2.5.42)
\end{aligned}$$

In the exponent in (2.5.42) we can pass to the limit $\varepsilon \to 0$, i.e. we can replace the symbolic integrals by ordinary ones in (2.5.10a–d). Thus, the finite value of ε will be absolutely eliminated from the result.

Now consider the case (2.5.10a–d) for arbitrary values of λ ($0 \leqslant \lambda \leqslant 1$). The term $\varepsilon f_1''[x]$ has no influence on (2.5.20) for small ε. The second derivative $\partial^2 f_1[x]/\partial x_{\alpha m}\,\partial x_{\beta n}$ is calculated by analogy with preceding discussion. Now additional terms appear in (2.5.28), namely,

$$\begin{aligned}
\varepsilon^{-1}\left[\frac{\partial^2 f_1[x]}{\partial x_{\alpha m}\,\partial x_{\beta n}}\right]_{x^1} = Q_{\alpha m,\beta n} + P_{\alpha m,\beta n} + S_{\alpha m,\beta n} \\
+ \varepsilon U_{\alpha m,\beta n} + \varepsilon V_{\alpha m,\beta n}, \qquad (2.5.43)
\end{aligned}$$

where

$$U_{\alpha m,\beta n} = -(\lambda - \tfrac{1}{2})\varepsilon^{-2}[D_{\alpha\gamma}g_{\gamma,\beta}(\tilde{x}_n^1) + D_{\beta\gamma}g_{\gamma,\alpha}(\tilde{x}_n^1)]\gamma_{mn}. \tag{2.5.44}$$

Here $V_{\alpha m,\beta n}$ contains the matrices of the same type as $P_{\alpha m,\beta n}$ and $S_{\alpha m,\beta n}$ do, and we need not consider them here.

The determinant of matrix (2.5.43) can be transformed as

$$\mathrm{Det}(\hat{Q} + \hat{P} + \hat{S} + \varepsilon\hat{U} + \varepsilon\hat{V}) = \mathrm{Det}(\hat{Q} + \hat{P} + \hat{S})$$
$$\times \mathrm{Det}\,[\hat{I} + \varepsilon(\hat{Q} + \hat{P} + \hat{S})^{-1}(\hat{U} + \hat{V})], \tag{2.5.45}$$

where $\hat{I}$ is the identity matrix ($I_{\alpha m,\beta n} = \delta_{\alpha\beta}\delta_{mn}$). Furthermore,

$$\mathrm{Det}[\hat{I} + \varepsilon(\hat{Q} + \hat{P} + \hat{S})^{-1}(\hat{U} + \hat{V})] = \mathrm{Det}[\hat{I} + \varepsilon(\hat{Q} + \hat{P} + \hat{S})^{-1}$$
$$\cdot(\hat{Q}\hat{Z} + \hat{P}\hat{Z} + \hat{S}\hat{Z} - \hat{P}\hat{Z} - \hat{S}\hat{Z} + \hat{V})], \tag{2.5.46}$$

where $\hat{Z} = \hat{Q}^{-1}\hat{U}$. It is obvious that

$$(2.5.46) = \mathrm{Det}[\hat{I} + \varepsilon\hat{Q}^{-1}\hat{U}$$
$$+ \varepsilon(\hat{Q} + \hat{P} + \hat{S})^{-1}(\hat{V} - \hat{P}\hat{Z} - \hat{S}\hat{Z})]. \tag{2.5.47}$$

Applying formulas of the type (2.4.34) to determinant (2.5.47) yields

$$\mathrm{Det}(\hat{Q} + \hat{P} + \hat{S} + \mu\hat{U} + \mu\hat{V}) = \mathrm{Det}(\hat{Q} + \hat{P} + \hat{S})\exp\{\varepsilon\,\mathrm{Tr}\,[\hat{Q}^{-1}\hat{U}]$$
$$+ \varepsilon\,\mathrm{Tr}[(\hat{Q} + \hat{P} + \hat{S})^{-1}(\hat{V} - \hat{P}\hat{Z} - \hat{S}\hat{Z})]\}. \tag{2.5.48}$$

We first consider the first term in the exponent on the right-hand side of (2.5.48). Using (2.5.31) and (2.5.44) gives

$$\varepsilon\,\mathrm{Tr}\{\hat{Q}^{-1}\hat{U}\} = -(\lambda - \tfrac{1}{2})\varepsilon\,\mathrm{Tr}\{\|D_{\delta\alpha}^{-1}[D_{\alpha\gamma}g_{\gamma,\beta} + D_{\beta\gamma}g_{\gamma,\alpha}]\gamma_{ml}^{-1}\gamma_{ln}\|\}$$
$$= -(\lambda - \tfrac{1}{2})2\varepsilon\sum_n g_{\beta,\beta}(x_n^1)$$
$$= -2(\lambda - \tfrac{1}{2})\int_0^T *\mathrm{d}t\,g_{\beta,\beta}(x^1(t)). \tag{2.5.49}$$

Let us now consider the second term in the exponent in (2.5.48). Let $G_{\alpha\beta}(t, t')$ be the kernel inverse of the kernel $Q_{\alpha\beta}(t, t') + P_{\alpha\beta}(t, t') + S_{\alpha\beta}(t, t')$, i.e. these kernels are connected by equations of the type (2.4.14) and (2.4.15). Using formulas of the type (2.4.20) gives

$$(\hat{Q} + \hat{P} + \hat{S})_{\alpha m,\beta n}^{-1} = \varepsilon^2 G_{\alpha\beta}(t_m, t_n). \tag{2.5.50}$$

Therefore
$$\mathrm{Tr}\{(\hat{Q} + \hat{P} + \hat{S})^{-1}(\hat{V} - \hat{P}\hat{Z} - \hat{S}\hat{Z})\}$$
$$= \int_0^T\int_0^T *\mathrm{d}t\,\mathrm{d}t'\,G_{\alpha\beta}(t, t')\{V_{\beta\alpha}(t', t) - (\lambda - \tfrac{1}{2})[P_{\beta\gamma}(t', t)$$
$$+ S_{\beta\gamma}(t', t)]D_{\gamma\rho}^{-1}[D_{\rho\sigma}g_{\sigma,\alpha}(x^1(t))]_{\mathrm{sym}}\}. \tag{2.5.51}$$

Here the matrix $\hat{Z} = \hat{Q}^{-1}\hat{U}$, diagonal in m and n, which we have explicitly

53

found earlier, is used. Substituting (2.5.51) into (2.5.48), we pass to the limit $\varepsilon \to 0$. In so doing the integral $\int\int *$ becomes the ordinary integral. The second term (2.5.51) in the exponent (2.5.48) vanishes since ε precedes the integral (2.5.51) in (2.5.48). Thus, using (2.5.48) and (2.5.49) gives

$$\mathrm{Det}^{-1/2}(\hat{Q} + \hat{P} + \hat{S} + \varepsilon\hat{U} + \varepsilon\hat{V}) = \mathrm{Det}^{-1/2}(\hat{Q} + \hat{P} + \hat{S})$$

$$\times \exp\left\{(\lambda - \tfrac{1}{2}) \int_0^T *\mathrm{d}t\, g_{\alpha,\alpha}(x^1(t))\right\}. \tag{2.5.52}$$

The integral in the exponent is the only term caused by $\varepsilon f_1''[x]$. It and the term $(\lambda - \tfrac{1}{2})\int_0^T *\mathrm{d}t\, g_{\alpha,\alpha}$ in (2.5.10d) cancel out. Therefore the result (2.5.37) is independent of λ. Analogously we can prove that in (2.4.50) the terms of the integral

$$-\frac{1}{8} \int_0^T *h_{\alpha i,\beta}h_{\beta j,\alpha}[h_{iy}^{-1}h_{j\delta}^{-1}\,\mathrm{d}x_y\,\mathrm{d}x_\delta - N_{ij}\mathrm{d}t]$$

give contributions that eventually cancel each other out. Here the terms with $\mathrm{d}x_y\,\mathrm{d}x_\delta$ should be included in $f_1[x]$, and the terms with $N_{ij}\mathrm{d}t$ should be included in $f_2[x]$, as we have done above.

The independence of the result of path integration from the mode of discretization was proved in Langouche, Roekaerts and Tirapegui (1979a–c) by using a diagram technique.

2.5.3   *Proof of the independence of the conditional probability density from the discretization parameter λ in the non-Markov case*

By virtue of (2.4.82) and (2.4.83a, b), (2.5.7) is valid for

$$f_1[x(t)] = f_1'[x(t)] + \varepsilon f_1''[x(t)], \tag{2.5.53a}$$

$$f_1'[x(t)] = \frac{1}{2} \int_0^T\int_0^T *\mathrm{d}t\,\mathrm{d}t'\, s_{ij}^0(t, t')h_{i\alpha}^{-1}(x)h_{j\beta}^{-1}(x')$$
$$\times [\dot{x}_\alpha - g_\alpha(x)][\dot{x}_\beta - g_\beta(x')], \tag{2.5.53b}$$

$$\varepsilon f_1''[x(t)] = -(\lambda - \tfrac{1}{2})\varepsilon \int_0^T\int_0^T *\mathrm{d}t\,\mathrm{d}t'\, s_{ij}^0(t, t')[m_{i,\alpha}(x)h_{j\beta}^{-1}(x')\dot{x}_\alpha\dot{x}_\beta$$
$$+ h_{i\delta}^{-1}(x)h_{\delta k,y}(x)h_{k\alpha}^{-1}(x)h_{j\beta}^{-1}(x')\dot{x}_\alpha\dot{x}_y\dot{x}_\beta], \tag{2.5.53c}$$

$$f_2'[x(t)] = \frac{1}{2} \int_0^T *\mathrm{d}t\,[h_{\alpha i}m_{i\alpha} + (h_{\alpha i,\alpha}h_{iy}^{-1} - h_{\alpha i,y}h_{i\alpha}^{-1})\dot{x}_y$$
$$- (\lambda - \tfrac{1}{2})Z_1]. \tag{2.5.53d}$$

Finding $\varepsilon^{-2}\partial^2 f_1'/\partial x_{\alpha m}\partial x_{\beta n}$ by using (2.5.53b), we find that this derivative contains the term in ε^{-2}, which we denote by $Q_{\alpha m, \beta n}$, and also the terms with

ε^{-1} and terms without ε, i.e.

$$\varepsilon^{-2}\left[\frac{\partial^2 f_1'[x]}{\partial x_{\alpha m}\,\partial x_{\beta n}}\right]_{x^1} = Q_{\alpha m,\beta n} + P_{\alpha m,\beta n}, \qquad (2.5.54)$$

where $P_{\alpha m,\beta n}$ denotes the sum of terms other than Q. We have

$$\sum Q_{\alpha m,\beta n} y_{\alpha m} y_{\beta n} = \varepsilon^{-2}\sum s_{ij}^0(t_m,t_n) h_{i\alpha}^{-1}(x_m^1) h_{j\beta}^{-1}(x_n^1)$$
$$\times (y_{\alpha m} - y_{\alpha,m-1})(y_{\beta n} - y_{\beta,n-1}). \qquad (2.5.55)$$

Introducing the matrix

$$\psi_{mn} = \begin{cases} 1 \text{ for } 1 \leqslant m = n \leqslant N-1 \\ -1 \text{ for } 1 \leqslant m = n+1 \leqslant N-1 \\ 0 \text{ for other values,} \end{cases} \qquad (2.5.56)$$

from (2.5.55) we find

$$Q_{\alpha m,\beta n} = \varepsilon^{-2}\psi_{mp} s_{ij}^0(t_p,t_q) h_{i\alpha}^{-1}(x_p^1) h_{j\beta}^{-1}(x_q^1)\psi_{qn}^{\mathrm{T}}$$
$$\approx \varepsilon^{-2} h_{i\alpha}^{-1}(x_m^1)\psi_{mp} s_{ij}^0(t_p,t_q)\psi_{qn}^{\mathrm{T}} h_{j\beta}^{-1}(x_n^1). \qquad (2.5.57)$$

We need not write down the terms appearing in $P_{\alpha m,\beta n}$. By analogy with (2.5.54) we represent the matrix $\varepsilon^{-2}\partial^2 f_1''[x]/\partial x_{\alpha m}\,\partial x_{\beta n}$ in the form

$$\varepsilon^{-2}\left[\frac{\partial^2 f_1''[x]}{\partial x_{\alpha m}\,\partial x_{\beta n}}\right]_{x^1} = U_{\alpha m,\beta n} + V_{\alpha m,\beta n}. \qquad (2.5.58)$$

The terms with ε^{-2} enter into $U_{\alpha m,\beta n}$. Selecting the quadratic terms in y that are proportional to ε^{-2} from $\varepsilon^{-2} f_1''[x^1 + y]$ yields

$$\tfrac{1}{2} U_{\alpha m,\beta n} y_{\alpha m} y_{\beta n} = -(\lambda - \tfrac{1}{2})\varepsilon^{-2} s_{ij}^0(t_m,t_n)\{h_{i\delta}^{-1}(x) h_{\delta k,\gamma}(x) h_{k\alpha}^{-1}(x)$$
$$\times h_{j\beta}^{-1}(x')[(y_{\alpha m} - y_{\alpha,m-1})(y_{\beta n} - y_{\beta,n-1})\dot{x}_{\gamma m}^1$$
$$+ (y_{\gamma m} - y_{\gamma,m-1})(y_{\beta n} - y_{\beta,n-1})\dot{x}_{\alpha m}^1$$
$$+ (y_{\alpha m} - y_{\alpha,m-1})(y_{\gamma m} - y_{\gamma,m-1})\dot{x}_{\beta n}^1]$$
$$+ m_{i\alpha}(x_m^1) h_{j\beta}^{-1}(x_n^1)(y_{\alpha m} - y_{\alpha,m-1})$$
$$\times (y_{\beta n} - y_{\beta,n-1})\}. \qquad (2.5.59)$$

Consequently

$$U_{\alpha m,\beta n} = U_{\alpha m,\beta n}' + U_{\alpha m,\beta n}'', \qquad (2.5.60a)$$

$$U_{\alpha m,\beta n}' = -(2\lambda - 1)\varepsilon^{-2}\psi_{mp} s_{ij}^0(t_p,t_q)\{h_{i\delta}^{-1}(x_p^1)[h_{k\alpha}^{-1}(x_p^1)$$
$$\times h_{\delta k,\gamma}(x_p^1) + h_{\delta k,\alpha}(x_p^1) h_{k\gamma}^{-1}(x_p^1)] h_{j\beta}^{-1}(x_q^1)\dot{x}_{\gamma p}^1$$
$$+ m_{\alpha i}(x_p^1) h_{j\beta}^{-1}(x_q^1)\}_{\mathrm{sym}}\psi_{qn}^{\mathrm{T}}. \qquad (2.5.60b)$$

Here $\{\cdots\}_{\text{sym}}$ denotes the symmetrization procedure of the type

$$\{m_{i\alpha}(x_p^1)h_{j\beta}^{-1}(x_q^1)\}_{\text{sym}} = \tfrac{1}{2}[m_{i\alpha}(x_p^1)h_{j\beta}^{-1}(x_q^1)$$
$$+ m_{i\beta}(x_q^1)h_{j\alpha}^{-1}(x_p^1)]. \tag{2.5.61}$$

$U''_{\alpha m, \beta n}$ in (2.5.60a) denotes the expression

$$U''_{\alpha m, \beta n} = -(\lambda - \tfrac{1}{2})\varepsilon^{-2}\sum_p s_{ij}^0(t_m, t_p)h_{j\gamma}^{-1}(x_p^1)\dot{x}_{\gamma p}h_{i\delta}^{-1}(x_m^1)$$
$$\times [h_{\delta k,\beta}(x_m^1)h_{k\alpha}^{-1}(x_m^1) + h_{\delta k,\alpha}(x_m^1)h_{k\beta}^{-1}(x_m^1)]\gamma_{mn}, \tag{2.5.62}$$

where γ_{mn} is the matrix (2.5.30). Expression (2.5.62) has a structure completely different from that of (2.5.60b), and we will ignore this expression for reasons that will be made clear later. Matrix (2.5.62) can be included in $V_{\alpha m, \beta n}$.

Now we consider the determinant entering into (2.5.13):

$$\text{Det} \left\| h_{i\alpha}^{\text{T}}(x_m^1)\left(\frac{\partial^2 f_1[x]}{\partial x_{\alpha m}\, \partial x_{\beta n}}\right)_{x^1} h_{\beta j}(x_n^1) \right\|$$
$$= \text{const}\cdot\text{Det} \|\tilde{Q}_{im,jn} + \tilde{P}_{im,jn} + \varepsilon\tilde{U}'_{im,jn} + \varepsilon\tilde{V}_{im,jn}\|, \tag{2.5.63}$$

where

$$\tilde{Q}_{im,jn} = Q_{\alpha m, \beta n}h_{\alpha i}(x_m^1)h_{\beta j}(x_n^1),$$
$$\tilde{U}'_{im,jn} = U'_{\alpha m, \beta n}h_{\alpha i}(x_m^1)h_{\beta j}(x_n^1), \tag{2.5.64}$$

etc. Using (2.5.57) and (2.5.60b) gives

$$\tilde{Q}_{im,jn} = \varepsilon^{-2}\psi_{mp}s_{ij}^0(t_p, t_q)\psi_{qn}^{\text{T}}, \tag{2.5.65}$$
$$\tilde{U}_{im,jn} = -(2\lambda - 1)\varepsilon^{-2}\psi_{mp}\{s_{kj}^0(t_p, t_q)[h_{k\delta}^{-1}(x_p^1)(h_{\delta i,\gamma}(x_p^1)$$
$$+ h_{\delta l,\alpha}(x_p^1)h_{\alpha i}(x_p^1)h_{l\gamma}^{-1}(x_p^1))\dot{x}_{\gamma p}$$
$$+ m_{k\alpha}(x_p^1)h_{\alpha i}(x_p^1)]\}_{\text{sym}}\psi_{qn}^{\text{T}}. \tag{2.5.66}$$

By analogy with (2.5.45)–(2.5.47), the multiplier $\text{Det}\|\tilde{Q}_{im,jn} + \tilde{P}_{im,jn}\|$ independent of λ is singled out from (2.5.63):

$$(2.5.63) = \text{const}\cdot\text{Det}\,(\hat{\tilde{Q}} + \hat{\tilde{P}})\cdot\text{Det}[\hat{I} + \varepsilon\hat{\tilde{Q}}^{-1}\hat{\tilde{U}}' + \varepsilon\hat{X}], \tag{2.5.67}$$

where

$$\hat{X} = (\hat{\tilde{Q}} + \hat{\tilde{P}})^{-1}(\hat{\tilde{V}} - \hat{\tilde{P}}\hat{\tilde{Q}}^{-1}\hat{\tilde{U}}'). \tag{2.5.68}$$

Applying (2.4.34) yields

$$\text{Det}^{-1/2}\left\| h_{i\alpha}^{\text{T}}(x_m^1)\left(\frac{\partial^2 f_1[x]}{\partial x_{\alpha m}\, \partial x_{\beta n}}\right)_{x^1} h_{\beta j}(x_n^1) \right\|$$
$$= \text{const}\cdot\text{Det}^{-1/2}(\hat{\tilde{Q}} + \hat{\tilde{P}})\cdot\exp\left[-\frac{\varepsilon}{2}\text{Tr}\{\hat{\tilde{Q}}^{-1}\hat{\tilde{U}}' + \hat{X}\}\right]. \tag{2.5.69}$$

The term $\varepsilon\,\text{Tr}\{\hat{\tilde{Q}}^{-1}\hat{\tilde{U}}'\}$ in the exponent is of special interest. Using (2.5.65) and

(2.5.66), we see that the matrices ψ_{mp}, ψ_{qn}^{T}, $s_{ij}^0(t_m, t_n)$ fall out, so that

$$\varepsilon \operatorname{Tr}\{\hat{Q}^{-1}\hat{U}'\} = -(2\lambda - 1)\varepsilon \sum_n \{h_{i\delta}^{-1}(x_n^1)[h_{\delta i,\gamma}(x_n^1)$$

$$+ h_{\delta l,\alpha}(x_n^1)h_{\alpha i}(x_n^1)h_{l\gamma}^{-1}(x_n^1)]\dot{x}_{\gamma n}^1 + m_{i,\alpha}(x_n^1)h_{\alpha i}(x_n^1)\}$$

$$= -(2\lambda - 1)\int_0^T *dt\{[h_{i\delta}^{-1}(x^1)h_{\delta i,\gamma}(x^1)$$

$$+ h_{\alpha l,\alpha}(x^1)h_{l\gamma}^{-1}(x^1)]\dot{x}_\gamma^1 + m_{i,\alpha}(x^1)h_{\alpha i}(x^1)\}. \tag{2.5.70}$$

Substituting (2.5.70) into (2.5.69) and applying (2.4.83a), (2.5.13) and (2.5.53d), we can see that the recently obtained terms with $\lambda - \frac{1}{2}$ and terms entering into $-(\lambda - \frac{1}{2})Z_1$ cancel out. Thus, the value of λ has no influence on the result. Other terms entering into (2.5.68) will be of lower order in ε and will vanish in the limit $\varepsilon \to 0$, as the expression (2.5.51) multiplied by ε does for $\varepsilon \to 0$. Matrix (2.5.62) has no influence on the result in the limit $\varepsilon \to 1$ since in the exponent in (2.5.69) $\operatorname{Tr}\{\hat{Q}^{-1}\hat{U}''\}$ is not the trace of the matrix diagonal in the subscripts m and n. Only the diagonal matrix is of importance since its trace is relatively large.

Thus, although the functional probability density (2.4.82) has the 'three-storeyed' form of (2.1.1), where

$$F_0[x(t)] = \int_0^T *dt \operatorname{tr}\{\ln \|h_{\alpha i}(x(t))\|\},$$

$$F_2[x(t)] = (\lambda - \frac{1}{2})\int_0^T\int_0^T *dt\,dt'\, s_{ij}(t,t')[m_{i,\alpha}(x)h_{i\beta}^{-1}(x')\dot{x}_\alpha\dot{x}_\beta$$

$$+ h_{i\delta}^{-1}(x)h_{\delta k,\gamma}(x)h_{k\alpha}^{-1}(x)h_{j\beta}^{-1}(x')\dot{x}_\alpha\dot{x}_\gamma\dot{x}_\beta'], \tag{2.5.71}$$

and the other terms enter into $F_1[x(t)]$, all 'storeys' correctly influence the result (2.5.13) that has a reasonable limit for $\varepsilon \to 0$.

2.6 Stationary probability density and approximate determination of the mean first-passage time

2.6.1 The solution of the stationary Fokker–Planck equation

For the broad class of stochastic processes the conditional probability density $p_{T0}(x_T|x_0)$ tends to the stationary probability density

$$\lim_{T \to \infty} p_{T0}(x_T|x_0) = p_0(x_T), \tag{2.6.1}$$

which is independent of x_0. These are known as mixing processes. Since $\partial p_0(x)/\partial T = 0$, from (2.4.43) we obtain the stationary Fokker–Planck equation for $p_0(x)$:

$$\frac{1}{2}\frac{\partial^2}{\partial x_\alpha \partial x_\beta}[b_{\alpha\beta}(x)p_0(x)] - \frac{\partial}{\partial x_\alpha}[a_\alpha(x)p_0(x)] = 0. \tag{2.6.2}$$

Sometimes we can find the explicit form of the stationary probability density $p_0(x)$ simply by solving (2.6.2). This simple solution is not always possible, only in the special case when the potentiality condition is fulfilled. To explain this, we denote

$$G_\alpha(x) = a_\alpha p_0 - \frac{1}{2}\frac{\partial}{\partial x_\beta}(b_{\alpha\beta}p_0), \tag{2.6.3}$$

and then (2.6.2) can be written in the form

$$\frac{\partial G_\alpha(x)}{\partial x_\alpha} = 0. \tag{2.6.4}$$

This equation certainly holds if $G_\alpha(x) = 0$ for all α and x, i.e. if

$$\frac{\partial(b_{\alpha\beta}p_0)}{\partial x_\beta} = 2a_\alpha p_0 \tag{2.6.5}$$

or

$$b_{\alpha\beta}\frac{\partial p_0}{\partial x_\beta} = \left(2a_\alpha - \frac{\partial b_{\alpha\beta}}{\partial x_\beta}\right)p_0. \tag{2.6.6}$$

Consequently,

$$\frac{\partial \ln p_0}{\partial x_\beta} = b_{\beta\alpha}^{-1}\left(2a_\alpha - \frac{\partial b_{\alpha\gamma}}{\partial x_\gamma}\right). \tag{2.6.7}$$

The potentiality condition amounts to the requirement that the right-hand side of (2.6.7) be the derivative of some function:

$$b_{\beta\alpha}^{-1}\left(2a_\alpha - \frac{\partial b_{\alpha\gamma}}{\partial x_\gamma}\right) = -\frac{\partial \Phi}{\partial x_\beta}. \tag{2.6.8}$$

Then from (2.6.7)

$$p_0(x) = \text{const} \cdot e^{-\Phi(x)}. \tag{2.6.9}$$

From (2.6.8) and (2.6.9), when the potentiality condition is fulfilled, the solution of (2.6.2) is in the form of a contour integral

$$p_0(x) = \text{const} \cdot \exp\left\{\int_{x^0}^{x} dx_\beta'\, b_{\beta\alpha}^{-1}(x')\left[2a_\alpha(x') - \frac{\partial b_{\alpha\gamma}}{\partial x_\gamma}(x')\right]\right\}. \tag{2.6.10}$$

The integral is independent of the integration contour. The choice of x^0 only influences the constant. Since the constant is determined from the normalization condition, the result is independent of x^0.

We easily verify that the solution (2.6.8)–(2.6.10) has the correct transformational properties. We represent the probability density in the form

$$p_0(x) = \det^{-1}\|h_{\alpha i}(x)\|e^{-\Psi(x)}, \tag{2.6.11}$$

where $\Psi = \Phi - \ln \det \hat{h}$. Substituting (2.2.22) into (2.6.8) and using

$$d(\ln \det \hat{h}) = \ln \det(\hat{h} + d\hat{h}) - \ln \det \hat{h} = \ln \det(\hat{I} + \hat{h}^{-1}\,d\hat{h})$$

$$= \mathrm{Tr}\{\ln(\hat{I} + \hat{h}^{-1}\,d\hat{h})\} = \mathrm{Tr}\{\hat{h}^{-1}\,d\hat{h}\},$$

we obtain from (2.6.9) and (2.6.11)

$$\frac{\partial \Psi}{\partial x_\alpha} = \frac{\partial \Phi}{\partial x_\alpha} - h_{iy}^{-1}\frac{\partial h_{yi}}{\partial x_\alpha} = -2b_{\alpha\beta}^{-1}g_\beta + h_{i\alpha}^{-1}\frac{\partial h_{yi}}{\partial x_y} - h_{iy}^{-1}\frac{\partial h_{yi}}{\partial x_\alpha}. \quad (2.6.12)$$

Since g_β is the vector and $b_{\alpha\beta}$ is the tensor with respect to curvilinear transformations, the expression $b_{\alpha\beta}^{-1}g_\beta$ has vector properties. The remaining expression on the right-hand side of (2.6.12) has already been considered in Section 2.4.3, where it was shown (formulas (2.4.70)–(2.4.72)) that it is transformed as a vector. Since the right-hand side of (2.6.12) is a vector, the derivative $\partial\Psi/\partial x_\alpha$ is also a vector. Consequently, $\Psi(x)$ is scalar. But $\det^{-1}\hat{h}$ is the scalar density, and therefore the probability density (2.6.11) must have the transformational properties of scalar density.

In the case of small fluctuations we have

$$b_{\alpha\beta}(x) = \kappa B_{\alpha\beta}(x), \quad (2.6.13)$$

where κ is a small parameter, and $B_{\alpha\beta}(x)$ are independent of κ. In this case (2.6.9) is of the form

$$p_0(x) = \mathrm{const}\cdot e^{-U(x)/\kappa}. \quad (2.6.14)$$

From (2.6.8) we get

$$\frac{\partial U}{\partial x_\beta} = -B_{\beta\alpha}^{-1}\left(2a_\alpha - \kappa\frac{\partial B_{\alpha y}}{\partial x_y}\right) \quad (2.6.15)$$

or, if we omit the term with κ,

$$\frac{\partial U}{\partial x_\beta} = -2B_{\beta\alpha}^{-1}a_\alpha. \quad (2.6.16)$$

Now the potentiality condition can be written as

$$\frac{\partial}{\partial x_y}(B_{\beta\alpha}^{-1}a_\alpha) = \frac{\partial}{\partial x_\beta}(B_{y\alpha}^{-1}a_\alpha). \quad (2.6.17)$$

When the potentiality condition is not fulfilled, one should use a more complex method. Substituting (2.6.14) into (2.6.2) (where we must put $b_{\alpha\beta} = \kappa B_{\alpha\beta}$) and discarding the terms of higher orders in κ, we obtain

$$B_{\alpha\beta}\frac{\partial U}{\partial x_\alpha}\frac{\partial U}{\partial x_\beta} - 2a_\alpha\frac{\partial U}{\partial x_\alpha} = 0. \quad (2.6.18)$$

Strictly speaking, (2.6.18) defines the function $U(x)$ in the equation

$$p_0(x) = \mathrm{const}\cdot\exp[-\kappa^{-1}(U + \kappa V + \cdots)] \quad (2.6.19)$$

We can solve (2.6.18) in the non-potential case finding several lower derivatives

$$u_{\alpha\beta} = \left(\frac{\partial^2 U}{\partial x_\alpha \partial x_\beta}\right)_{x=x_{st}}, \quad u_{\alpha\beta\gamma} = \left(\frac{\partial^3 U}{\partial x_\alpha \partial x_\beta \partial x_\gamma}\right)_{x_{st}}, \dots \tag{2.6.20}$$

taken at stable stationary point x_{st}, which is assumed to be unique. It is determined by the set of equations

$$a_\alpha(x) = 0, \quad \alpha = 1, \dots, r \tag{2.6.21}$$

and by the stability condition. To obtain linear algebraic equations determining the derivatives (2.6.20) one should represent functions U, a_α and $B_{\alpha\beta}$ in (2.6.18) by their Taylor series in $x - x_{st}$ and separately equate to zero the coefficients of quadratic, cubic, etc., powers of deviation $x - x_{st}$. Quadratic terms yield the equation

$$a_{\alpha,\sigma} v_{\sigma\beta} + a_{\beta,\sigma} v_{\sigma\alpha} = -B_{\alpha\beta}(x_{st}), \tag{2.6.22}$$

where $v_{\alpha\beta} = u_{\alpha\beta}^{-1}$. Cubic terms give

$$3\{a_{\alpha,\sigma} v_{\sigma\rho} u_{\rho\pi\eta} v_{\tau\beta} v_{\pi\gamma}\}_{\text{sym}} = 3\{a_{\alpha,\sigma\tau} v_{\sigma\beta} v_{\tau\gamma}\}_{\text{sym}} + 3\{B_{\alpha\beta,\sigma} v_{\sigma\gamma}\}_{\text{sym}}, \tag{2.6.23}$$

where $a_{\alpha,\sigma} = (\partial a_\alpha/\partial x_\sigma)_{x_{st}}$, etc.; symmetrization operations contain three terms. When we begin to solve (2.6.23), we note that the matrix $\hat{v}$ has been already found from (2.6.22), so that the right-hand side of (2.6.23) is known, and we have to solve linear equations. The analogous situation also takes place for other equations. This method of obtaining the derivatives in (2.6.20) is given in a number of works, for instance in Stratonovich (1985).

2.6.2   Using the functional probability density to find the stationary probability density

When the potentiality condition (2.6.17) is not fulfilled, besides the method of successive determination of derivatives (2.6.20), one can apply the method based on using the functional probability densities.

By virtue of (2.6.19), (2.6.1) assumes the form

$$\lim_{T\to\infty} p_{T0}(x_T|x_0) = \text{const} \cdot \exp\{-\kappa^{-1} U(x_T) - V(x_T) - \cdots\}. \tag{2.6.24}$$

Substituting (2.5.13) into the left-hand side of (2.6.24) gives

$$U(x_T) = \lim_{T\to\infty} f_1[x^1(t)]. \tag{2.6.25}$$

Here $f_1[x(t)]$ is the functional determined by (2.5.8a). In the Markov case $s_{ij}^0(t,t')h_{i\alpha}^{-1}(x)h_{j\beta}^{-1}(x) = B_{\alpha\beta}(x)\delta(t-t')$, and we have

$$f_1[x(t)] = \frac{1}{2}\int_0^T *dt\, B_{\alpha\beta}^{-1}(x)(\dot{x}_\alpha - g_\alpha)(\dot{x}_\beta - g_\beta). \tag{2.6.26}$$

After the substitution $x(t) = x^1(t)$ into (2.6.26) we can pass to the limit $\varepsilon \to 0$ and interpret the integral in the conventional sense. Consequently,

$$U(x_T) = \lim_{T \to \infty} \frac{1}{2} \int_0^T dt\, B_{\alpha\beta}^{-1}(x^1)(\dot{x}_\alpha^1 - g_\alpha(x^1))(\dot{x}_\beta^1 - g_\beta(x^1)). \quad (2.6.27)$$

The function g_α is associated with a_α by (2.2.21) (or by (2.2.22)). In the small noise approximation considered here we can take a_α instead of g_α. Using the Lagrange function (2.5.23), (2.6.27) can be written as

$$U(x_T) = \lim_{T \to \infty} \int_0^T L(x^1(t), \dot{x}^1(t))\, dt.$$

As well as the Lagrange function, it is convenient to introduce the Hamilton function

$$H(x, p) = p_\alpha \dot{x}_\alpha - L(x, \dot{x}). \quad (2.6.28)$$

On the right-hand side one should express $\dot{x}$ in terms of p and x by using the definition of momenta

$$p_\alpha = \frac{\partial L(x, \dot{x})}{\partial \dot{x}_\alpha}. \quad (2.6.29)$$

Applying the Hamilton function, instead of (2.5.24) we can consider the Hamilton equations

$$\dot{x}_\alpha = \frac{\partial H}{\partial p_\alpha}, \quad \dot{p}_\alpha = -\frac{\partial H}{\partial x_\alpha}. \quad (2.6.30)$$

Using these we can easily verify that $H(x, p)$ is constant in time in the stationary case, i.e. when L and H do not explicitly depend on time (in our case $g_\alpha(x)$ and $B_{\alpha\beta}(x)$ must be independent of time). Using (2.6.28) and the constancy of the Hamilton function, we transform (2.6.27) to

$$U(x_T) = \lim_{T \to \infty} \left\{ \int_0^T p_\alpha^1 \dot{x}_\alpha^1 - T H(x^1, p^1) \right\}. \quad (2.6.31)$$

Subsequently we will drop the superscript 1 on x and p.

Substituting (2.5.23) into (2.6.29), in our case we find

$$p_\alpha = B_{\alpha\beta}^{-1}(\dot{x}_\beta - g_\beta), \quad (2.6.32)$$

hence

$$\dot{x}_\beta = B_{\beta\alpha} p_\alpha + g_\beta. \quad (2.6.33)$$

Inserting (2.5.23) and (2.6.33) into (2.6.28) yields

$$H(x, p) = \tfrac{1}{2} B_{\alpha\beta} p_\alpha p_\beta + p_\alpha g_\alpha. \quad (2.6.34)$$

The boundary conditions for (2.6.30) are rather unusual for mechanics:

$$x(0) = x_0, \quad x(T) = x_T. \quad (2.6.35)$$

However, the limit (2.6.31) or (2.6.27) must be independent of x_0; therefore we must discard the condition $x(0) = x_0$, replacing it by the requirement that the time of motion must be equal to infinity. Let v be the modulus of velocity vector $\dot{x}_\alpha$, and dx be the modulus of vector dx_α. We have $dt = dx/v$. Integrating this equation gives

$$\int_{x_0}^{x_T} \frac{dx}{v} = T. \tag{2.6.36}$$

In order that this quantity is infinite for finite x_0 and x_T, it is necessary that the velocity $\dot{x}_\alpha$ (and modulus v) become zero at some point or points x_{rt} of the extremal function

$$\dot{x}_\alpha = 0 \quad \text{for} \quad x = x_{rt}. \tag{2.6.37}$$

Point (or points) x_{rt} is (are) called retarding point(s). The time of staying in the vicinity of the point x_{rt}, $x(t)$, must be infinite. Consider what contribution the vicinity of the point x_{rt} makes to the right-hand side of (2.6.27), i.e. in the integral

$$\frac{1}{2} \int_0^\infty dt\, B_{\alpha\beta}^{-1} (\dot{x}_\alpha - g_\alpha)(\dot{x}_\beta - g_\beta). \tag{2.6.38}$$

Because the system stays for an infinite length of time in the vicinity of x_{rt} and also by (2.6.37), this contribution will be equal to

$$\infty \cdot B_{\alpha\beta}^{-1} (0 - g_\alpha(x_{rt}))(0 - g_\beta(x_{rt})). \tag{2.6.39}$$

In order that this quantity is finite, the equation

$$g_\alpha(x_{rt}) = 0 \tag{2.6.40}$$

is necessary. We suppose that $B_{\alpha\beta}^{-1}$ is non-zero everywhere. By virtue of (2.6.37) and (2.6.40), from (2.6.32) we find that

$$p_\alpha = 0 \quad \text{for} \quad x = x_{rt}. \tag{2.6.41}$$

Equation (2.6.41) implies that the value of the Hamiltonian (2.6.34) is equal to zero at the retarding point. But the value of the Hamiltonian is constant along the path, therefore, for all moments of time:

$$H(x, p) = 0. \tag{2.6.42}$$

Substituting (2.6.42) into (2.6.31) and taking $-\infty$ as the initial moment of time gives

$$U(x_T) = \int_{-\infty}^T p_\alpha \dot{x}_\alpha\, dt = \int_{x_{st}}^{x_T} p_\alpha\, dx_\alpha. \tag{2.6.43}$$

Here we carry out the integration along the contour found by using (2.6.30). The solution of (2.6.30) yields the functions $x_\alpha(t)$, $p_\alpha(t)$, which serve as a parametric representation of the contour.

62

The initial point in phase space

$$x(-\infty) = x_{\text{st}}, \quad p(-\infty) = 0 \tag{2.6.44}$$

is a retarding point (other retarding points may also exist). In its vicinity the equations of motion (2.6.30), i.e. equations

$$\dot{x}_\alpha = B_{\alpha\beta}(x)p_\beta + g_\alpha, \quad \dot{p}_\alpha = -p_\beta \frac{\partial g_\beta(x)}{\partial x_\alpha} - \frac{1}{2}\frac{\partial B_{\beta\gamma}}{\partial x_\alpha} p_\beta p_\gamma \tag{2.6.45}$$

can be replaced by the equations

$$\dot{p} = \hat{A}^{\mathrm{T}} p, \quad \dot{\xi} = -\hat{A}\xi + \hat{D}p \tag{2.6.46}$$

linearized with respect to p and $\xi = x - x_{\text{st}}$. Here $\hat{D} = \|B_{\alpha\beta}(x_{\text{st}})\|$, $\hat{A} = -\|(\partial g_\alpha/\partial x_\beta)_{x=x_{\text{st}}}\|$. The real parts of the eigenvalues of matrix $\hat{A}$ are positive. The solution of (2.6.46) with initial conditions (2.6.44) is

$$p(t) = \varepsilon \exp(\hat{A}^{\mathrm{T}} t)n, \quad \xi(t) = \varepsilon\hat{M} \exp(\hat{A}^{\mathrm{T}} t)n, \tag{2.6.47}$$

where ε is a small quantity, $n = \{n_\alpha\}$ is the unit vector and $\hat{M}$ is the matrix satisfying equation $\hat{A}\hat{M} + \hat{M}\hat{A}^{\mathrm{T}} = \hat{D}$. The values (2.6.47) taken at $t = 0$ may be regarded as initial conditions for the non-linear equations (2.6.45). The unit vector must be selected so that path $x(t)$ passes through x_T.

We can also find the integration contour in (2.6.43) by solving the equations of motion in reverse time, i.e. equations

$$\dot{x}_\alpha = -B_{\alpha\beta}p_\beta - g_\beta, \quad \dot{p}_\alpha = \frac{\partial g_\beta}{\partial x_\alpha}p_\beta + \frac{1}{2}\frac{\partial B_{\beta\gamma}}{\partial x_\alpha} p_\beta p_\gamma. \tag{2.6.48}$$

If the point x_T is not the retarding point, we can take

$$p(0) = an, \quad x(0) = x_T \tag{2.6.49}$$

as the initial conditions for (2.6.48), where a is found from (2.6.42). Substituting (2.6.49) into (2.6.34) and (2.6.42) yields

$$a = -2n_\alpha g_\alpha(x_T)/B_{\alpha\beta}(x_T)n_\alpha n_\beta. \tag{2.6.50}$$

The unit vector n must be chosen such that $x(t) \to x_{\text{st}}$ for $t \to \infty$.

2.6.3 Approximate calculation of mean first-passage time

It is well known (see, for example, Stratonovich, 1963) that in the multicomponent Fokker–Planck case the mean time $m(x)$ of the first reaching of boundary Γ, when the initial point is x, satisfies the equation

$$\frac{1}{2}b_{\alpha\beta}(x)\frac{\partial^2 m(x)}{\partial x_\alpha \partial x_\beta} + a_\alpha(x)\frac{\partial m(x)}{\partial x_\alpha} = -1, \tag{2.6.51}$$

where

$$m(x) = 0 \quad \text{for } x \in \Gamma. \tag{2.6.52}$$

Equation (2.6.51) has a simple solution in the one-component case when it assumes the form

$$m'' + 2a/(\kappa B)m' = -2\kappa^{-1}B^{-1}. \tag{2.6.53}$$

Here $m' = \mathrm{d}m/\mathrm{d}x$, $m'' = \mathrm{d}^2m/\mathrm{d}x^2$, and we also set $b = \kappa B$. It is convenient to use the function

$$U(x) = -2\int_{x_{\mathrm{st}}}^{x}\frac{a(x')}{B(x')}\,\mathrm{d}x' \tag{2.6.54}$$

defined by analogy with (2.6.16). Here x_{st} is the root of the equation $\mathrm{d}a/\mathrm{d}x = 0$. Using (2.6.54), we can write (2.6.53) in the form

$$m'' - \kappa^{-1}U'm' = -2\kappa^{-1}B^{-1}. \tag{2.6.55}$$

We first consider the case when the point b plays the role of Γ, and the reflecting boundary, in whose approach we are not interested, is placed at the point $a < b$. Then for (2.6.55) the following boundary conditions

$$m'(a) = 0, \quad m(b) = 0 \tag{2.6.56}$$

are valid. Multiplying (2.6.55) by $\exp[-U(x)/\kappa]$ and performing integration, we get

$$m'(x) = -2\kappa^{-1}\exp[U(x)/\kappa]$$
$$\times \int_{a}^{x}\exp[-U(z)/\kappa]B^{-1}(z)\,\mathrm{d}z. \tag{2.6.57}$$

The integration limits are chosen so that the first condition in (2.6.56) is fulfilled. The right-hand side of (2.6.57) can be expressed by the stationary distribution

$$p_0(x) = CB^{-1}(x)e^{-U(x)/\kappa}, \tag{2.6.58}$$

which, as we easily verify, satisfies the stationary Fokker–Planck equation

$$\frac{\varepsilon}{2}(B(x)p_0)'' - (a(x)p_0)' = 0. \tag{2.6.59}$$

Using (2.6.58), we transform (2.6.57) to

$$m'(x) = -2(\kappa B(x)p_0(x))^{-1}\int_{a}^{x}\mathrm{d}z\,p_0(z). \tag{2.6.60}$$

Taking into account the second condition of (2.6.56), we perform the repeated integration:

$$m(x) = 2\kappa^{-1}\int_{x}^{b}\frac{\mathrm{d}y}{B(y)p_0(y)}\int_{a}^{y}\mathrm{d}z\,p_0(z). \tag{2.6.61}$$

From this precise formula we obtain the simpler approximate formula.

Suppose that $a(x) < 0$, i.e. $U'(x) > 0$ for $x_{st} < x < b$ and that $x_{st} - a$ and $b - x_{st}$ are much more than root-mean-square deviation

$$\sigma = \left[\int (x - x_{st})^2 p_0(x) \, dx \right]^{1/2}, \tag{2.6.62}$$

which is of the order of $\kappa^{1/2}$. The integral over y in (2.6.61) is mainly influenced by the region where the values of $b - y$ are very small (of the order of $\kappa^{1/2}$) and where the function

$$[B(y)p_0(y)]^{-1} = C^{-1} e^{U(x)/\kappa} \tag{2.6.63}$$

has a maximum. But

$$\int_a^y dz \, p_0(z) \approx 1 \tag{2.6.64}$$

for $y - x_{st} \gg \sigma$, $x_{st} - a \gg \sigma$ since $p_0(x)$ markedly differs from zero only at $x - x_{st} \sim \sigma$. Therefore from (2.6.61) we have

$$m(x) \approx 2\kappa^{-1} \int_x^b \frac{dy}{B(y)p_0(y)}. \tag{2.6.65}$$

This formula can be also derived by the method used in Kramers (1940). It was obtained by another method, for $B(x) = \text{const}$ and $x = x_{st}$, in Stratonovich (1967, formula (1.43)).

The point $x = x_{st}$ is of most interest in (2.6.65), since for small κ $x(t)$ stays in this vicinity almost all the time. The path $x(t)$ seldom reaches point b. The first passage through the point b is immediately followed by a series of other frequent passages. Then the path returns to the stable point x_{st}, this return being similar to the return according to equation $\dot{x}_\alpha = a_\alpha(x)$.

Under the condition $b - x_{st} \gg \sigma$, the appearances of the series of passages almost form the Poisson process. Let us consider a relatively large period of time T and denote the number of series appearing during T by n. The probabilities of various values of n are equal to

$$P_n = \frac{(\beta_1 T)^n}{n!} e^{-\beta_1 T}. \tag{2.6.66}$$

Using this probability, we easily find the mean number of series for a given time as

$$\langle n \rangle = \beta_1 T. \tag{2.6.67}$$

Thus, β_1 is the mean number of events per unit time. From (2.6.67) we find the mean time between the events as

$$\frac{T}{\langle n \rangle} = \frac{1}{\beta_1}.$$

This mean time is simply the mean first-passage time

$$\frac{1}{\beta_1} = m(x_{st}), \tag{2.6.68}$$

the initial point being x_{st}. By (2.6.65) we can write (2.6.68) in the form

$$\beta_1 = \kappa C \bigg/ 2 \int_{x_{st}}^{b} e^{U(y)/\kappa} \, dy, \tag{2.6.69}$$

where C is the normalization constant

$$C^{-1} = \int_a^c e^{-U(x)/\kappa} \frac{dx}{B(x)}. \tag{2.6.70}$$

It is sufficient to approximate the integral in (2.6.70). Expanding the function $- U(x)/\kappa - \ln B(x)$ into a series in $x - x_{st}$ and omitting cubic and higher order terms, we readily obtain an integral that is easily evaluated, whence

$$C = B(x_{st})e^{U(x_{st})/\kappa} \left[\frac{U''(x_{st})}{2\pi\kappa} \right]^{1/2} [1 + O(\kappa)]. \tag{2.6.71}$$

If the point a, not b, plays the role of boundary Γ, and the reflecting barrier is placed at b, then, as is easily seen, instead of (2.6.69) we will have the analogous formula

$$\beta_2 = \kappa C \bigg/ 2 \int_a^{x_{st}} e^{U(x)/\kappa} \, dy, \tag{2.6.72}$$

where C is defined by (2.6.71).

We now consider the combined case when both points a and b constitute the boundary Γ. The passes of $x(t)$ through a and b should be regarded as independent events when they are rare. Therefore the total event density must be equal to the sum of densities (2.6.69) and (2.6.72):

$$\beta = \beta_1 + \beta_2. \tag{2.6.73}$$

The function $U(x)$ increases for $x_{st} < x < b$ by virtue of the fact that $a(x) < 0$ for $x_{st} < x < b$, and $\exp[U(x)/\kappa]$ quickly increases when x increases. Therefore the integral

$$\int_{x_{st}}^{b} e^{U(y)/\kappa} \, dy \tag{2.6.74}$$

in (2.6.69) is chiefly influenced by values of the expression $\exp[U(y)/\kappa]$ in the vicinity of the point b. We substitute

$$U(x) = U(b) + 2\alpha(x - b) - \gamma(x - b)^2 + \cdots,$$

where

$$2\alpha = U'(b) = - 2a(b)/B(b) > 0, \quad \gamma = -\tfrac{1}{2}U''(b), \tag{2.6.75}$$

into (2.6.74), and assume that $\gamma > 0$. Discarding terms of higher orders than $(x - b)^2$, we can replace the lower limit by $-\infty$ because of the presence of the small parameter κ. We easily obtain

$$(2.6.74) = \frac{1}{2}\left(\frac{\pi\kappa}{\gamma}\right)^{1/2} \exp\left[\frac{U(b)}{\kappa} + \frac{\alpha^2}{\kappa\gamma}\right] \mathrm{erfc}[\alpha(\kappa\gamma)^{-1/2}]. \qquad (2.6.76)$$

By virtue of (2.6.71) and (2.6.76), (2.6.69) gives

$$\beta_1 = (2\pi)^{-1}[U''(x_{st})|U''(b)|]^{1/2} B(x_{st})$$

$$\times \frac{\exp(-\alpha^2(\kappa\gamma)^{-1})}{\mathrm{erfc}(\alpha(\kappa\gamma)^{-1/2})} \exp\left[-\frac{U(b) - U(x_{st})}{\kappa}\right]. \qquad (2.6.77)$$

In particular, if $\alpha/(\kappa\gamma) \gg 1$, then

$$\beta_1 = \left[\frac{U''(x_{st})}{2\pi\kappa}\right]^{1/2} \alpha B(x_{st}) \exp\left[-\frac{U(b) - U(x_{st})}{\kappa}\right]. \qquad (2.6.78)$$

Besides, for $\alpha = 0$ from (2.6.77) we have

$$\beta_1 = (2\pi)^{-1}[U''(x_{st})|U''(b)|]^{1/2} B(x_{st})$$

$$\times \exp\left[-\frac{U(b) - U(x_{st})}{\kappa}\right]. \qquad (2.6.79)$$

When $\alpha = 0$, $U''(b) < 0$, the upper point of the potential barrier separating the metastable state from the other states that correspond to minima of $U(x)$ has coordinate b. If the path $x(t)$ reaches the upper point b, then it descends in the previous potential well with probability $\frac{1}{2}$, and it descends the other side of the barrier with the same probability. Therefore the mean lifetime $\bar{T}_0$ of the metastable state is twice as large as $m(x_{st})$:

$$\bar{T}_0 = \frac{2}{\beta_1} = 2\pi \left[\frac{B(b)}{B(x_{st})}\right]^{1/2} [|a'(x_{st})|a'(b)]^{-1/2}$$

$$\times \exp\left[\frac{U(b) - U(x_{st})}{\kappa}\right]. \qquad (2.6.80)$$

β_2 can be written in the form analogous with (2.6.77)–(2.6.79). Now we consider the total frequency (2.6.73). If $|U(b) - U(a)| \gg \kappa$, then because of the factor $\exp\{[U - U(x_{st})]/\kappa\}$ one of the quantities β_1 or β_2 is much greater than the other. Discarding the smaller quantity, we arrive at

$$\beta = \max(\beta_1, \beta_2), \qquad (2.6.81a)$$

$$\bar{T} = m(x_{st}) = \frac{1}{\beta} = \exp\left\{\frac{U_{min} - U(x_{st})}{\kappa}[1 + O(\kappa)]\right\}, \qquad (2.6.81b)$$

where

$$U_{min} = \min(U(a), U(b)). \qquad (2.6.82)$$

Formula (2.6.81b) is not only valid in the one-component case, but it is more difficult to substantiate it in the multicomponent case. The function $U(x)$ in (2.6.81b) is defined by (2.6.16) or (2.6.43) in the multicomponent case. In the multicomponent case, the value of U in (2.6.81b) at the most accessible point of the boundary whose attainment is of interest is taken as U_{min}, i.e.

$$U_{min} = \inf_{x \in \Gamma} U(x). \tag{2.6.83}$$

Formula (2.6.81b) for the multicomponent case does not contradict the non-fully substantiated result of Landauer and Swanson (1961) and the results of Ryter (1985). Formula (2.6.81b) can also be written as

$$\ln \bar{T} = \ln \left[\frac{p_0(x_{st})}{(p_0)_{max}} \right] (1 + O(\kappa)), \tag{2.6.84}$$

where

$$(p_0)_{max} = \sup_{x \in \Gamma} p_0(x). \tag{2.6.85}$$

If we want to find the mean lifetime of the metastable state, then the hypersurface with the property

$$v_\alpha(x) \frac{\partial U(x)}{\partial x_\alpha} = 0 \quad \text{for } x \in \Gamma \tag{2.6.86}$$

should be taken as the boundary Γ, where $v_\alpha(x)$ is the unit vector normal to Γ at point x.

The 'lowest', i.e. saddle, point x_{sd} of the hypersurface Γ plays an important role since $U_{min} = U(x_{sd})$. The saddle point satisfies the same equations (2.6.40) as the stable point x_{st} and retarding point (x_{sd} is one of the retarding points).

We note that the presence of the small parameter κ is not necessary in the initial stochastic differential equations. In order that (2.6.81b) and (2.6.84) hold, it is only important that boundary Γ be sufficiently remote, so that it will be quite a rare event for the system to reach the boundary (i.e. the ratio $p_0(x_{st})/(p_0)_{max}$ should be very large). It is possible that (2.6.84) can be used in the non-Markov case.

2.7   Appendix. Evaluation of the M-fold multiple integral by the saddle-point method

Let us evaluate the integral

$$I = \int \prod_{a=1}^{M} dx_a \, \varphi(x) e^{-f(x)/\kappa}. \tag{2.7.1}$$

Here $x = (x_a, a = 1, \ldots, M)$. Applying the formulas derived below to integral (2.5.3), we should treat x_a as $x_\alpha(t_n)$, i.e. the pair (α, n) must play the role of subscript a; $M = (N - 1)r$. Let $\varphi(x)$ in (2.7.1) be the scalar density with respect to the transformation $y = y(x)$, and let $f(x)$ be scalar. Then the integral I is invariant under these transformations.

Suppose that the function $f(x)$ is differentiable a necessary number of times and has a unique minimum at the point x^1. The appropriate condition of extremum

$$\frac{\partial f(x)}{\partial x_a} = 0, \quad a = 1, \ldots, M \tag{2.7.2}$$

unambiguously determines x^1. We expand the function $\ln \varphi - f(x)/\kappa$ into its Taylor series at the minimum point. Applying (2.7.2) gives

$$\ln \varphi(x) - \kappa^{-1} f(x) = \ln \varphi(x^1) - \kappa^{-1} f(x^1) + \left(\frac{\varphi_a}{\varphi}\right)_{x^1} y_a$$

$$- \frac{1}{2\kappa}(f_{ab})_{x^1} y_a y_b + \frac{1}{2}\left(\frac{\varphi_{ab}}{\varphi} - \frac{\varphi_a \varphi_b}{\varphi^2}\right) y_a y_b$$

$$- \frac{1}{6\kappa}(f_{abc})_{x^1} y_a y_b y_c - \frac{1}{24\kappa}(f_{abcd})_{x^1} y_a y_b y_c y_d + \cdots, \tag{2.7.3}$$

where $y_\alpha = x_\alpha - x^1_\alpha$, and $(\cdots)_{x^1}$ means that the functions are taken at the point x^1. Subscripts of f and φ denote the corresponding derivatives. Equation (2.7.3) can be written as

$$\ln \varphi(x) - \kappa^{-1} f(x) = \ln \varphi(x^1) - \kappa^{-1} f(x^1)$$

$$- \frac{1}{2\kappa}(f_{ab})_{x^1} y_a y_b + D, \tag{2.7.4}$$

where D designates the sum of other terms in (2.7.3). Substituting (2.7.4) into (2.7.1) and expanding the exponent into a series in D yields

$$I = \varphi(x^1)e^{-f(x^1)/\kappa} \int \prod_{a=1}^{M} dx_a \exp\left[-\frac{1}{2\kappa}(f_{ab})_{x^1} y_a y_b\right]$$

$$\times [1 + D + \tfrac{1}{2}D^2 + \tfrac{1}{6}D^3 + \cdots]. \tag{2.7.5}$$

We introduce the Gaussian probability density

$$p(y) = (2\pi\kappa)^{-M/2} \det^{1/2} \|f_{ab}(x^1)\| \exp\left[-\frac{1}{2\kappa}f_{ab}(x^1)y_a y_b\right]. \tag{2.7.6}$$

The matrix $f_{ab}(x^1)$ is assumed to be positive definite. If the exponent in (2.7.5) is expressed by $p(y)$, the integral in (2.7.5) will simply be the integral of averaging (with weight (2.7.6)):

$$I = \varphi(x^1)e^{-f(x^1)/\kappa}(2\pi\kappa)^{M/2} \det^{-1/2} \|f_{ab}(x^1)\|$$

$$\cdot [1 + \langle D \rangle + \tfrac{1}{2}\langle D^2 \rangle + \tfrac{1}{6}\langle D^3 \rangle + \cdots]. \tag{2.7.7}$$

R. L. STRATONOVICH

We repeat here that D is the sum of terms in y in (2.7.3), excluding the term in f_{ab}. Polynomials of the Gaussian random variables with zero mean are averaged by using well-known simple rules, and all the odd powers fall out. As a result we obtain the series in progressive powers of the small parameter κ. If we do not include the terms of the order of κ and of higher orders, we will have

$$I = (2\pi\kappa)^{M/2} \det^{-1/2} \| f_{ab}(x^1) \| \varphi(x^1) e^{-f(x^1)/\kappa} [1 + O(\kappa)]. \qquad (2.7.8)$$

It is easy to write down more precise formulas.

Let us now verify the invariance of result (2.7.8) under curvilinear transformations. We can easily prove that $f_{ab}(x^1)$ is the tensor with respect to these transformations. Using formulas of the type (2.4.63), in new variables $y(x)$ we obtain

$$\bar{f}_{pq} = \partial[\partial f/\partial y_q]/\partial y_p = \partial[f_b C_{bq}^{-1}]/\partial x_a \, C_{ap}^{-1}$$
$$= f_{ab} C_{bq}^{-1} C_{ap}^{-1} + f_b \partial(C_{bq}^{-1})/\partial x_a, \qquad (2.7.9)$$

where $C_{ap}^{-1} = \partial x_a/\partial y_p$. By (2.7.2), the term $f_b \partial(C_{bq}^{-1})/\partial x_a$ falls out at the point x^1, and we have

$$\bar{f}_{pq} = f_{ab} C_{ap}^{-1} C_{bq}^{-1}. \qquad (2.7.10)$$

This proves the correct tensor properties of martix f_{ab}.
From (2.7.10) we get

$$\det^{-1/2} \| \bar{f}_{pq}(y^1) \| = \det^{-1/2} \| f_{ab}(x^1) \| \det \| C_{pa} \|. \qquad (2.7.11)$$

Since φ is the scalar density, it is transformed as

$$\bar{\varphi} = \varphi \det \left\| \frac{\partial x_a}{\partial y_q} \right\| = \varphi \det^{-1} \| C_{pa} \|. \qquad (2.7.12)$$

From (2.7.11) and (2.7.12) we see that the product $\varphi(x^1)\det^{-1/2} \| f_{ab}(x^1) \|$ is invariant. Consequently, (2.7.8) is invariant. If (2.7.8) is used for calculating (2.5.3), then $C_{pa} = C_{pm,an}$ should be taken in the form $c_{p\alpha}\delta_{mn}$ in (2.7.9)–(2.7.12).

If $f(x)$ consists of two terms of different orders in κ,

$$f(x) = f_1(x) + \kappa f_2(x), \qquad (2.7.13)$$

the easiest thing to do when evaluating integral (2.7.1) is to use (2.7.2)–(2.7.8) taking $f_1(x)$ as $f(x)$ and $\varphi(x)e^{-f_2(x)}$ as $\varphi(x)$. Then from (2.7.8) we get

$$I = (2\pi\kappa)^{M/2} \det^{-1/2} \| f_{1,ab}(x^1) \| \varphi(x^1)$$
$$\times \exp[-\kappa^{-1} f_1(x^1) - f_2(x^1)][1 + O(\kappa)], \qquad (2.7.14)$$

x^1 being determined from the equation

$$\partial f_1(x)/\partial x_a = 0. \qquad (2.7.15)$$

70

The function $\varphi(x)$ has the correct transformational properties (2.7.12) if

$$\varphi(x) = \det^{-1} \| G_{aJ}(x) \|, \qquad (2.7.16)$$

where $G_{aJ}(x)$, $J = 1, \ldots, M$ is the set of vectors being transformed as $\bar{G}_{pJ} = C_{pa} G_a$. In this case, substituting (2.7.16) into (2.7.14) and combining both determinants yields

$$I = (2\pi\kappa)^{M/2} \det^{-1/2} \| (G_{Ja}^{\mathrm{T}} f_{1,ab} G_{bL})_{x^1} \|$$

$$\times \exp[-\kappa^{-1} f_1(x) - f_2(x)][1 + O(\kappa)]. \qquad (2.7.17)$$

Here the determinant of the matrix whose elements are scalars is taken.

References

Ghirsanov, I. V. 1960. *Theory of Probability and Its Applications* **5**, 285–301.

Graham, R. 1973. In *Quantum Statistics in Optics and Solid-State Physics* (G. Höhler, ed.), Springer Tracts in Modern Physics 66, pp. 1–97. Berlin: Springer.

Graham, R. 1977. *Z. Phys. B* **26**, 281–90.

Graham, R. 1978. In *Stochastic Processes in Nonequilibrium Systems*, Lecture Notes in Physics 84, pp. 82–138. Berlin: Springer.

Ito, K. 1944. *Proc. Jpn. Acad.* **20**, 519–24.

Ito, K. 1946. *Proc. Jpn. Acad.* **22**, 32–5.

Ito, K. 1951. *Memoirs Am. Math. Soc.* **4**, 51–89.

Kramers, H. A. 1940. *Physica* 7, 284–304.

Landauer, R. and Swanson, J. A. 1961. *Phys. Rev.* **121**, 1668.

Langouche, F., Roekaerts, D. and Tirapegui, E. 1979a. *Physica A* **95**, 252–74.

Langouche, F., Roekaerts, D. and Tirapegui, E. 1979b. *Nuovo Cimento B* **53**, 135–59.

Langouche, F., Roekaerts, D. and Tirapegui, E. 1979c. *Phys. Rev. D* **20**, 433–8.

Onsager, L. and Machlup, S. 1953. *Phys. Rev.* **91**, 1505–15.

Phythian, R. 1977. *J. Phys. A* **10**, 777–89.

Ryter, D. 1985. *Physica A* **130**, (1/2), 205.

Stratonovich, R. L. 1962. *Transactions of Sixth All-Union Conference on the Theory of Probability and Mathematical Statistics*. Vilnius. (In Russian.) (English translation in *Selected Translations in Mathematical Statistics and Probability*. Providence, RI: Am. Math. Soc. 1972, p. 273.)

Stratonovich, R. L. 1963. *Introduction to the Theory of Random Noise*, Vol. 1. New York: Gordon and Breach.

Stratonovich, R. L. 1964. *Vestn. Mosk. Univ. ser. Math. Mech.* 1, 3–11. (In Russian.)

Stratonovich, R. L. 1967. *Introduction to the Theory of Random Noise*, Vol. 2. New York: Gordon and Breach.

Stratonovich, R. L. 1985. *Nonlinear Nonequilibrium Thermodynamics*. Moscow: Nauka. (In Russian.)

Wiener, N. 1923. *J. Math. Phys.* **2**, 131–74.

Wiener, N. 1924. *Proc. Lond. Math. Soc.* **22**, 454–67.

3   Langevin equations with colored noise

J.M. SANCHO and M. SAN MIGUEL

3.1   Introduction

This chapter is devoted to the study of Langevin equations with colored noise. These are stochastic differential equations which, for a single variable q and a single noise source, take the form

$$d_t q = f(q) + g(q)\xi(t), \tag{3.1.1}$$

where $f(q)$ and $g(q)$ are general functions of q and the driving noise $\xi(t)$ is a colored noise. We use the term 'colored noise' in a broad sense, meaning a nonwhite noise, that is a noise $\xi(t)$ such that its correlation at different times are not proportional to delta functions. When $\xi(t)$ is a white noise the process $q(t)$ is a Markovian process (Arnold, 1974). In particular, if $\xi(t)$ is a Gaussian white noise, $q(t)$ is a diffusion process. There exist well known techniques to deal with these Markovian processes. They have been reviewed in recent textbooks (Gardiner, 1983; Risken, 1984; Van Kampen, 1981). When $\xi(t)$ is a colored noise, $q(t)$ is a non-Markovian process. Techniques and approximation methods for non-Markovian processes are not so well known. Although this is by no means a new problem in the literature (Kubo, 1962b; Stratonovich, 1963), a renovated interest and new developments have appeared in recent years. In particular, one aims to describe the behavior of different statistical quantities characterizing the process $q(t)$ as a function of the correlation time of the noise. In the limit of vanishing correlation time, $q(t)$ becomes Markovian. In this chapter we review our work in this field, restricting ourselves to the case in which $\xi(t)$ is an Ornstein–Uhlenbeck process*.

Recent strong motivation to study Langevin equations with colored noise has arisen mostly in connection with studies of systems under the influence of external noise (Horsthemke and Lefever, 1984). In these situations the statistical properties of the noise are unrelated to the intrinsic dynamics of the system under consideration. As a consequence, the usual justification of white

---

* We will often loosely refer in the following to a 'non-Markovian process' meaning the special class of non-Markovian processes defined by (3.1.1) when $\xi(t)$ is an Ornstein–Uhlenbeck noise.

noise based on the separation of time scales cannot be invoked. A classical situation of this type in which colored noise is known to be of importance is in the problem of magnetic resonance in a fluctuating magnetic field (Kubo, 1962b, 1963). More recently, dye lasers have become laboratory systems of interest for which colored noise effects seem to be important (see Chapter 4, Volume 3). Other types of experimental studies of relevance in the study of colored noise effects are those in which one studies the response of a system to a noise source which can be controlled. In particular, it is possible to test stability properties when varying noise parameters as, for example, the correlation time measuring deviations from white noise (see Volume 3). It should be mentioned that Langevin equations with colored noise also appear in a different context, namely in the adiabatic elimination of variables: the reduction of the number of variables used in the description of a system usually leads to a colored noise problem in the contracted space of variables. A well known example is the description of Brownian motion in configuration space starting from a Markovian description in phase space (San Miguel and Sancho, 1980b).

Our approach to the handling of non-Markovian problems of the type defined by (3.1.1) aims to satisfy two key ideas. The first is the necessity of a general context with systematic approximation schemes of potential use in different problems and situations. The second idea is a pragmatic point of view: the techniques summarized in this chapter are intended as a unified calculational machinery to obtain concrete results in each specific situation. Our approximations are mostly based on expansions in the characteristic parameters of the noise, essentially its correlation time and intensity. As a general statement it is fair to say that some of the approximations discussed in this chapter, although systematic, may not always be completely justified from a rigorous mathematical point of view. These are the same type of problems that occur, for example, in the truncation of cumulant expansions, or Kramers–Moyal type of expansion, for the equation obeyed by a probability density. The approximations are used to calculate time independent properties (stationary distributions and moments) as well as time dependent properties (correlation functions and characteristic relaxation times). We generally obtain rather simple final expressions for quantities of interest. Avoiding pathological situations and taking into account specific features of the particular situation, these expressions compare successfully with digital or analogical simulations (see Section 3.4). We note that important dynamical non-Markovian features are displayed by time dependent properties which, however, have not yet been discussed at length by other alternative approaches.

The application of the methods summarized in this chapter to physical systems is not reviewed systematically here. However, we would like to mention the application to instabilities in nematic liquid crystals (Sagués and San Miguel, 1985; San Miguel, 1985; San Miguel and Sagués, 1986. See also

Volumes 2 and 3 for other contributions). The effect of colored noise in dye lasers has been studied in stationary solutions (Aguado and San Miguel, 1988; San Miguel and Sancho, 1981b), correlation functions (Hernández-Machado, San Miguel and Katz, 1985, 1986; San Miguel, Pesquera, Rodriguez and Hernández-Machado, 1987) and transient dynamics (De Pasquale, Sancho, San Miguel and Tartaglia, 1986a, b). We refer to the paper by Roy, Yu and Zhu (Chapter 4, Volume 3) for more details and references of other related work in the dye-laser problem. The spectrum of transmitted light in optical bistability has also been analyzed from this point of view (Hernández-Machado and San Miguel, 1986).

This chapter is divided in six sections. Section 3.2 is devoted to the problem of the calculation of the probability density associated with (3.1.1). Fokker–Planck approximations for the equation satisfied by the probability distribution are discussed and the calculation of the stationary distribution is addressed. We also indicate connections with other related approaches to the same problem. Section 3.3 is concerned with the study of non-Markovian dynamics in the steady state. Topics considered are the equation satisfied by the joint probability distribution and the calculation of correlation functions and their characteristic relaxation times. It is seen that the decay of the correlation function involves essential non-Markovian dynamical effects. Section 3.4 reviews a few examples in which the general results of Sections 3.2 and 3.3 are made more explicit. The final section, 3.5, addresses questions related to transient dynamics, especially decay of unstable states.

Finally we would like to include a warning on the use of the equations which are the subject of this chapter. Within the context of external noise problems (3.1.1) appears as a phenomenological equation, whose origin and validity deserve further analysis. Indeed, external noise is usually modelled replacing a control parameter of a deterministic description by a stochastic process. This procedure bypasses the difficulty of a joint description of external noise and internal fluctuations. In addition, the statistical properties assumed for the process modelling the external noise often do not respect physical characteristics of the control parameters. For instance, Gaussian noise does not respect intrinsic positivity of some parameters. Besides Gaussian noise (white or Ornstein–Uhlenbeck) other models of external noise used in the literature are dichotomic Markov processes and Poisson noise. An analysis of positivity requirements and their implications on stability properties and changes in the stationary distribution has been recently reported (Sancho, San Miguel, Pesquera and Rodriguez, 1987). This analysis is based on the use of Poisson white noise to model bounded external fluctuations. The joint description of internal and external fluctuation has been addressed considering master equations with fluctuating parameters (Sancho and San Miguel, 1984c; San Miguel and Sancho, 1982). Positivity of the fluctuating transition probabilities is guaranteed if a dichotomic Markov process (Sancho and San Miguel, 1984b) or a Poisson white noise (Rodriguez, Pesquera, San Miguel and Sancho, 1985) is used for the fluctuating parameters. As a general result one finds coupling effects between external and internal fluctuations.

3.2 Equation for the probability density: $P(q, t)$. Steady state distribution

3.2.1 Fokker–Planck equation

We start by considering stochastic differential equations (for a single variable q) of the form (3.1.1). When $g(q)$ is independent of q, the noise term is called additive. In the general case, in which the fluctuating term in (3.1.1) depends on the instantaneous value of q, it is called multiplicative noise. Equation (3.1.1), and the given initial conditions, define the process $q(t)$ as a functional of $\xi(t)$. In the case of Gaussian white noise of zero mean the probability density $P(q, t)$ for the process $q(t)$ is known to satisfy the following Fokker–Planck equation* (Stratonovich, 1963):

$$\partial_t P = - \partial_q f(q) P + D \partial_q g(q) \partial_q g(q) P, \tag{3.2.1}$$

where D is the noise intensity

$$\langle \xi(t) \xi(t') \rangle = 2D\delta(t - t') \tag{3.2.2}$$

We address here the situation in which $\xi(t)$ is not a white noise. The process $q(t)$ is then non-Markovian, and no exact simple equation for $P(q, t)$ is known in general. We take $\xi(t)$ to be an Ornstein–Uhlenbeck process, that is a Gaussian process of zero mean and correlation

$$\langle \xi(t)\xi(t') \rangle = \frac{D}{\tau} e^{-|t-t'|/\tau}, \tag{3.2.3}$$

where τ is the correlation time of the noise. In the limit $\tau = 0$, $\xi(t)$ becomes the white noise defined by (3.2.2). One can benefit from well established Markovian techniques by enlarging the state space considering $\xi(t)$ as an auxiliary variable driven by white noise and studying a two-variable (q, ξ) Markovian process. The starting point to study the process $q(t)$ is then a Fokker–Planck equation for the probability density $P(q, \xi, t)$. Information on the process $q(t)$ can be obtained subsequently by several methods. We mention, for example, the perturbative expansion in τ of Horsthemke and Lefever (1984) and continued fraction expansions extensively used by Risken (1984, 1985) and coworkers. As an alternative method we will focus here on the description of the process $q(t)$ in its natural state space finding an approximated Fokker–Planck equation for the probability density $P(q, t)$. We

---

* This equation corresponds to the Stratonovich interpretation of (3.1.1) (see, for example, Horsthemke and Lefever, 1984). This is the correct interpretation when $\xi(t)$ is obtained as the limit of noise with a very small correlation time.

believe that in this approach it is possible to obtain a more intuitive picture of the process $q(t)$. In addition, the mathematical calculations are much simpler once the general results have been established, since it is possible to deal with a single variable equation. As an example we mention that the stationary solution of a Fokker–Planck equation for a single variable is known, while for two variables this is generally not the case.

The probability density $P(q, t)$ is generally given by

$$P(q, t) = \langle \delta(q(t) - q) \rangle, \tag{3.2.4}$$

where $q(t)$ is a solution of (3.1.1) for a given realization of $\xi(t)$ and for a given initial condition. q is a point in the state space and the average is taken over initial conditions and over the realizations of $\xi(t)$. From (3.2.4) and (3.1.1) one gets (Sancho and San Miguel, 1980)

$$\partial_t P(q, t) = -\partial_q f(q) P(q, t) - \partial_q g(q) \langle \xi(t) \delta(q(t) - q) \rangle. \tag{3.2.5}$$

To proceed further we make use of functional methods (Hänggi, 1985; Klyatskin, 1974, 1986; Novikov, 1965) to transform the average left in (3.2.5). Taking advantage of the Gaussian character of $\xi(t)$, (3.2.5) can be written as (Sancho and San Miguel, 1980; Sancho, San Miguel, Katz and Gunton, 1982a)

$$\partial_t P(q, t) = -\partial_q f(q) P(q, t) + \partial_q g(q) \partial_q \int_0^t dt' (D/\tau) e^{-(t-t')/\tau}$$

$$\times \left\langle \frac{\delta q(t)}{\delta \xi(t')} \bigg|_{q(t) = q} \delta(q(t) - q) \right\rangle \tag{3.2.6}$$

Equation (3.2.6) is not a closed equation for $P(q, t)$ because in general the response function $\delta q(t)/\delta \xi(t')$ depends explicitly on $\xi(t)$ so that it cannot be taken out of the average within the angled brackets. The problem has then been reduced to the evaluation of this response function. Note that in the white noise limit the response function only contributes when $t' = t$. Recalling that

$$\frac{\delta q(t)}{\delta \xi(t')} \bigg|_{t = t'} = g(q),$$

then the equation (3.2.6) becomes the Markovian Fokker–Planck equation (3.2.1). In general the response function cannot be calculated explicitly. An exception (Sancho and San Miguel, 1980; Sancho et al., 1982a; San Miguel and Sancho, 1980a) are linear equations or equations which can be transformed into linear models by changes of variables (San Miguel, 1979). For $f(q) = aq$, $g(q) = 1$, we have $\delta q(t)/\delta \xi(t') = e^{a(t-t')}$ and (3.2.6) becomes

$$\partial_t P(q, t) = [-\partial_q aq + D(t)\partial_q^2] P(q, t), \tag{3.2.7}$$

$$D(t) = \int_0^t dt' (D/\tau) e^{-(t-t')/\tau} e^{a(t-t')}. \tag{3.2.8}$$

Equation (3.2.7) is a Fokker–Planck equation local in time for a non-Markovian process. It is valid for all times and involves no approximation. It should be stressed that the same equation would be obeyed by the probability density of a Markovian process driven by a Gaussian white noise with correlation $\langle \xi(t)\xi(t') \rangle = 2D(t)\delta(t - t')$. The difference between these two processes appears when considering multitime probabilities (see Section 3.3).

In more general cases some approximation has to be made to calculate the response function. Most known approximations lead to an equation of the Fokker–Planck type for $P(q, t)$. This is achieved by approximating the response function by a function independent of $\xi(t)$:

$$\frac{\delta q(t)}{\delta \xi(t')} = R(q(t), t, t'). \tag{3.2.9}$$

Equation (3.2.6) then becomes

$$\partial_t P(q, t) = \left[-\partial_q f(q) + D\partial_q g(q)\partial_q H(q, t) \right] P(q, t), \tag{3.2.10}$$

where

$$H(q, t) = \int_0^t dt' \frac{e^{-(t - t')/\tau}}{\tau} R(q, t, t'). \tag{3.2.11}$$

An important point in connection with (3.2.10) is that the diffusion term vanishes at $t = 0$. This implies a transient behavior qualitatively different from that of the associated Markov process obtained when $\tau = 0$. In the latter process the diffusion term is time independent. Other important differences persist for long time scales, in which the limit $t \to \infty$ is taken in (3.2.11).

We now address the question of the explicit calculation of $R(q(t), t, t')$. A general framework for calculating the response function is the operational approach of Martin, Siggia and Rose (1973) (see also Garrido and San Miguel, 1978; Sancho and San Miguel, 1980; San Miguel and Sancho, 1980a). As a more convenient method we use here a functional calculus approach (Sancho *et al.*, 1982a). From (3.1.1) it is immediate to obtain that

$$\frac{\delta q(t)}{\delta \xi(t')} = g(q(t'))\exp \int_{t'}^t ds[f'(q(s)) + g'(q(s))\xi(s)]\frac{\delta q(s)}{\delta \xi(t')} \tag{3.2.12}$$

for $t > t'$. If τ is small it is clear from (3.2.6) that the main contribution of the response function comes from t' close to t. An expansion around the white noise limit is obtained by expanding the response function in (3.2.6) in powers of $(t - t')$:

$$\frac{\delta q(t)}{\delta \xi(t')} = \sum_n \frac{(-1)^n}{n!} \frac{d^n}{dt'^n} \frac{\delta q(t)}{\delta \xi(t')}\bigg|_{t'=t} (t - t')^n. \tag{3.2.13}$$

The different terms in the expansion (3.2.13) can be calculated from (3.2.12). Substitution of (3.2.13) in (3.2.6) gives rise to an expansion in powers of τ often called the 'τ-expansion' of (3.2.6). In this expansion the coefficients of the terms are of the form $D^l\tau^m$, $l \leqslant m$. It is easy to see that for $n \geqslant 2$ in (3.2.13) the

77

coefficients of $(t - t')^n$ have contributions which explicitly contain $\xi(t)$. These are the contributions that when substituted in (3.2.6) break the Fokker–Planck form introducing extra ∂_q-derivatives. In addition these are the contributions which lead to terms with $l > 1$ in the coefficient $D^l \tau^m$. Neglecting these contributions one obtains an approximation of the form (3.2.9). From what we have said, it is clear that such an approximation contains all the terms which have a coefficient linear in D. It can be understood as an approximation valid for small noise intensity ('small D-approximation') containing contributions to all orders in τ. This subseries of terms of (3.2.13) can be formally summed up. It corresponds to the calculation of the response function as (San Miguel and Sancho, 1981b)

$$\frac{\delta q(t)}{\delta \xi(t')} = g(q(t')) \exp \int_{t'}^{t} ds \, f'(q(s)), \tag{3.2.14}$$

with

$$\dot{q}(t) = f(q(t)). \tag{3.2.15}$$

Equations (3.2.14) and (3.2.15) are obviously the lowest order approximations in the noise $\xi(t)$ to (3.2.12). The final result for $H(q, t)$ from (3.2.11) in this approximation is (Sancho and Sagués, 1985; Sancho et al., 1982a)

$$H(q, t) = f(q)[(1 + \tau f(q)\partial_q)]$$

$$\times \{1 - \exp[-(1/\tau + f(q)\partial_q)t]\} \frac{g(q)}{f(q)}. \tag{3.2.16}$$

If we also take τ as a small parameter we can approximate (3.2.16) keeping only the first order contribution in τ

$$H(q, t) \simeq h(q, t) \equiv g(q)(1 - e^{-t/\tau})$$

$$+ \tau g(q)\left(\frac{f(q)}{g(q)}\right)'[(1 - e^{-t/\tau}) - (t/\tau)e^{-t/\tau}]. \tag{3.2.17}$$

This approximation contains all the terms which are of first order in τ in the τ-expansion, and it is therefore often called the 'small τ-approximation'.

Equation (3.2.10) with (3.2.16) or (3.2.17) is an equation of the Fokker–Planck type for the probability density. It includes transient terms and is valid from $t = 0$ onwards. At $t = 0$ we have assumed statistical independence of $\xi(t)$ and $q(0)$. If we are only interested in the steady state distribution for the process $q(t)$ we can take the limit $t = \infty$ in (3.2.16) or (3.2.17). This gives a Fokker–Planck equation with time independent coefficients. The function $H(q)$ satisfies an ordinary differential equation (Sancho et al., 1982a):

$$H(q) + \tau[H'(q)f(q) - H(q)f'(q)] = g(q) \tag{3.2.18}$$

and

$$h(q) = g(q)(1 + \tau g(q)(f/g)'). \tag{3.2.19}$$

A general solution for (3.2.18) has been given by Masoliver, West and Lindenberg (1987).

For future reference we rewrite (3.2.10) with $H(q, \infty)$ in terms of a Fokker–Planck operator $L_q(\tau)$

$$\partial_t P = L_q(\tau) P \tag{3.2.20}$$

$$L_q(\tau) = -\partial_q f(q) + D\partial_q g(q)\partial_q H(q). \tag{3.2.21}$$

The original non-Markovian process defined by (3.1.1) is often approximated by a Markovian process characterized by the Fokker–Planck equation (3.2.20). We will refer to this approximation as the quasi-Markovian approximation to (3.1.1).

There are several alternative ways of deriving the Fokker–Planck equations (3.2.10) and (3.2.20). A derivation often overlooked of (3.2.20) is that of Stratonovich (1963, eq. (4.180) and the supplement to chapter 4 of Stratonovich's monograph). Fokker–Planck approximations can also be obtained using cumulant techniques (Fox, 1979; Kubo, 1962a; Lax, 1966). The second cumulant approximation of Van Kampen (1976) is equivalent to our small D-approximation equation (3.2.20) (Garrido and Sancho, 1982; Lindenberg and West, 1983; Sancho and Sagués, 1985). The explicit connection between our approximation for the response function and the second cumulant approximation was discussed by San Miguel and Sancho (1981b). Different techniques of adiabatic elimination of variables can also be used to obtain a Fokker–Planck equation. These derivations start from a Markovian Fokker–Planck equation in an extended space of variables. A contraction of the description leads to closed equations for $P(q, t)$. Examples of these procedures are the derivation of the small D-approximation by Grigolini (1986) (see also Chapter 5 of this volume).

The variety of possible derivations of (3.2.20) have in common a truncation of a Kramers–Moyal expansion. The shortcomings of such truncations are well known. They are related to the fact that the τ-expansion is a singular perturbation theory, so that the relative importance of the neglected terms is not easy to estimate. A major problem of the approximation is that it is not homogeneous for all values of q. This introduces in many cases artificial boundaries at the points where $H(q)/g(q) = 0$. In these cases the approximation has to be handled with care, taking into account the particular situation under study. Several authors have recently tried to patch up these deficiencies, modifying the form of $H(q)$ but remaining in a Fokker–Planck approximation. Using the same functional methods as discussed above Fox (1986b) has proposed an approximation in which $H(q)$ is given by*

$$\bar{H}(q) = g(q)/(1 - \tau g(f/g')). \tag{3.2.22}$$

$\bar{H}(q)$ is an approximation to $H(q)$ which can be alternatively understood as an extrapolation of the value obtained for $h(q)$ in the approximation to first order

---

* In an earlier paper Fox introduced a different and more involved approximation to $H(q)$ which was later corrected (Fox, 1986a). The particular terms retained in the approximation have been discussed by Masoliver, West and Lindenberg (1987).

in τ. For small τ

$$h(q) = g(q)(1 + \tau g(f/g)') \simeq g(q)/(1 - \tau g(f/g)').$$ (3.2.23)

Obviously, when expanding $\bar{H}(q)$ in powers of τ one obtains a series of terms which are all contained in the τ-expansion of $H(q)$, but $H(q)$ contains additional terms. However, in particular situations the use of $\bar{H}(q)$ may avoid the appearance of artificial boundaries. In addition, $\bar{H}(q)$ may also be a reasonable approximation to $H(q)$ in situations in which $H(q)$ cannot be explicitly obtained. A different approximation proposed by Hänggi, Mroczkowski, Moss and McClintock (1985) for distribution with small width consists in replacing the denominator in (3.2.23) by its mean value:

$$H(q) = g(q)/(1 - \tau \langle g(f/g)' \rangle).$$ (3.2.24)

In this approximation the equation for $P(q, t)$ is the same as in the white noise limit but with a renormalized diffusion constant $D(\tau) = D/(1 - \tau \langle g(f/g)' \rangle)$. In this way the approximation is forced to be homogeneous for all values of q, and the problem of artificial boundaries is circumvented. It is important to realize that in this case the equation for $P(q, t)$ is not closed and it has to be solved self-consistently with $D(\tau)$. We finally mention that the Fokker–Planck approximations (3.2.22) and (3.2.24) have been recovered as particular cases of (3.2.18) by Grigolini (1986) within the context of adiabatic elimination techniques. In the same context it has also been considered as a way of taking into account, within a Fokker–Planck approximation, terms obtained beyond the approximation leading to (3.2.20) (Faetti, Fronzoni, Grigolini and Mannella, 1987a).

3.2.2 Steady state distribution

The formal solution for the steady state distribution $P_{st}(q)$ of (3.2.20) is given by

$$P_{st}(q) \sim H(q)^{-1} \exp\left[\int^q dq' \, f(q')/Dg(q')H(q') \right].$$ (3.2.25)

We have already mentioned that artificial boundaries appear at points where $H(q)/g(q) = 0$. A positive definite solution for $P_{st}(q)$ given by (3.2.25) exists in the interval (q_1, q_2) where q_1 and q_2 are points at which $g(q)H(q) = 0$. From a practical point of view this solution makes sense as a reasonable approximation if $P_{st}(q) = 0$ at the possible artificial boundaries. A divergence of $P_{st}(q)$ given by (3.2.25) at an artificial boundary indicates the necessity of a more careful analysis. A possible way out in these circumstances is to construct a perturbative solution of the equation obtained in the small τ-approximation. One writes

$$P_{st}(q) = P_0(q) + \tau P_1(q),$$ (3.2.26)

where $P_0(q)$ is the normalized white noise solution. Substituting in (3.2.20) one obtains

$$P_{st}(q) = P_0(q)[1 - \tau(G(q) - \langle G(q) \rangle_0)], \tag{3.2.27}$$

where

$$G(q) = g(f/g)' + (1/2D)(f/g)^2, \tag{3.2.28}$$

and $\langle \cdots \rangle_0$ means that the average is taken with $P_0(q)$.

P_{st} in (3.2.27) is normalized but may become negative for certain values of q. An *ad hoc* exponentiation of the terms in the square brackets in (3.2.27) gives a distribution which is positive definite, free of artificial boundaries and which is a solution of (3.2.20) for small values of τ:

$$P_{st}(q) = N P_0(q) e^{-\tau G(q)} \tag{3.2.29}$$

where N is a normalization constant. We will see in the examples in Section 3.4 that (3.2.29) gives a good approximation in several cases. An expansion like (3.2.26) but in powers of D cannot be constructed since $D = 0$ is a singular limit. However, in some cases, the difference between the small τ-approximation and the small D-approximation is just a renormalization of the parameter τ: $H(q, \tau) = h(q, \tau_R)$. In such cases an ansatz of the form (3.2.29) with τ changed by τ_R is still useful beyond a first order in τ-approximation (see Section 3.4).

Other approaches to the calculation of $P_{st}(q)$ have as a starting point the Markovian Fokker–Planck equation for $P(q, \xi, t)$. Continued fraction expansions (Risken, 1984) are essentially numerical procedures giving very reliable results. The perturbation expansion of Horsthemke and Lefever (1984) is also a τ-expansion but performed in the Markovian Fokker–Planck equation. Their result for the probability density reproduces (3.2.27). A recent result by Jung and Hänggi (1987) for $P_{st}(q)$ based on adiabatic elimination techniques can be formally reobtained from (3.2.27)–(3.2.29) if only the second contribution to $G(q)$ in (3.2.28) is exponentiated. This result coincides with (3.2.25) when $H(q)$ is replaced by its approximation (3.2.22). However, the derivation by Jung and Hänggi is based on a different Fokker–Planck equation.

An important characteristic of the probability distribution is the number and location of its extrema. The equation for the extrema is obtained from (3.2.25):

$$f(q) - Dg(q)H'(q) = 0. \tag{3.2.30}$$

It is clear from (3.2.30) that not only the position but also the number of extrema can be changed with respect to the white noise limit when τ becomes nonzero. This can happen also for additive noise. It is important here to note that approximations like (3.2.24) which only produce a renormalization of the diffusion constant are unable to account for the appearance of new extrema in $P_{st}(q)$.

3.2.3 Multivariable and nonlinear noise problems

The extension of the above developments to N-variable problems is straightforward but cumbersome. For the process defined by

$$d_t q^\mu(t) = f^\mu(q(t) + g_k^\mu(q(t)\xi_k(t),$$

$$\mu = 1, \ldots, n, \quad k = 1, \ldots, m, \tag{3.2.31a}$$

$$\langle \xi_k(t)\xi_l(t') \rangle = \frac{D_{kl}}{\tau_{kl}} \exp(-|t - t'|/\tau_{kl}), \tag{3.2.31b}$$

an equation for the probability density to first order in τ has been derived by Hernández-Machado, Sancho, San Miguel and Pesquera (1983) completing earlier calculations (Dekker, 1982; San Miguel and Sancho, 1980a). It is found that there is no general consistent Fokker–Planck equation to first order in τ. Particular cases in which such an equation exists are for additive noise or when the problem can be transformed to an additive noise problem (San Miguel, 1979) and also when D_{kl} is diagonal and all the relaxation times τ_{kl} are equal. The derivation of the equation for $P(q,t)$ has also been considered using cumulant techniques (Fox, 1983; Garrido and Sancho, 1982; Lindenberg and West, 1984). The fourth cumulant contains the terms of order τ which break the Fokker–Planck form. The result of San Miguel and Sancho (1980a) for additive noise has been reproduced by Schenzle and Tél (1985) using an adiabatic elimination method. These authors also discuss a method to calculate the stationary distribution for weak noise.

A problem of particular interest is defined by a Langevin equation for a single variable like (3.1.1) supplemented with an extra additive independent white noise

$$d_t q = f(q) + g_1(q)\xi(t) + g_2(q)\zeta(t), \tag{3.2.32a}$$

where

$$\langle \xi(t)\zeta(t') \rangle = 0 \tag{3.2.32b}$$

$$\langle \zeta(t)\zeta(t') \rangle = 2\varepsilon\delta(t - t'). \tag{3.2.32c}$$

This is a particular case of (3.2.31) with $n = 1$, $m = 2$. It is of special physical relevance because it often appears when $\xi(t)$ is an external noise modelling fluctuations of a control parameter in $f(q)$ and $\zeta(t)$ models small internal fluctuations of the system which can be usually taken to be white. In some cases $\zeta(t)$ is additive $(g_2(q) = \text{constant})$, but in other cases $\zeta(t)$ must be multiplicative to respect boundaries of the problem, for example when q is a positive definite variable. The equation for the probability density of (3.2.32a–c) to first order in τ is:

$$\partial_t P = -\partial_q f(q)P + \varepsilon\partial_q g_2(q)\,\partial_q g_2(q)P + D\partial_q g_1(q)\,\partial_q h_1(q)P(q,t)$$

$$+ D\varepsilon\tau\partial_q g_1(q)\partial_q g_1^2(q)\left(\frac{f(q)}{g_1(q)}\right)'\partial_q g_2(q)P(q,t), \tag{3.2.33}$$

where $h_1(q)$ is $h(q)$ defined in (3.2.19) for $g_1(q)$. The last term in (3.2.33) is a coupling term between ξ and ζ which breaks the Fokker–Planck form of the equation for $P(q,t)$. Therefore, the recipe for constructing a Fokker–Planck equation, adding the diffusion coefficients associated with $\xi(t)$ and $\zeta(t)$, which is correct in the white noise limit, is not strictly correct here. However, the physical interpretation of $\zeta(t)$ implies that usually $\varepsilon \ll 1$, and the last term in (3.2.33) proportional to $\varepsilon\tau$ can be neglected, recovering a Fokker–Planck approximation. Within this approximation one can also extend (3.2.33) substituting $h_1(q)$ by $H_1(q)$. The final equation is then (3.2.20) supplemented with an independent diffusion term associated with ζ. We finally mention that situations more complicated than the one defined by (3.1.1) appear when considering equations of the form

$$d_t q = F(q, [\xi(t)]), \tag{3.2.34}$$

where F is a nonlinear function of $\xi(t)$ (Horsthemke and Lefever, 1984; Sagués, 1984; San Miguel and Sancho, 1981a; Wodckiewicz, 1982). In the particular case in which $F(q, [\zeta(t)]) = f(q) + g(q)\xi^2(t)$ and using D and τ as small parameters with D/τ finite, one obtains, after neglecting transients, the following equation for $P(q,t)$:

$$\partial_t P(q,t) = L_q P(q,t), \tag{3.2.35}$$

$$L_q = -\partial_q(f(q) + (D/\tau)g(q))$$
$$+ (D^2/\tau)\partial_q g(q)\partial_q g(q)(1 + (\tau/2)g(q)(f(q)/g(q))') \tag{3.2.36}$$

Equation (3.2.36) has been used to study nonlinear noise effects in the Freedericksz transition in nematic liquid crystals. Noise is associated with fluctuations in the magnetic field. Effects which have been considered among others are changes in the stationary distribution (Sagués and San Miguel, 1985; San Miguel and Sancho, 1981a) and shifts in the point where a correlation length diverges (San Miguel, 1985).

3.3   Equation for $P(q, t; q', t')$ and steady state dynamics

3.3.1   Basic equations

The knowledge of $P(q,t)$ is not enough to characterize a non-Markovian process. Quantities like the correlation functions and relaxation times, which give a characterization of the peculiarities of non-Markovian dynamics, cannot be obtained from $P(q,t)$. We also need to know the joint probability distribution $P(q,t;q',t')$ (Hernández-Machado and San Miguel, 1984; Hernández-Machado *et al.*, 1983; Sancho and San Miguel, 1984a). For Markovian processes, $P(q,t;q',t')$ obeys the same equation as $P(q,t)$ but with a different initial condition. This is not the case for the non-Markovian process defined by (3.1.1) and (3.2.3). Our first task is then the derivation of an equation for $P(q,t;q',t')$. The joint probability distribution is given by

$$P(q,t;q',t') = \langle \delta(q-q(t))\delta(q'-q(t')) \rangle; \quad t > t', \tag{3.3.1}$$

where $q(t)$ is a solution of (3.1.1) for a given realization of $\xi(t)$ and with initial conditions specified at $t = 0$, $q(t=0) = q_0$. Taking the time derivative in (3.3.1) we get

$$\partial_t P(q,t;q',t') = -\partial_q f(q)P(q,t;q',t') + \partial_q g(q)\partial_q \int_0^t dt_1 - \frac{D}{\tau}e^{-(t-t_1)/\tau}$$

$$\times \left\langle \frac{\delta q(t)}{\delta \zeta(t_1)}\bigg|_{q(t)=q} \delta(q-q(t))\delta(q'-q(t')) \right\rangle + \partial_q g(q)\partial_{q'} \int_0^{t'} dt_1 - \frac{D}{\tau}e^{-(t-t_1)/\tau}$$

$$\times \left\langle \frac{\delta q(t')}{\delta \xi(t_1)}\bigg|_{q(t')=q'} \delta(q-q(t))\delta(q'-q'(t')) \right\rangle. \tag{3.3.2}$$

The last term in (3.3.2) vanishes in the Markovian limit and it does not appear in the equation for $P(q,t)$. Assuming the standard approximation (3.2.11), (3.3.2) transforms into

$$\partial_t P(q,t;q',t') = [-\partial_q f(q) + D\partial_q g(q)\partial_q H(q,t)$$

$$+ D\partial_q g(q)\partial_{q'} H(q',t,t')]P(q,t;q',t'), \tag{3.3.3}$$

where $H(q,t,t')$ is a generalization of (3.2.11):

$$H(q,t;t') = \int_0^{t'} dt_1 [e^{-(t-t_1)/\tau}/\tau]R(q',t',t_1)$$

$$= e^{-(t-t')/\tau}H(q',t'). \tag{3.3.4}$$

The equation satisfied by $P(q,t;q',t')$ in the steady state is obtained by the following limiting procedure:

$$P_{st}(q,q',s) = \lim_{t\to\infty;t'\to\infty} P(q,t;q',t'); \quad t-t' = s. \tag{3.3.5}$$

Equation (3.3.3) becomes in this limit (Hernández-Machado et al., 1983):

$$\partial_s P_{st}(q,q',s) = [L_q(\tau) + De^{-s/\tau}\partial_q g(q)\partial_{q'} H(q')]$$

$$\times P_{st}(q,q',s), \tag{3.3.6}$$

where $L_q(\tau)$ is the Fokker–Planck operator (3.2.21).

The last term of (3.3.6) is characteristic of the non-Markovian dynamics and it vanishes in the white noise limit $\tau = 0$. This term does not appear in quasi-Markovian Fokker–Planck theories, but it is very important in the initial decay (small s) of the correlation function as we will see below. The modification of the initial decay also influences the long time dynamics. For $s \gg \tau$ one can neglect the last term in (3.3.6) provided one introduces an effective initial condition (Haake and Lewenstein, 1983). The generalization of

(3.3.6) to multivariable and nonlinear noise problems has been considered by Hernández-Machado *et al.* (1983) and Sagués, San Miguel and Sancho (1984), respectively.

Once the equation satisfied by $P_{st}(q, q', s)$ is known it is simple to write down the equation satisfied by the correlation function in the steady state defined by

$$\langle q(s)q \rangle_{st} = \lim_{t \to \infty; t' \to \infty} \langle q(t)q(t') \rangle$$

$$= \int dq \, dq' \, qq' P_{st}(q, q'; s); t' - t = s. \tag{3.3.7}$$

From (3.3.6) one obtains*

$$d_s \langle q(s)q \rangle_{st} = \langle f(q(s))q \rangle_{st} + D \langle g'(q(s))H(q(s))q \rangle_{st}$$
$$+ D e^{-s/\tau} \langle g(q(s))H(q) \rangle_{st}. \tag{3.3.8}$$

There are two important results implied by (3.3.8). First we note that the equation for the second moment is (for $t \gg \tau$)

$$d_t \langle q^2 \rangle = 2 \langle (f(q) + Dg'(q)H(q))q \rangle + 2D \langle g(q)H(q) \rangle, \tag{3.3.9}$$

and in the steady state ($t = \infty$) it reduces to

$$\langle (f(q) + Dg'(q)H(q))q \rangle_{st} + D \langle g(q)H(q) \rangle_{st} = 0. \tag{3.3.10}$$

Making use of this equation and of (3.3.8) it is seen that the time derivative of the correlation function is identically zero at $s = 0$:

$$d_s \langle q(s)q \rangle_{st}|_{s=0} = 0. \tag{3.3.11}$$

This is an important and general result independent of the kind of approximations involved (Pesquera and Rodriguez, 1985) which characterizes clearly the initial time regime of the non-Markovian process defined by (3.1.1).

The second result is the following. A formal integration of (3.3.8) is possible in the lowest order in D. After some algebra (Sancho and San Miguel, 1984a) an expression for $\langle q(s)q \rangle_{st}$ is obtained:

$$\langle q(s)q \rangle_{st} = \langle q(s)q \rangle_{st}^0 - \tau(1 - e^{-s/\tau})$$
$$\times \langle q(s)(f(q) + Dg'(q)g(q)) \rangle_{st}^0 \tag{3.3.12}$$

where $\langle \cdots \rangle_{st}^0$ indicates the steady state correlation function associated with the Markovian diffusion process defined by $L_q(\tau)$ (quasi-Markovian approximation). Note that (3.3.12) is exact for $s = 0$ and $s = \infty$. The usefulness of (3.3.12) is that it reduces the problem of the calculation of $\langle q(s)q \rangle_{st}$ to the calculation of the correlation functions of a Markovian diffusion process for

---

* A direct derivation of (3.3.8) from the Langevin equation (3.1.1) is given by Hernández-Machado and San Miguel (1984). This derivation makes clear the implications of stationarity.

which standard methods are available. We will profit from this fact in the study of relaxation times.

We finally mention that Sancho and Sagués (1985) have reobtained (3.3.3) and (3.3.6), also in its backward form, using Van Kampen cumulants. Roerdink (1982, 1984), using cumulant techniques, has also obtained some results for $P(q, t; q', t')$ and correlation functions. Effects of colored noise in the behavior of correlation functions using the techniques discussed here have been studied in the short time behavior of intensity correlation functions of dye lasers (Hernández-Machado, San Miguel and Katz, 1985, 1986; San Miguel et al., 1987) and in the spectrum of transmitted light in optical bistability (Hernández-Machado and San Miguel, 1986).

To discuss and clarify some important facts which appear in the steady state dynamics due to the non-Markovian character of the process we present in the following the study of a pedagogical example for which exact results can be obtained, a decoupling ansatz for more general cases, and methods to evaluate the relaxation time and effective eigenvalue associated with the correlation function.

3.3.2   A linear model

We consider the linear model

$$\dot{q} = -\alpha q + \xi(t) \tag{3.3.13}$$

for which non-Markovian properties can be clearly displayed. For this model $R(q, t, t_1)$, $H(q, t)$ are known exactly (see (3.2.7) and (3.2.8)), and (3.3.6) is easily obtained. The solution for the steady state correlation reads*

$$\langle q(s)q \rangle_{st} = D/(1 - \alpha^2 \tau^2)[(1/\alpha)e^{-\alpha s} - \tau e^{-s/\tau}], \tag{3.3.14}$$

which shows the characteristic zero initial slope (Figure 3.1). Different approximations to (3.3.14) are also shown in Figure 3.1 (Sancho, 1985b). The white noise limit ($\tau = 0$) gives a decaying exponential from $s = 0$. In the quasi-Markovian approximation the process is completely characterized by $L_q(\tau)$, so that the last term in (3.3.8) is neglected. It gives a decaying exponential but with correct value at $s = 0$

$$\langle q(s)q \rangle_{st} = (D/\alpha(1 + \alpha\tau))e^{-\alpha s}. \tag{3.3.15}$$

A last possible approximation is the quasi-Markovian approximation with effective initial conditions (Haake and Lewenstein, 1983) giving the correct behavior for $s \gg \tau$. This is obtained here neglecting the last term in (3.3.14): $e^{-s/\tau} = 0$.

---

* It is obvious that this is also directly obtained by integrating (3.3.13).

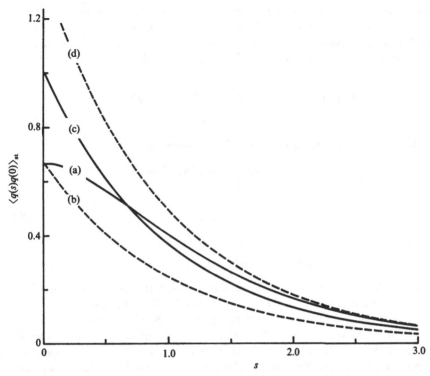

Figure 3.1. The correlation function of the linear model (3.3.13). (a) Exact result (3.3.14); (b) quasi-Markovian approximation (3.3.15); (c) white noise limit ($\tau = 0$); (d) quasi-Markovian approximation with effective initial conditions ($D = \alpha = 1; \tau = 0.5$).

3.3.3 The decoupling ansatz of Stratonovich

Stratonovich (1967) introduced, for a Markovian problem, a decoupling ansatz to truncate the infinite hierarchy of equations for the correlation functions. A generalized correlation function is defined by

$$C_n(s) = \langle q^{n-1}(s)q \rangle_{\text{st}} - \langle q^{n-1} \rangle_{\text{st}} \langle q \rangle_{\text{st}}. \tag{3.3.16}$$

The decoupling ansatz states that all these correlation functions decay in the same form. This is formulated explicitly assuming that

$$\frac{C_n(s)}{C_n(0)} = \frac{C_2(s)}{C_2(0)}, \tag{3.3.17}$$

where

$$C_2(s) = \langle \Delta q(s) \Delta q \rangle_{\text{st}} = \langle q(s)q \rangle_{\text{st}} - \langle q \rangle_{\text{st}}^2. \tag{3.3.18}$$

When the hypothesis (3.3.17) is used in a Markovian process, $C_2(s)$ obeys a linear equation of motion and it decays exponentially from $s = 0$.

Stratonovich's ansatz is equivalent to the lowest order truncation of a continued fraction expansion (Hernández-Machado, San Miguel and Sancho, 1984) and gives correct results for short times. We now extend this ansatz for a non-Markovian process. From (3.3.8) and (3.3.10) the equation satisfied by $C_2(s)$ is

$$d_s C_2(s) = \langle [f(g(s)) + Dg'(q(s))H(q(s))] \Delta q \rangle_{st}$$

$$+ De^{-s/\tau} \langle g(q(s))H(q) \rangle_{st}. \qquad (3.3.19)$$

Introducing now the ansatz (3.3.17) it reduces to

$$d_s C_2(s) = \frac{\langle (f(q) + Dg'(q)H(q))\Delta q \rangle_{st} C_2(s)}{C_2(0)}$$

$$+ De^{-s/\tau} \langle g(q)H(q) \rangle_{st}, \qquad (3.3.20)$$

where in the last term we have made the approximation

$$\langle g(q(s))H(q) \rangle_{st} = \langle g(q)H(q) \rangle_{st} \qquad (3.3.21)$$

because the dominant contribution is included in the factor $e^{-s/\tau}$. Equation (3.3.20) is linear and has a similar solution to (3.3.14). The dynamical behavior of $C_2(s)$ is now described by two decaying exponentials giving a vanishing slope at $s = 0$. Equation (3.3.20) includes the correct initial condition for $s = 0$, and it is very useful in the study of short time behavior. The Stratonovich decoupling can be understood as a linearization of the dynamics, taking into account nonlinearities through correct statics. Approximations for the correlation functions beyond the Stratonovich decoupling are discussed by Hernández-Machado, San Miguel and Sancho (1984) and Sancho and San Miguel (1984a).

3.3.4   Relaxation times

We have seen in (3.3.11) that a characteristic consequence of having $\tau \neq 0$ in (3.3.8) is the fact that the correlation function has a vanishing slope at $s = 0$. We have found this slow initial decay in the linear model of Section 3.3.2 and also when analyzing the decoupling ansatz of Stratonovich. To address these questions with more generality we have to characterize the relaxation of the steady state correlation function and the time scales which appear in the relaxation. Quite generally the correlation function can be expressed as a superposition of decaying exponentials:

$$C_2(s) = \sum_k C_k e^{-\lambda_k s}.$$

The index k might include a continuous part. In the Markovian limit the decay rates λ_k are the eigenvalues associated with the Fokker–Planck equation. Knowledge of all the λ_k characterize the decay of the correlation function.

A first characterization of the relaxation of $C_2(s)$ is given by the relaxation time defined by

$$T = \frac{1}{C_2(0)} \int_0^\infty C_2(s)\,\mathrm{d}s = \sum_k (C_k/\lambda_k)/\sum_k C_k. \qquad (3.3.22)$$

The relaxation time is the area under the normalized correlation function and therefore gives a global characterization of the relaxation of the correlation function. The calculation of this quantity for non-Markovian processes will be analyzed below. We will find that it contains special features which give clear information on the non-Markovian nature of the process. The relaxation time T includes, in principle, contributions from all the λ_k; however, it gives no information on possible different time regimes of evolution. An alternative partial characterization of the decay of $C_2(s)$ is given by an effective eigenvalue λ_{eff} which is a weighted decay rate defined by

$$\lambda_{\mathrm{eff}} = \sum_k (C_k\lambda_k)/\sum_k C_k. \qquad (3.3.23)$$

λ_{eff} also includes contributions from all the λ_k. It is the reciprocal time constant associated with the initial slope

$$\lambda_{\mathrm{eff}} = \lim_{t=0^+} -\frac{\mathrm{d}_t C_2(s)}{C_2(s=0)}. \qquad (3.3.24)$$

The calculation of λ_{eff} for non-Markovian processes is discussed in the following section. The behavior of T and λ_{eff} could be similar. In fact they coincide when a single exponential dominates the relaxation of $C_2(s)$. In the Markovian limit the Stratonovich ansatz gives for $C_2(s)$ a single decaying exponential whose decay rate is λ_{eff}. This coincidence would be accidental for non-Markovian processes, since, even in the linear case, $C_2(s)$ is the superposition of two exponentials.

The relaxation time T can be calculated exactly for diffusion Markovian processes (Jung and Risken, 1985b; Sancho, Mannella, McClintock and Moss, 1985). Before extending these techniques to non-Markovian processes we comment on a relevant aspect in this context of the linear model (3.3.13) and of the Stratonovich decoupling. For the linear model (3.3.13), the exact value of T is easily obtained from its solution (3.3.14):

$$T = \alpha^{-1} + \tau, \qquad (3.3.25)$$

where α is the relaxation time in the Markovian limit and τ is the new time scale introduced by the non-Markovian character of the process. The relaxation time obtained by integrating (3.3.20) coming from the Stratonovich decoupling is

$$T = [\gamma^0(\tau)]^{-1} + \tau, \qquad (3.3.26)$$

where

$$\gamma^0(\tau) = \frac{-\langle (f(q) + Dg'(q)H(q))\Delta q \rangle_{st}}{C_2(0)}, \tag{3.3.27}$$

which is the coefficient of the linear term in (3.3.20). We see that T has the same structure as in the linear case (3.3.25): $[\gamma^0(\tau)]^{-1}$ is the relaxation time in the same approximation for the quasi-Markovian approximation to the process. The second contribution, τ, is of a pure non-Markovian dynamical character, and arises from the last term in (3.3.19). The relative importance of these two contributions depends on the properties of $P_{st}(q)$, and it will be discussed below. If we wish to go beyond the Stratonovich decoupling to get better results for T, we need to find a different approach giving exact results for the white noise case.

Recently, Nadler and Schulten (1985) and Jung and Risken (1985b) presented independently for the white noise case a theoretical formalism which allows us to write an explicit expression for the relaxation time in terms of integrals which can be solved at least numerically. Following this method one can evaluate time integrals for generalized correlation functions of the form

$$\int_0^\infty \langle f_1(q(s))f_2(q)\rangle_{st}^0 \, ds, \tag{3.3.28}$$

where $f_1(q)$ and $f_2(q)$ are arbitrary nonlinear functions and the average, $\langle \cdots \rangle_{st}^0$, was introduced in (3.3.12). We summarize now the main steps of the use of this method to calculate T for a non-Markovian process. The solution (3.3.12) can be simplified if we discard the $e^{-s/\tau}$ term, whose contribution to T is of a higher order in τ. Then (3.3.12) can be rewritten as (Casademunt et al., 1987; Sancho and San Miguel, 1984a)

$$C_2(s) = \langle \Delta q(s)\Delta q \rangle_{st}^0 - \tau \langle \Delta q(s)(f(q) + Dg(q)g'(q))\rangle_{st}^0. \tag{3.3.29}$$

Dividing by $C_2(0)$ and integrating we obtain

$$T = T_0(\tau) + \tau T_1, \tag{3.3.30}$$

where T_0 and T_1 correspond to the time integrals of the first and second terms of the right-hand side of (3.3.29). These integrals are of the general type of (3.3.28). The calculation of T_1 gives $T_1 = 1$, and hence (3.3.30) has the same basic structure as (3.3.25) and (3.3.26). $T_0(\tau)$ is the relaxation time of the quasi-Markovian approximation, and takes the explicit form (Casademunt et al., 1987)

$$T_0(\tau) = \frac{1}{C_2(0)} \int_a^b \frac{F^2(q)}{Dg(q)H(q)P_{st}(q)} \, dq, \tag{3.3.31}$$

$$F(q) = -\int_a^q (q' - \langle q \rangle_{st})P_{st}(q') \, dq', \tag{3.3.32}$$

and $[a, b]$ is the domain of q. Taking the limit $\tau = 0$ in (3.3.31), we obtain the

Langevin equations with colored noise

exact result for the Markovian limit (Jung and Risken, 1985b). Non-Markovian effects enter in two different ways in (3.3.30). The first term, $T_0(\tau)$, incorporates non-Markovian contributions which come from $P_{st}(q)$. These are the only non-Markovian contributions taken into account in the quasi-Markovian approximation. The second contribution in (3.3.30), τ, is a non-Markovian dynamical effect and it comes from the extra term in the equation for $P(q,t;q',t')$.

Casademunt *et al.* (1987) have studied the relative relevance of the two contributions mentioned above. They manifest the importance of two different physical mechanisms. When the system in the steady state has a bimodal distribution, the relevant dynamical mechanism of relaxation is the diffusion through the corresponding barrier. The dominant time scale in the Markovian limit follows an Arrhenius law (Kramers, 1940)

$$T \simeq e^{\Delta\phi/D}. \tag{3.3.33}$$

Hence, small changes in D are magnified by the exponential. Since the colored noise reduces diffusion for values of q close to the maximum of $P_{st}(q)$ ($H(q) \langle g(q)$ in these points) then T increases exponentially with τ. In this situation $T_0(\tau)$ is dominant. Thus the relaxation time and the mean first passage time (MFPT), which measures lifetimes of metastable states, are closely related. They represent the unique relevant time scale of relaxation. This fact justifies the widely used quasi-Markovian approximation to evaluate MFPTs. For the same reason the approach of Hänggi *et al.* (1985) with a renormalized value of D is also valid in this special situation (Fox, 1986b; Masoliver, West and Lindenberg, 1987). On the other hand, when the system has a monomodal steady state distribution and D is small enough, the linear or Gaussian approximation gives correct results. The relaxation time is given in this situation by (Hernández-Machado, San Miguel and Sancho, 1984)

$$T = \text{const.} + \tau + O(D). \tag{3.3.34}$$

Hence a small variation of D is not of special relevance and the effect of the dynamical non-Markovian contribution, τ, is important. In the examples studied in Section 3.4 this contribution is of the same order as the quasi-Markovian one.

3.3.5 Effective decay rate

The effective decay rate which was defined in (3.3.24) characterizes the initial decay of the correlation function. We will now show that it is possible to obtain exact formulas for λ_{eff} in terms of stationary moments. For a general equation of the type (3.1.1) λ_{eff} vanishes when $\xi(t)$ is a colored noise, as in (3.2.3). It becomes nonzero in the white noise limit $\tau = 0$. In order to discuss how λ_{eff} becomes nonzero when introducing a white noise we analyze the model (3.2.32a–c) of Section 3.2.3. The main idea involved in the calculation of λ_{eff}

91

(San Miguel *et al.*, 1987) is that $C_2(s)$ is an even function of s, so, if it has a continuous derivative at $s = 0$, it must be $\dot{C}_2(s = 0) = 0$. The time derivative of $C_2(s)$ is continuous for a nonwhite noise, but white noise introduces a discontinuity of $\dot{C}_2(s)$ at $s = 0$. A straightforward calculation of $C_2(s)$ from (3.2.32a–c) gives

$$\dot{C}_2(s) = \langle q(t)f(q(t+s))\rangle_{\rm st} + \langle q(t)g_1(q(t+s))\xi(t+s)\rangle_{\rm st}$$
$$+ \langle q(t)g_2(q(t+s))\zeta(t+s)\rangle_{\rm st}. \tag{3.3.35}$$

The first two terms on the right-hand side of (3.3.35) are continuous at $s = 0$, but

$$\lim_{s=0^+} \langle q(t)g_2(q(t+s))\zeta(t+s)\rangle_{\rm st} = \varepsilon\langle qg_2g_2'\rangle_{\rm st} \tag{3.3.36}$$

$$\lim_{s=0^-} \langle q(t)g_2(q(t+s))\zeta(t+s)\rangle_{\rm st} = \varepsilon\langle qg_2g_2'\rangle_{\rm st} + \varepsilon\langle g_2^2\rangle_{\rm st}. \tag{3.3.37}$$

Equations (3.3.36) and (3.3.37) are obtained using Novikov's theorem (Novikov, 1965) and causality requirements for the response function. To calculate $\dot{C}_2(0^+)$ we have to use (3.3.36). Substituting (3.3.36) in (3.3.35) and using the relation obtained imposing $d_t\langle q^2\rangle_{\rm st} = 0$ we immediately find (San Miguel *et al.*, 1987)

$$\dot{C}_2(0^+) = -\varepsilon\langle g_2^2\rangle_{\rm st}, \tag{3.3.38}$$

so that

$$\lambda_{\rm eff} = \frac{-\dot{C}_2(0^+)}{C(0)} = \frac{\varepsilon\langle g_2^2\rangle_{\rm st}}{\langle q^2\rangle_{\rm st} - \langle q\rangle^2}. \tag{3.3.39}$$

Therefore, in the limit in which white noise disappears ($\varepsilon = 0$), $\lambda_{\rm eff} = 0$ and we find a zero initial slope. A nonvanishing initial slope indicates the presence of a noise source whose correlation time can be completely neglected in comparison with the time scale of evolution of the relevant variable. When (3.2.32a–c) model a system under the influence of an external noise $\xi(t)$, and $\zeta(t)$ stands for a small internal noise, a large value of $\lambda_{\rm eff}$ indicates a domain of parameters in which the effect of $\zeta(t)$ cannot be neglected.

3.4 Selected examples

3.4.1 *Brownian motion*

There are several illustrative situations in Brownian motion problems in which non-Markovian effects can be discussed using the techniques presented in the previous sections. We give here a short list of such situations and refer the reader to the literature for details. A first problem concerns the description in configuration space of a Brownian particle which follows Markovian dynamics in phase space. This is a problem of elimination of variables which can be discussed by different methods. Using the methods of this chapter the exact equation for the probability density in configuration space of a free Brownian particle and a Brownian particle in a harmonic potential were derived by San Miguel and Sancho (1980b). They also derived the Smoluchowski equation for a general anharmonic potential to third order in the inverse of the damping coefficient. Equations for the joint probability distribution and steady state correlations in configuration space for this problem were discussed by Hernández-Machado and Sagués (1983). Nonconstant damping coefficients were considered in this context by Sancho, San Miguel and Dürr (1982). A second problem is defined by the equation for the velocity of a free Brownian particle with a damping memory kernel. The joint probability density and correlation functions have been discussed for this problem by Hernández-Machado *et al.* (1983). For these two problems fluctuation–dissipation relations determine the correct equilibrium distribution. This is not the case for a third problem, which is of pedagogical interest. It considers Brownian motion in phase space with an arbitrary force potential and with a momentum driven by a phenomenological colored noise. The equation for the probability density in phase space to third order in τ was derived by San Miguel and Sancho (1980a). The solution of this equation (Ramirez de la Piscina and Sancho, 1988) for a bistable potential shows an asymmetry found in analogic simulations (Fronzoni *et al.*, 1986; Moss, Hänggi, Mannella and McClintock, 1986). This asymmetry does not appear in other approximations (Fronzoni *et al.*, 1986; Schenzle and Tél, 1985).

3.4.2   *A prototype model*

We next discuss a prototype model which has a wide applicability in different situations. The general model is defined by the following equation:

$$d_t q = \alpha q - q^3 + c + aq\xi(t) + b\zeta(t), \qquad (3.4.1)$$

where $\zeta(t)$ is a second independent noise of intensity ε and correlation time τ_1. Depending on the values of the parameters a, b and c, the prototype model reduces to particular situations of interest. The parameter c represents an external field which controls the shape of the potential

$$V(q) = \frac{q^4}{4} - \frac{\alpha q^2}{2} - cq. \qquad (3.4.2)$$

For $c > c_0 = (4\alpha^3/27)^{1/2}$, $V(q)$ is monostable and for $c < c_0$ it becomes bistable. In the following we will study the most relevant non-Markovian features in different cases of (3.4.1).

Additive colored noise ($a = 0$, $b = 1$)
This situation corresponds to the well known time dependent Landau mean

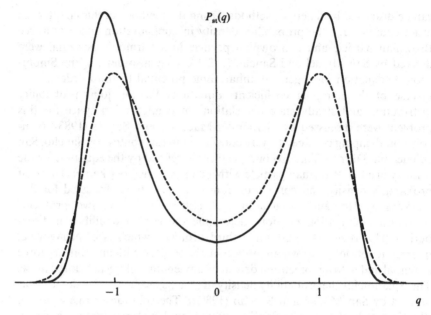

Figure 3.2. $P_{st}(q)$ for the model (3.4.3) ($c = 0$; $\varepsilon = 0.25$; $\tau_1 = 0.3$; $\alpha = 1$). The broken line is the white noise case ($\tau = 0$).

field model with colored fluctuations. Explicitly (3.4.1) becomes

$$\dot{q} = c + \alpha q - q^3 + \zeta(t). \tag{3.4.3}$$

Now the steady state properties can be obtained from (3.2.27) and (3.2.28). In particular for the simplest situation ($c = 0$) the stationary distribution (3.2.29) is

$$P_{st}(q) = N \exp \left\{ \frac{1}{\varepsilon} \left[(\alpha - \alpha\tau_1 + (6\tau_1\varepsilon)) \frac{q^2}{2} \right. \right.$$
$$\left. \left. -(1 - 4\alpha\tau_1)\frac{q^4}{4} - \frac{\tau_1}{2}q^6 \right] \right\}. \tag{3.4.4}$$

A transition from monostability to bistability is induced by the colored noise for fixed $\alpha < 0$ at $\tau_1 = -\alpha/6D$. (See also Faetti *et al.*, 1987b.) For $\alpha > 0$, $P_{st}(q)$ has two maxima at $q_{1,2} = \pm(\alpha + 6\varepsilon\tau_1)^{1/2}$ and a minima at $q = 0$ (Figure 3.2). The colored noise produces more pronounced and displaced maxima. This fact is important when considering the characteristic time scales. From a numerical integration of (3.3.32) we obtain the behavior of T shown in Figure 3.3 as a function of the external field c for fixed values of ε and τ_1. The comparison with the white noise case ($\tau_1 = 0$) is also shown. The general discussion of Section 3.3.4 is clearly visualized here: first, the effect of $\tau_1 \neq 0$ is much more important in the bistable situation. Second, the relative error of

Langevin equations with colored noise

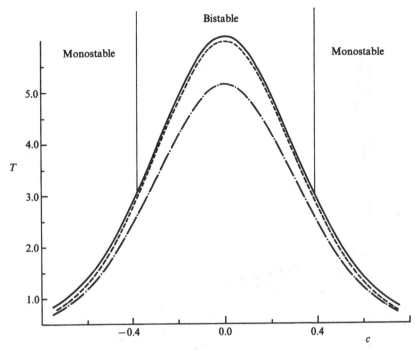

Figure 3.3. Linear relaxation time T versus the parameter $c(\varepsilon = 0.25; \tau_1 = 0.1;$ $\alpha = 1)$. The solid line is the numerical evaluation of (3.3.30) and (3.3.31). The broken line is the quasi-Markovian prediction $T_0(\tau_1)$. The broken dotted line is the white noise case.

taking $T = T_0(\tau_1)$ is much smaller for the bistable situation than for the monostable one.

Multiplicative noise $(b = 0; c = 0; a = \alpha = 1)$

This model was introduced by Stratonovich (1967) in the context of fluctuations in electric circuits. Explicitly the model is:

$$\dot{q} = q - q^3 + q\xi(t). \tag{3.4.5}$$

It can be understood as a Landau model in which the coefficient of the linear term fluctuates around its mean value. This model has been studied analytically and numerically by Sancho *et al.* (1982a) and with electronic simulations by Sancho, San Miguel, Yamazaki and Kawakubo (1982b). Here we summarize the main results obtained using the methods of Sections 3.2 and 3.3.

The stationary solution obtained from (3.2.29) is

$$P_{\text{st}}(q) = Nq^{1/D-1}\exp\left[\left(\tau\frac{2D+1}{D} - \frac{1}{2D}\right)q^2 - \frac{\tau}{2D}q^4\right], \tag{3.4.6}$$

95

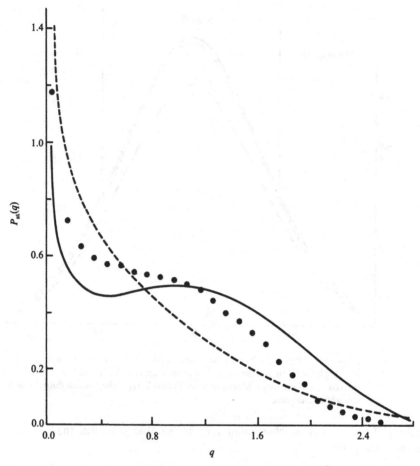

Figure 3.4. $P_{st}(q)$ for the model (3.4.5) (Sancho *et al.*, 1982a) ($D = 1.5$; $\tau = \frac{1}{3}$). The circles represent the result of the simulation, the full line represents the theoretical prediction (3.4.6), and the broken line is the white noise result.

whose extrema are given by

$$q = 0 \qquad\qquad (3.4.7)$$

$$2\tau q^4 - (2(2D + 1)\tau - 1)q^2 + D - 1 = 0. \qquad\qquad (3.4.8)$$

The solution is restricted to positive values of q. For $D < 1$, $q = 0$ is a minimum and (3.4.8) gives a unique positive solution for q. For $D > 1$ and small values of τ there is a unique maximum at $q = 0$. Therefore the general form of $P_{st}(q)$ is the same as in the white noise limit. Nevertheless, for large enough τ, (3.4.8) has positive solutions for q, which indicates the appearance of a new relative minimum and a relative maximum of $P_{st}(q)$. These new extrema do not exist in the white noise limit.

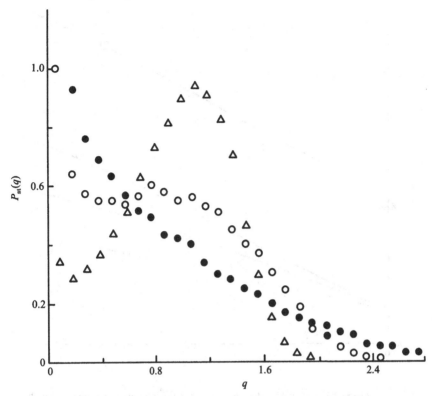

Figure 3.5. $P_{st}(q)$ for the model (3.4.5) obtained from numerical simulation (Sancho *et al.*, 1982a). $P_{st}(q)$ changes when increasing τ for fixed $D = 1.5$. ● $\tau = 1/18$; ○ $\tau = 1/2.1$; △ $\tau = 2.0$.

These results for $P_{st}(q)$ are based on an extrapolation of the result for small τ to larger values of τ. In particular, the appearance of the new maxima occurs for relatively large τ. Nevertheless this prediction has been confirmed by digital (Sancho *et al.*, 1982a) and analog (Sancho *et al.*, 1982b) simulations, and also by a continued matrix expansion calculation (Jung and Risken, 1984). Figure 3.4 shows a comparison of the white noise distribution, the solution (3.4.6) and simulation for values of the parameters for which new extrema are predicted. Figure 3.5 exhibits the emergence of these relative extrema when increasing τ. In this model the small D approximation only amounts to a renormalization of the parameter τ such that $H(q)$ in (3.2.21) is just $h(q)$ with τ replaced by τ_R:

$$\tau_R = \tau/(1 + 2\tau). \tag{3.4.9}$$

The substitution of τ by τ_R in the results for small τ should be a way of taking into account larger values of τ. In this case the theoretical results for the

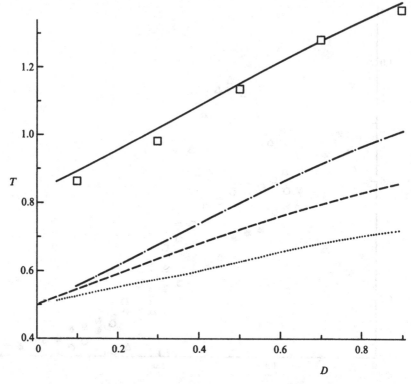

Figure 3.6. Linear relaxation time T versus the intensity of the noise D for the model (3.4.5). —— Result of (3.3.30) and (3.3.31); ---- white noise result; $\cdots$ quasi-Markovian approximation with $H(q)$ given by (3.2.22); $\ldots$ quasi-Markovian approximation with $H(q)$ given by (3.2.24). The squares depict the result of a digital simulation by Casademunt *et al.* (1987) (for $\tau = \frac{1}{3}$).

moments of the stationary distribution give a better agreement with numerical simulation (Sancho *et al.*, 1982a).

Concerning the relaxation time Hernández-Machado, San Miguel and Sancho, (1984) found the correct qualitative dependence of T on D and τ for small values of D using Stratonovich's decoupling ansatz and other more elaborate approximations. However, the agreement with simulations for large D is not satisfactory because the approximation is already unreliable in the white noise limit. This deficiency is overcome using the general method discussed in Section 3.4. The result of (3.3.30) and (3.3.31) for this model is shown in Figure 3.6. It shows perfect agreement with simulation. Figure 3.6 also shows the results that would be obtained using quasi-Markovian approximations defined by $L_q(\tau)$ with $H(q)$ in the approximations proposed by Fox, (3.2.22), or Hänggi *et al.*, (3.2.24). These approximations were not proposed for this sort of dynamical calculation and they do not incorporate the explicit τ-contribution in (3.3.30). However, if this contribution is added to the result following from (3.2.22) the result is quite satisfactory.

3.4.3   Dye lasers

As a relevant example reviewed elsewhere (see Chapter 4, Volume 3) we mention here that through a simple change of variables, which reduces the cubic nonlinearity to a quadratic one, (3.4.5) becomes a model equation widely used to describe the behavior of the intensity of a single mode dye laser in which nonwhite pump noise fluctuations are known to be important. The predictions of (3.4.6) regarding the appearance of a stationary distribution with two relative extrema have been confirmed for this system by experimental measurements (Lett, Gage and Chyba, 1987) and by other calculations (Jung and Hänggi, 1987; Jung and Risken, 1984). The steady state intensity correlation function has also been calculated for this model using the methods of Section 3.3 (Hernández-Machado, San Miguel and Katz, 1985, 1986). This function exhibits an initial plateau associated with the vanishing initial slope (3.3.11). Comparison with simulations shows that the Stratonovich decoupling (Section 3.3.3) gives a good description of the initial decay. Such behavior of the correlation function was reproduced by other calculations (Jung and Risken, 1985a) and confirmed by experiment (P. Lett, 1986, PhD thesis).

The discussion above for a dye laser neglects quantum noise, which is particularly important below threshold. The inclusion of quantum white noise has an important consequence in that the steady state solution can then be calculated for negative values of the pump parameter, whereas (3.4.5) has no steady state solution if the coefficient of the linear term (α in (3.4.1)) is negative. The equation for the light intensity including quantum noise is of the form (3.2.32a–c) with $g_2(q) \sim q^{1/2}$, imposing a boundary such that the light intensity remains positive. A study of this problem based on (3.2.33) and (3.2.29) has been recently reported (Aguado and San Miguel, 1988). The relative fluctuations of the intensity exhibit a maximum and tend to zero as the mean intensity goes to zero. This quantity diverges in the absence of quantum noise. The maximum is much less pronounced than in the white noise limit. The calculation of this quantity reproduces satisfactorily experimental and simulation data by Lett, Short and Mandel (1984). Other approximations (Fox and Roy, 1987) based on (3.2.24) do not reproduce the emergence of relative extrema which can also occur for negative values of the pump parameter. The distribution obtained by simulation by Lett, Gage and Chyba (1987) for positive pump parameters are well reproduced extending the result by Jung and Hänggi (1987) to include quantum noise (Aguado and San Miguel, 1988). However, this approximation fails for negative pump parameters. We finally mention that quantum white noise invalidates (3.3.11) and λ_{eff} becomes very large for small values of $\langle I \rangle$ (San Miguel *et al.*, 1987).

3.5   Transient dynamics

3.5.1   Generalities

The study of transient dynamics concerns the evolution of a stochastic process from an initial condition far from the steady state. Two particularly interesting cases from the physical point of view are the relaxation of metastable and unstable states. For Markovian processes defined by Langevin equations (3.1.1) driven by white noise there are well known methods to describe transient dynamics (Gardiner, 1983; Risken, 1984; Stratonovich, 1963; Van Kampen, 1981). Metastability is usually characterized by its lifetime defined as a mean first passage time (MFPT) to leave the vicinity of a locally stable state. The description of the decay of an unstable state is generally more delicate. The time evolution of the variance in the decay of an unstable state exhibits anomalously large transient fluctuations. This behavior is satisfactorily described by the scaling theory of Suzuki (1977) or alternative and more refined elaborations (De Pasquale and Tombesi, 1979). An alternative description of the early stages of the decay of an unstable state is given by the FPT distribution to leave the vicinity of the unstable state (Haake, Haus and Glauber, 1981). We will address the question of extending these ideas to non-Markovian situations.

It is first fair to say that transient problems have, up till now, been analyzed in much less detail than the steady state properties and that few well established results are available. In this context it is important to identify which are the difficulties intrinsically associated with a non-Markovian dynamics besides other difficulties associated with the nonlinearities already present in Markovian dynamics. A point to be stressed is that the use of the approximate Fokker–Planck equation (3.2.20) to study transient dynamics, if possible at all, has to be justified in each particular case. We have already seen that the non-Markovian character of the process implies that steady state dynamics cannot be studied using (3.2.20) but that a different equation for $P_{st}(q, t; q', t')$ is needed (Section 3.3). For transient processes the evolution of time dependent quantities cannot be calculated from (3.2.20), even if they depend on a single time, since transient terms $\simeq e^{(-1/t)/\tau}$ have been neglected. It is precisely in the initial regime where the effect of a colored noise is usually more important. For long time dynamics it is necessary to remember that (3.2.20) is not a bona fide Fokker–Planck equation. For example, MFPT are calculated straightforwardly for Markovian processes using the backward equation associated with (3.2.20). The calculation of MFPT for non-Markovian processes is much more subtle. It suffices to say that an exact calculation of MFPT seems not to be known for the pure diffusion process $\dot{q} = \xi(t)$ when $\xi(t)$ is an Ornstein–Uhlenbeck noise, while the probability distribution for this process follows immediately from (3.2.7).

100

3.5.2 Metastability and escape times

It has already been mentioned that metastability is usually characterized by an MFPT which is difficult to calculate for non-Markovian processes. The problem is simpler when the noise $\xi(t)$ is a dichotomic Markov process instead of the Ornstein–Uhlenbeck noise. In this case exact results are known (Hänggi and Talkner, 1985; Masoliver, Lindenberg and West, 1986; Rodriguez and Pesquera, 1986; Sancho, 1985a). For the Ornstein–Uhlenbeck noise the key problem is to find the equation which is satisfied by the MFPT incorporating the correct boundary conditions. An attempt to obtain a genuine non-Markovian equation for the MFPT was done by Sancho, Sagués and San Miguel (1986). However, that equation already incorporates Markovian boundary conditions. Most other approaches to the problem are based on quasi-Markovian approximations (Fox, 1986a, b; Hänggi, Marchesoni and Grigolini, 1984; Hänggi *et al.*, 1985; Masoliver, West and Lindenberg, 1987). We recall that in quasi-Markovian approximations, the non-Markovian process is approximated by a Markovian one obeying a Fokker–Planck type equation like (3.2.20). Masoliver, West and Lindenberg, (1987) have shown that all these approaches give the same result up to first order in τ. A physical reason to understand the relatively good results of quasi-Markovian approximations when dealing with metastability was outlined in Section 3.3.4 in connection with the calculation of relaxation times. For a more detailed discussion of the calculation of MFPTs for non-Markovian processes the reader is referred to Chapter 4 of this volume by Lindenberg, West and Masoliver.

3.5.3 Decay of unstable states

We will discuss the main effects caused by a colored driving noise in the decay of unstable states through the study of two particular relevant models. We also mention that the techniques discussed below have been successfully applied to describe the transient dynamics of a dye laser (De Pasquale *et al.*, 1986b) and that the effect of nonlinear noise, (3.2.34), in the decay of unstable states, has been considered in the context of the Freedericksz transition in nematics (Sagués and San Miguel, 1985; San Miguel and Sagués, 1986).

Additive colored noise

We first consider the model (3.4.3) $(c = 0)$ already discussed in Section 3.4.2. We are now interested in the relaxation of the unstable state $q_0 = 0$ when $\zeta(t)$ is the Ornstein–Uhlenbeck noise. Fluctuations are crucial to trigger the decay process. The amplification of these initial fluctuations by the deterministic dynamics gives rise to the phenomenon of anomalous fluctuations. With this picture in mind our approximation consists in replacing the process $q(t)$ by a deterministic process with random initial conditions. The meaning of such effective random initial conditions is clearly seen considering the linearized version of (3.4.3) $(c = 0)$ whose solution can be written as

$$q^2(t) = h^2(t)e^{2\alpha t}, \tag{3.5.1}$$

where

$$h(t) = \int_0^t dt' e^{\alpha t'} \zeta(t'). \tag{3.5.2}$$

For times $t \gg \alpha^{-1}$, τ_1, $h(t)$ can be replaced by $h(\infty)$, which is a Gaussian variable of zero mean and variance

$$\langle h^2(\infty) \rangle = \frac{\varepsilon}{\alpha(1 + \alpha\tau_1)} = \varepsilon'. \tag{3.5.3}$$

$h^2(t)$ in (3.5.1) plays the role of a random initial condition for each time t. The deterministic solution of (3.4.3) with this random initial condition gives an approximation to the process $q(t)$ for which different moments $\langle q^n(t) \rangle$ can be calculated. In this approximation the effect of the finite correlation time of the noise only enters the problem by modifying the variance of the effective random initial conditions. For $t \gg \alpha^{-1}$, τ_1 this reduces to a renormalization of the noise intensity which is now ε', (3.5.3).

Using the above approximation we wish to characterize the decay process in terms of an FPT distribution. Our approach here (De Pasquale *et al.*, 1986b) is a reformulation and extension to colored noise problems of the work by Haake, Haus and Glauber (1981). Taking into account the symmetry of the problem we look for the time t_I at which $q^2(t)$ takes a prescribed value I:

$$I = q^2(t_I). \tag{3.5.4}$$

This time is a stochastic variable, and we want to calculate the moments of its distribution. To this end we will consider the generating function $W(\lambda)$:

$$W(\lambda) = \langle e^{-\lambda t} \rangle. \tag{3.5.5}$$

The MPT T and the variance σ_T^2 are calculated in terms of the cumulant generating function $\Omega(\lambda)$:

$$\Omega(\lambda) = \ln W(\lambda) \tag{3.5.6a}$$

$$T = \langle t_I \rangle = -\frac{d\Omega}{d\lambda}\bigg|_{\lambda = 0} \tag{3.5.6b}$$

$$\sigma_T^2 = \langle t_I^2 \rangle - \langle t_I \rangle^2 = -\frac{d^2\Omega}{d\lambda^2}\bigg|_{\lambda = 0}. \tag{3.5.6c}$$

The PT distribution is well approximated using the linearized solution (3.5.1) with $h^2(t)$ replaced by $h^2(\infty)$. This approximation is reliable for small values of the noise intensity and provided that the reference value I is not too close to the

Langevin equations with colored noise

final equilibrium value of $q^2(t)$. From (3.5.1) we have

$$t_I = \frac{1}{2\alpha}\ln\frac{I}{h^2}, \tag{3.5.7}$$

so that

$$W(\lambda) = \langle e^{-\lambda t}\rangle, \tag{3.5.8}$$

$$\Omega(\lambda) = -\frac{\lambda}{2\alpha}\ln\left(\frac{\alpha I}{\varepsilon'}\right) + \ln\Gamma\left(1 + \frac{\lambda}{2\alpha}\right). \tag{3.5.9}$$

It should be noted that strictly (3.5.7) gives any passage time and not the first passage time. Multiple crossing of the reference value I is not expected to be important for small noise intensities.

From (3.5.6) and (3.5.9) we obtain

$$T = \frac{1}{2\alpha}\ln\left(\frac{\alpha I}{\varepsilon'}\right) - \frac{\Psi(1)}{2\alpha} = T_0 - \frac{\Psi(1)}{2\alpha} + \frac{1}{2\alpha}\ln(1 + \alpha\tau_1), \tag{3.5.10}$$

$$\sigma_T^2 = \frac{\Psi'(1)}{4\alpha^2}, \tag{3.5.11}$$

where $\Psi(1)$ and $\Psi'(1)$ are, respectively, the digamma function and its derivative, and

$$T_0 = \frac{1}{2\alpha}\ln\left(\frac{\alpha I}{\varepsilon}\right) \tag{3.5.12}$$

is the dominant contribution to T in the Markovian limit. The last term in (3.5.10) is the non-Markovian contribution coming from the renormalization of ε. It gives an increase of T in a quantity independent of ε and typically very small. The time T gives a measure of the lifetime of the unstable state. It essentially coincides with the 'matching time' used in scaling theories (Suzuki, 1977) and gives the time scale at which nonlinear contributions start to become important. Our basic result for T has also been obtained for a 'matching time' by Suzuki, Liu and Tsuno (1986) and Valsakamar (1985).

Nonlinear contributions in the calculation of T can be included following the same procedure of Haake, Haus and Glauber (1981). They only give rise to the substitution of I by $I/1 - (I/\alpha)$ in (3.5.12). This only changes the value of T_0 in (3.5.12) when I is very close to α, and it does not introduce any important new qualitative feature.

Additive and multiplicative noise

We next consider the model ($c = 0$; $a = b = 1$). Physically $\xi(t)$ models an external colored noise while $\zeta(t)$ models internal fluctuations of much smaller intensity which are taken to be a white noise ($\tau_1 = 0$). This is a particular case of (3.2.32a–c). Given the multiplicative character of $\xi(t)$, the decay of the unstable

103

state is initiated by $\zeta(t)$, which is therefore an essential ingredient to describe the process. We approximate the process $q(t)$ following the same basic idea as in the previous case; that is, the additive noise responsible for the decay is replaced by effective random initial conditions. The difference is that now, instead of a deterministic evolution from the random initial conditions, we have stochastic evolution given by (3.4.1) with $\varepsilon = 0$. (De Pasquale *et al.*, 1986a, b). The approximation is then given by the following process $q(t)$:

$$q(t) = \frac{h(t)\exp[\alpha t + w(t)]}{[1 + 2h^2(t)\int_0^t dt' \exp[2(\alpha t' + w(t'))]]^{1/2}} \quad (3.5.13)$$

where

$$w(t) = \int_0^t dt' \, \xi(t'), \quad (3.5.14)$$

and $h(t)$ is defined in (3.5.2). Equation (3.5.13) gives the formal solution of (3.4.1) for $\varepsilon = 0$ and initial condition $h(\infty)$. The Gaussian process $w(t)$ has zero mean and variance

$$\langle w^2(t) \rangle = D(t) = D(t - \tau) + D\tau e^{-t/\tau} \approx D(t - \tau). \quad (3.5.15)$$

We note that the values taken by the process $w(t)$ at time t can be obtained by replacing $w(t)$ by $vD(t)$ where v is a Gaussian variable of zero mean and unit variance.

The approximation (3.5.13) gives the passage time statistics as well as the time dependence of the moments. As we discussed in the previous case the passage time statistics can be obtained within a linear regime, while the nonlinear regime is better described in terms of the time evolution of the variance which exhibits the anomalous fluctuation phenomenon. For the calculation of passage time statistics we neglect the denominator in (3.5.13) and look for the times t_I at which $q^2(t)$ takes a reference value I. As we did before we also approximate $h(t)$ by $h(\infty)$. We get

$$t - \tau + vD(t - \tau) = T_0 - \tau + \frac{1}{2\alpha}\ln\frac{\varepsilon\alpha}{h^2(\infty)}, \quad (3.5.16)$$

where T_0 is given in (3.5.12). The generating function $W(\lambda)$ can be calculated by expanding the solution of (3.5.16) for $t - \tau$ using T_0 as a large parameter and taking a double average over v and h. We obtain (De Pasquale *et al.*, 1986b)

$$W(\lambda) = W_{D=0}(\lambda)\left(1 + \frac{\lambda D}{\alpha^2}\right)^{-1/2} \exp\left[\frac{\lambda^2 D(T_0 - \tau)}{2\alpha^2\left(1 + \frac{\lambda D}{\alpha^2}\right)}\right], \quad (3.5.17)$$

where

$$W_{D=0}(\lambda) = \Gamma\left(\frac{\lambda}{2\alpha} + 1\right)e^{-\lambda T_0} \quad (3.5.18)$$

is the generating function for $D = 0$. The MFPT and variance are obtained

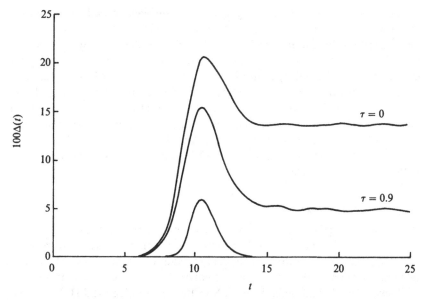

Figure 3.7. Transient fluctuations for the model (3.4.1) ($c = 0$; $a = b = 1$) calculated from (3.5.13) for $\varepsilon = 2 \times 10^{-9}$, $D = 0.14$ (De Pasquale *et al.*, 1986b). Bottom curve corresponds to $D = 0$.

from (3.5.17):

$$T = T_0 - \frac{\Psi(1)}{2\alpha} + \frac{D}{2\alpha^2} \tag{3.5.19}$$

$$\sigma_T^2 = \frac{\Psi'(1)}{4\alpha^2} + \frac{D(T_0 - \tau)}{\alpha^2} + \frac{D^2}{2\alpha^4}. \tag{3.5.20}$$

The results show the effect of a multiplicative colored noise in T and σ_T^2 (compare also with (3.5.10) and (3.5.11)). It is seen that T is independent of τ and essentially unaffected by the multiplicative noise in the usual case in which $T_0 \gg 1$. However, the variance has an important enhancement due to the multiplicative noise through the term $D(T_0 - \tau)/\alpha^2$. The enhancement is partially reduced by the finite correlation time τ.

The nonlinear regime is well described by $\Delta(t) = \langle q^4(t) \rangle - \langle q^2(t) \rangle^2$, which has a large maximum at a time slightly larger than T (Figure 3.7). Numerical simulations (De Pasquale *et al.*, 1986a, b) indicate that the behavior of $\Delta(t)$ is reproduced accurately by (3.5.13). The physical mechanism which causes the peak of $\Delta(t)$ for $D = 0$ is well understood. The peak is caused by the spread of stochastic trajectories at intermediate times due to the random times at which they leave the unstable state (linear regime). Multiplicative noise starts to be effective once the system leaves the vicinity of $q = 0$. This causes an additional spread, which becomes dominant for larger values of q. Close to the final state

105

the additive noise becomes negligible. This additional spread gives rise to a much larger value of $\Delta(t)$ at its maximum and also at its final value $\Delta(\infty)$. Detailed results for $\Delta(t)$ in the white noise case ($\tau = 0$) are given by De Pasquale *et al.* (1986a). These authors also show that the position of the maximum of $\Delta(t)$ is essentially unaffected by multiplicative noise and that for large values of D the anomalous fluctuation defined as the difference between peak and final values of $\Delta(t)$ tends to disappear. The novel feature that appears for $\tau \neq 0$ is a large decrease in $\Delta(t)$ with respect to the white noise case both at the maximum value and at $t = \infty$ (Figure 3.7). This fact and the result for σ_T^2 indicate that a finite τ compensates, in general, for the effects caused by a multiplicative noise with respect to the case of $D = 0$.

Acknowledgements

We acknowledge the collaboration in the work reviewed here of our colleagues and friends, F. De Pasquale, J. D. Gunton, A. Hernández-Machado, S. Katz, L. Pesquera, M. A. Rodriguez, F. Sagués and P. Tartaglia. We have also benefited from collaborations involving analogic simulations by T. Kawakubo, R. Mannella, P. V. E. McClintock, F. Moss and H. Yamazaki. We finally acknowledge the collaboration of M. Aguado, J. Casademunt and L. Ramirez de la Piscina in several points of this chapter. Financial support is acknowledged from Comisión Asesora para la Investigación Científica y Técnica (Project. 361/84) and Dirección General de Investigación Científica y Técnica (Project. PB-86-0534) (Spain).

References

Aguado, M. and San Miguel, M. 1988. *Phys. Rev. A* **37**, 450.
Arnold, L. 1974. *Stochastic Differential Equations*, New York: Wiley and Sons.
Casademunt, J., Mannella, R., McClintock, P. V. E., Moss, F. E. and Sancho, J. M. 1987. *Phys. Rev. A* **35**, 5183.
Dekker, H. 1982. *Phys. Lett. A* **90**, 26.
De Pasquale, F. and Tombesi, P. 1979. *Phys. Lett. A* **72**, 7.
De Pasquale, F., Sancho, J. M., San Miguel, M. and Tartaglia, P. 1986a. *Phys. Rev. A* **33**, 4360.
De Pasquale, F., Sancho, J. M., San Miguel, M. and Tartaglia, P. 1986b. *Phys. Rev. Lett.* **56**, 2473.
Faetti, S., Fronzoni, L., Grigolini, P. and Mannella, R. 1987a. Preprint.
Faetti, S., Fronzoni, L., Grigolini, P., Palleschi, V. and Tropiano, G., 1987b. Preprint.
Fox, R. F. 1979. *Phys. Rep.* **486**, 171.
Fox, R. F. 1983. *Phys. Lett. A* **94**, 281.
Fox, R. F. 1986a. *Phys. Rev. A* **33**, 476.
Fox, R. F. 1986b. *Phys. Rev. A* **34**, 4525.
Fox, R. F. and Roy, R. 1987. *Phys. Rev. A* **35**, 1838.

Langevin equations with colored noise

Fronzoni, L., Grigolini, P., Hänggi, P., Moss, F., Mannella, R. and McClintock, P. V. E. 1986. *Phys. Rev. A* **33**, 3320.

Gardiner, C. W. 1983. *Handbook of Stochastic Methods.* Berlin: Springer-Verlag.

Garrido, L. and Sancho, J. M. 1982. *Physica A* **115**, 479.

Garrido, L. and San Miguel, M. 1978. *Progr. Theor. Phys.* **59**, 40, 55.

Grigolini, P. 1986. *Phys. Lett. A* **119**, 157.

Haake, F. and Lewenstein, M. 1983. *Phys. Rev. A* **28**, 3606.

Haake, F., Haus, J. W. and Glauber, R. 1981. *Phys. Rev. A* **23**, 3235.

Hänggi, P. 1985. In *Stochastic Processes Applied to Physics* (L. Pesquera and M. A. Rodriguez, eds), p. 69. Singapore: World Scientific.

Hänggi, P. and Talkner, P. 1985. *Phys. Rev. A* **32**, 1934.

Hänggi, P., Marchesoni, F. and Grigolini, P. 1984. *Z. Phys. B* **23**, 333.

Hänggi, P., Mroczkowski, T. J., Moss, F. and McClintock, P. V. E. 1985. *Phys. Rev. A* **32**, 695.

Hernández-Machado, A. and Sagués, F. 1983. *Phys. Lett.* **98A**, 98.

Hernández-Machado, A. and San Miguel, M. 1984. *J. Math. Phys.* **25**, 1066.

Hernández-Machado, A. and San Miguel, M. 1986. *Phys. Rev. A* **33**, 2481.

Hernández-Machado, A., San Miguel, M. and Katz, S. 1985. *Phys. Rev. A* **31**, 2362.

Hernández-Machado, A., San Miguel, M. and Katz, S. 1986. *Recent Developments in Nonequilibrium Thermodynamics: Fluids and Related Topics.* Lecture Notes in Physics 253, p. 368. Berlin: Springer-Verlag.

Hernández-Machado, A., San Miguel, M. and Sancho, J. M. 1984. *Phys. Rev. A* **29**, 3388.

Hernández-Machado, A., Sancho, J. M., San Miguel, M. and Pesquera, L. 1983. *Z. Phys. B* **52**, 335.

Horsthemke, W. and Lefever, R. 1984. *Noise Induced Transitions.* Berlin: Springer-Verlag.

Jung, P. and Hänggi, P. 1987. *Phys. Rev. A.* **35**, 4464.

Jung, P. and Risken, H. 1984. *Phys. Lett.* **103A**, 38.

Jung, P. and Risken, H. 1985a. *Z. Phys. B* **61**, 367.

Jung, P. and Risken, H. 1985b. *Z. Phys. B* **59**, 469.

Klyatskin, V. I. 1974. *Sov. Phys. JETP* **38**, 27.

Klyatskin, V. I. 1986. In *Ondes et Equations Stochastiques dans les Milieux Aleatoirement Non Homogenes.* Paris: Les Editions de Physique.

Kramers, H. A. 1940. *Physica* **7**, 284.

Kubo, R. 1962a. *J. Phys. Soc. Jpn.* **17**, 1100.

Kubo, R. 1962b. In *Fluctuations, Relaxation and Resonance in Magnetic Systems* (D. Ter Haar, ed.). Olyver and Boyd.

Kubo, R. 1963. *J. Math. Phys.* **4**, 174.

Lax, M. 1966. *Rev. Mod. Phys.* **38**, 541.

Lett, P., Gage, E. C. and Chyba, T. H. 1987. *Phys. Rev. A* **35**, 746.

Lett, P., Short, R. and Mandel, L. 1984. *Phys. Rev. Lett.* **52**, 341.

Lindenberg, K. and West, B. J. 1983. *Physica A* **119**, 485.

Lindenberg, K. and West, B. J. 1984. *Physica A* **128**, 25.

Martin, P. C., Siggia, E. D. and Rose, H. A. 1973. *Phys. Rev. A* **88**, 423.

Masoliver, J., Lindenberg, K. and West, B. J. 1986. *Phys. Rev. A* **34**, 2531.

Masoliver, J., West, B. J. and Lindenberg, K. 1987. *Phys. Rev. A* **35**, 3086.

Moss, F., Hänggi, P., Mannella, R. and McClintock, P. V. E. 1986. *Phys. Rev. A* **33**, 4459.

Nadler, W. and Schulten, K. 1985. *Z. Phys. B* **59**, 53.

Novikov, E. A. 1965. *Soviet Phys. JETP* **20**, 1290.

Pesquera, L. and Rodriguez, M. 1985. *Recent Developments in Nonequilibrium Thermodynamics: Fluids and Related Topics*: Lecture Notes in Physics 253, p. 372. Berlin: Springer-Verlag.

Ramirez de la Piscina, L. and Sancho, J. M. 1988. *Phys. Rev. A* **37**, 4469.

Risken, H. 1984. *The Fokker–Planck Equation.* Berlin: Springer-Verlag.

Risken, H. 1985. *Stochastic Processes Applied to Physics*, (L. Pesquera and M. A. Rodriguez, eds.), p. 29. Singapore: World Scientific.

Rodriguez, M. A. and Pesquera, L. 1986. *Phys. Rev. A* **34**, 4532.

Rodriguez, M. A., Pesquera, L., San Miguel, M. and Sancho, J. M. 1985. *J. Stat. Phys.* **40**, 669.

Roerdink, J. B. T. M. 1982. *Physica A* **112**, 557.

Roerdink, J. B. T. M. 1984. *Physica A* **123**, 369.

Sagués, F. 1984. *Phys. Lett. A* **104**, 1.

Sagués, F. and San Miguel, M. 1985. *Phys. Rev. A* **32**, 1843.

Sagués, F., San Miguel, M. and Sancho, J. M. 1984. *Z. Phys. B* **55**, 269.

Sancho, J. M. 1985a. *Phys. Rev. A* **31**, 3523.

Sancho, J. M. 1985b. *Stochastic Processes Applied to Physics.* (L. Pesquera and M. A. Rodriguez, eds.), p. 96. Singapore: World Scientific.

Sancho, J. M. and Sagués, F. 1985. *Physica A* **132**, 489.

Sancho, J. M. and San Miguel, M. 1980. *Z. Phys. B* **36**, 357.

Sancho, J. M. and San Miguel, M. 1984a. *Fluctuations and Sensitivity in Nonequilibrium Systems.* (W. Horsthemke and D. K. Kondepudi, eds.), p. 114. Berlin: Springer-Verlag.

Sancho, J. M. and San Miguel, M. 1984b. *J. Stat. Phys.* **37**, 151.

Sancho, J. M. and San Miguel, M. 1984c. *Recent Developments in Nonequilibrium Thermodynamics.* Lecture Notes in Physics 199, p. 337. Berlin: Springer-Verlag.

Sancho, J. M., Sagués, F. and San Miguel, M. 1986. *Phys. Rev. A* **33**, 3399.

Sancho, J. M., San Miguel, M. and Dürr, D. 1982. *J. Stat. Phys.* **28**, 291.

Sancho, J. M., Mannella, R., McClintock, P. V. E. and Moss, F. 1985. *Phys. Rev. A* **32**, 3639.

Sancho, J. M., San Miguel, M., Katz, S. and Gunton, J. D. 1982a. *Phys. Rev. A* **26**, 1589.

Sancho, J. M., San Miguel, M., Pesquera, L. and Rodriguez, M. A. 1987. *Physica A* **142**, 532.

Sancho, J. M., San Miguel, M., Yamazaki, H. and Kawakubo, T. 1982b. *Physica A* **116**, 560.

San Miguel, M. 1979. *Z. Physik B* **33**, 307.

San Miguel, M. 1985. *Phys. Rev. A* **32**, 3811.

San Miguel, M. and Sagués, F. 1986. *Recent Developments in Nonequilibrium Thermodynamics: Fluids and Related Topics.* Lecture Notes in Physics 253, p. 305. Berlin: Springer-Verlag.

San Miguel, M. and Sancho, J. M. 1980a. *Phys. Lett. A* **76**, 97.

San Miguel, M. and Sancho, J. M. 1980b. *J. Stat. Phys.* **22**, 605.

San Miguel, M. and Sancho, J. M. 1981a. *Z. Physik B* **43**, 361.

San Miguel, M. and Sancho, J. M. 1981b. *Stochastic Nonlinear Systems in Physics, Chemistry and Biology*, (L. Arnold and R. Lefever, eds.), Springer Series in Synergetics 8, p. 137. Berlin: Springer-Verlag.

San Miguel, M. and Sancho, J. M. 1982. *Phys. Lett.* **90A**, 455.

San Miguel, M., Pesquera, L., Rodriguez, M. and Hernández-Machado, A. 1987. *Phys. Rev. A* **35**, 208.

Schenzle, A. and Tél, T. 1985. *Phys. Rev. A* **32**, 596.

Stratonovich, R. L. 1963. *Introduction to the Theory of Random Noise*, vol. I. New York: Gordon and Breach.

Stratonovich, R. L. 1967. *Introduction to the Theory of Random Noise*, vol. II. New York: Gordon and Breach.

Suzuki, M. 1977. *J. Stat. Phys.* **16**, 11.

Suzuki, M., Liu, Y. and Tsuno, T. 1986. *Physica A* **138**, 433.

Valsakamar, M. C. 1985. *J. Stat. Phys.* **39**, 347.

Van Kampen, N. G. 1976. *Phys. Rep.* **24c**, 171.

Van Kampen, N. G. 1981. *Stochastic Processes in Physics and Chemistry*. Amsterdam: North-Holland.

Wodckiewicz, K. 1982. *J. Math. Phys.* **23**, 11.

4 First passage time problems for non-Markovian processes

KATJA LINDENBERG, BRUCE J. WEST and
JAUME MASOLIVER

4.1 Introduction

This book is a reflection of the recent interest in the correlation properties of fluctuations driving linear and nonlinear systems. The interest arises because fluctuations in physical systems are never truly white and because of the sensitive dependence of certain system properties on these correlations. In the last few years a number of theoretical techniques have been proposed for the proper treatment of colored noise in stochastic systems. Also, a number of experiments have been designed to explore the effects of color on system response.

In this chapter we concentrate on one class of properties that are particularly responsive to the color of the noise that drives a physical system. The first time $T(\mathbf{x}_p)$ that an evolving stochastic physical process $\mathbf{X}(t)$ reaches a preassigned value $\mathbf{x}_p$ is called the first passage time and depends sensitively on the correlation properties of the noise. Of course the first passage time $T(\mathbf{x}_p)$ is itself a statistical quantity and is characterized by a distribution function $f(t)$ as distinct from the distribution $P(\mathbf{x}, t)$ for the state variable.

The first passage time from an initial state $\mathbf{X}(0) = \mathbf{x}_0$ to $\mathbf{x}_p$ is sensitive to the distribution density $P(\mathbf{x}, t)$ for values of $\mathbf{x}$ covered by $\mathbf{X}(t)$ as it evolves from $\mathbf{x}_0$ to $\mathbf{x}_p$. The biggest contributions to $T(\mathbf{x}_p)$ arise from those portions of a trajectory for which $P(\mathbf{x}, t)$ is small. These improbable regions of phase space through which the trajectory passes are the most difficult to traverse and therefore contribute most to the first passage time. It is therefore particularly important to the accurate calculation of the first passage time distribution to have a faithful description of these low probability regions of phase space. The clearest example of this effect is in the asymptotic region of the distribution $P(\mathbf{x}, t)$ where all values of $\mathbf{x}$ are improbable. The calculation of the first passage time to one of these values requires accurate knowledge of the tail of the distribution. It should furthermore be noted that the connection between the distribution $P(\mathbf{x}, t)$ and the first passage time properties is, in general, quite complicated. Thus, even with complete knowledge of $P(\mathbf{x}, t)$ one may not be able to determine $f(t)$ or its moments. This difficulty has only recently been overcome in some one-dimensional problems (see below). In higher dimensions, theories exist for the calculation of 'exit times', i.e. the time that it takes to leave a closed region of phase space for the first time. These theories, which are mostly based on singular perturbation methods, are beyond the scope of this review (Schuss, 1980).

Historically, it has been notoriously difficult to faithfully determine the distribution for processes driven by arbitrarily correlated noise. Until quite recently, the only well-developed methods required the noise to be Gaussian and white. In this case $P(x, t)$ satisfies a Fokker–Planck equation (Wang and Uhlenbeck, 1945). The relaxation of these restrictions on the noise has stimulated a great deal of activity, leading to a number of important new results that will be discussed subsequently (see Sections 4.3 and 4.4). The introduction of (exponentially correlated) colored Gaussian noise has been dealt with in two ways. In one, the noise itself is treated as a dynamical process driven by white fluctuations. In this approach one extends the phase space of the dynamical variables to include the colored noise. In this extended phase space the Fokker–Planck formalism again applies. Even if the original process $X(t)$ is one-dimensional, the extended process is at least two-dimensional and thus all the difficulties mentioned earlier arise when one wants to calculate the first passage time to x_p. The second strategy for the introduction of Gaussian colored noise is to use one of a number of perturbative techniques involving the intensity and/or correlation time of the fluctuations. These perturbative techniques typically break down in the asymptotic (tail) region of the distribution $P(x, t)$. Thus, any calculation of first passage times to asymptotic regions based on these methods is suspect. Even the calculation of first passage times to the central region of the distribution (which may at times be accurately calculated via perturbation methods) must be handled with caution because the systematic relation between $P(x, t)$ and $f(t)$ has not been established. This remains true in one dimension when the noise is colored.

The difficulties in dealing with Gaussian colored noise have motivated studies of models of non-Gaussian colored noise. In particular, many investigators have found correlated dichotomous (i.e. two-valued) processes to be useful models of the noise statistics. For these models not only is it possible to calculate $P(x, t)$ exactly, but one can also develop independent exact methods for the calculation of the distribution of first passage times. These methods have recently been extended to more sophisticated distributions of the noise (see Section 4.4).

In this chapter we deal with a number of model systems to illustrate the approaches mentioned above and the resultant predictions. The variety of physically interesting problems may obscure the common features related to the introduction of color in the fluctuations. Thus it is useful to state the ubiquitous effect that has emerged from both the exact and the approximate treatment of color: the distribution $P(x, t)$ is usually narrowed when correlations are introduced into the noise and consequently the mean first passage time to a preassigned value is increased (an exception occurs in the Hongler model treated by Doering, 1986). This property has been observed in the treatment of colored noise whose statistics are Gaussian, dichotomous, discrete multivalued as well as shot noise, and we conjecture its even broader veracity.

The processes to be considered throughout this chapter evolve according to a dynamic equation of the form (West and Lindenberg, 1987)

$$\dot{X}(t) = G(X(t)) + g(X(t))\eta(t), \tag{4.1.1}$$

where $G(X)$ is the deterministic 'force' and $\eta(t)$ is a zero-centered fluctuating quantity of prescribed statistics. The two-time correlation function of this fluctuating quantity is written as

$$\langle \eta(t)\eta(\tau) \rangle = Q(t - \tau), \tag{4.1.2}$$

where the brackets $\langle \quad \rangle$ denote an average over an ensemble of realizations of $\eta(t)$. The fluctuations in (4.1.1) are 'additive' if $g(X) = $ constant and 'multiplicative' otherwise. The intensity D of the noise is the integrated value of the correlation function and the correlation time τ_c is its normalized first moment:

$$D = \int_0^\infty Q(t)\, dt, \tag{4.1.3}$$

$$\tau_c = \frac{1}{D} \int_0^\infty t Q(t)\, dt. \tag{4.1.4}$$

4.2 Gaussian white fluctuations

The theory of first passage times is well established for one-dimensional processes driven by delta-correlated Gaussian fluctuations $\eta(t)$. In this case the correlation function (4.1.2) is given by

$$Q(t - \tau) = 2D\delta(t - \tau) \tag{4.2.1}$$

and the correlation time $\tau_c \to 0$. It is useful to briefly review this theory so that its dependency on the correlation function (4.2.1) and on the Gaussian distribution of $\eta(t)$ become apparent (Lindenberg and West, 1986; Weiss, 1966).

When $\eta(t)$ in the dynamical equation (4.1.1) is Gaussian and white, then an equivalent description of the process is given by the Fokker–Planck equation (Wang and Uhlenbeck, 1945) (using the Stratonovich calculus)

$$\frac{\partial}{\partial t} P_t = -\frac{\partial}{\partial x} G(x) P_t + D \frac{\partial}{\partial x} g(x) \frac{\partial}{\partial x} g(x) P_t \tag{4.2.2}$$

together with the appropriate boundary conditions and the initial condition $P_0 = \delta(x - x_0)$. Here $P_t\, dx \equiv P(x, t|x_0)\, dx$ is the probability that the process

$X(t)$ lies in the range $x \leqslant X(t) \leqslant x + dx$ at time t given that initially $X(0) = x_0$ (in Section 4.1 we were less precise about the definition of the probability density). The steady-state solution of (4.2.2) is given by

$$P_{ss}(x) \equiv \lim_{t \to \infty} P(x,t|x_0) = \frac{N}{g(x)} \exp\left[\frac{1}{D}\int^x dx' \frac{G(x')}{g^2(x')}\right], \qquad (4.2.3)$$

where N is the normalization constant. A related function that is useful for the calculation of the first passage time to $x = x_p$ is the probability density $W_t \equiv W(x,t|x_0)$ for $X(t)$ to lie in the interval $(x, x + dx)$ but subject to the additional constraint that $X(\tau)$ has not crossed the value x_p for $0 \leqslant \tau \leqslant t$. This probability density also satisfies (4.2.2) together with the absorbing boundary condition

$$W(x_p, t|x_0) = 0 \qquad (4.2.4)$$

replacing one of the original boundary conditions. This distribution function can be used to define the conditional cumulative distribution $f(t, x_p|x_0)$ of first passage times,

$$f(t, x_p|x_0) = \int_{-\infty}^{x_p} dx \, W(x,t|x_0), \qquad (4.2.5)$$

where we have taken $x_0 < x_p$ (otherwise the limits on the integral are x_p and ∞). The moments of the first passage time distribution are often used to characterize the process, and are given by

$$T_n(x_p|x_0) \equiv \int_0^\infty dt \, nt^{n-1} f(t, x_p|x_0). \qquad (4.2.6)$$

The first moment, $T_1(x_p|x_0)$, is the mean first passage time from x_0 to x_p and is a frequently used measure of the transition of a process from one state to another. A hierarchy of equations for the moments T_n can be derived directly from the backward Kolmogorov equation (i.e. the adjoint equation of (4.2.2)):

$$\frac{\partial}{\partial t} W_t = [G(x_0) + Dg(x_0)g'(x_0)] \frac{\partial}{\partial x_0} W_t + Dg^2(x_0) \frac{\partial^2}{\partial x_0^2} W_t. \qquad (4.2.7)$$

Multiplying (4.2.7) by nt^{n-1} and integrating over $x < x_p$ and over t from 0 to ∞ gives the exact equation

$$Dg^2(x_0) \frac{\partial^2}{\partial x_0^2} T_n(x_p|x_0) + [G(x_0) + Dg(x_0)g'(x_0)] \frac{\partial}{\partial x_0} T_n(x_p|x_0)$$
$$= -nT_{n-1}(x_p|x_0), \qquad (4.2.8)$$

with $T_0(x_p|x_0) \equiv 1$ (Pontryagin, Andronov and Vitt, 1933). The boundary conditions for (4.2.8) follow directly from those for W_t. In particular, if the process begins at x_p then its first passage time to x_p vanishes and hence

$$T_n(x_p|x_p) = 0 \qquad (4.2.9)$$

113

for all n. This condition follows from the absorbing boundary condition (4.2.4). The other condition may, for instance, be a reflecting boundary at x_L:

$$\frac{\partial}{\partial x_0} T_n(x_p|x_0)\big|_{x_0 = x_L} = 0. \tag{4.2.10}$$

With these conditions, (4.2.8) can be solved exactly and one obtains for the mean first passage time

$$T_1(x_p|x_0) = \int_{x_0}^{x_p} dz \frac{1}{g^2(z)P_{ss}(z)} \int_{x_L}^{z} dy\, P_{ss}(y). \tag{4.2.11}$$

If one wishes to calculate the first passage time statistics to boundaries at, say, x_p and $-x_p$ then (4.2.10) is replaced by a second absorbing boundary condition which insures zero first passage time for a process that starts at the lower boundary. In this case

$$T_1(x_p| \pm x_p) = 0. \tag{4.2.12}$$

Equation (4.2.11) is then slightly modified:

$$T_1(x_p|x_0) = \int_{x_0}^{x_p} dz \frac{1}{g^2(z)P_{ss}(z)} \int_{0}^{z} dy\, P_{ss}(y), \tag{4.2.13}$$

where we have assumed $P_{ss}(z)$ and $g^2(z)$ to be even functions of z (other symmetries can be easily considered).

Two features of the preceding analysis should be emphasized. The first is the differential structure (Sturm–Liouville form) of the equation of evolution for the probability densities which leads to the differential equation for the first passage time moments. This structure is a direct consequence of the Gaussian and white nature of the noise. The second feature is the 'boundary value' nature of the problem, i.e. the fact that the first passage time to a boundary can be expressed as an absorbing boundary condition. Any continuous process must touch a boundary in order to cross it (in contrast with discontinuous processes of, e.g., the stable or Lévy form which can jump over points). Thus for any such process the first passage time distribution is constrained by boundary conditions. The process generated by (4.1.1) with Gaussian white noise is certainly continuous and hence leads to boundary conditions of this sort. However, the precise nature of the absorbing boundary condition is a direct consequence of the whiteness of the noise, which causes $\dot{X}(t)$ to change at each instant of time.

We thus anticipate that changes in the statistics and correlation properties of the noise will lead to changes in the equation of evolution and in the nature of the boundary conditions. However, the existence of such boundary conditions will persist as long as the process $X(t)$ is continuous.

4.3   Continuous colored (mostly Gaussian) fluctuations

Let us now return to (4.1.1) and consider the case of correlated fluctuations as in (4.1.2). The oldest procedure to deal with such problems is based on an extension of the space of variables so that $\eta(t)$ itself becomes a variable driven by white noise $\zeta(t)$. In particular, if $\eta(t)$ is exponentially correlated Gaussian noise then one can write the two-variable set of stochastic differential equations

$$\dot{X}(t) = G(X) + g(X)\eta(t) \tag{4.3.1}$$

$$\dot{\eta}(t) = -\frac{1}{\tau_c}\eta(t) + \zeta(t), \tag{4.3.2}$$

where $\zeta(t)$ is Gaussian white noise with correlation function

$$\langle \zeta(t)\zeta(\tau)\rangle = 2D\,\delta(t-\tau). \tag{4.3.3}$$

It can easily be seen that (4.3.2) then leads to the exponential correlation function

$$\langle \eta(t)\eta(\tau)\rangle = \frac{D}{\tau_c}e^{-|t-\tau|/\tau_c}. \tag{4.3.4}$$

The probability density $P(x, \eta, t|x_0, \eta_0)$ then obeys a Fokker–Planck equation. The difficulty lies in connecting this two-dimensional Fokker–Planck equation to the first passage time to x_p or to $|x_p|$ regardless of the values of η and η_0. We have already mentioned the difficulties that exist in the calculation of first passage time properties for systems of more than one dimension.

The next step that one may wish to consider is to re-express the evolution of the system as a one-dimensional process, i.e. to determine an evolution equation for $P(x, t|x_0)$. One can approach this task in a number of ways, including the integration of the two-dimensional Fokker–Planck equation for $P(x, \eta, t|x_0, \eta_0)$ over η and η_0. A formal expression of the resulting equation has the form

$$\frac{\partial}{\partial t}P(x, t|x_0) = -\frac{\partial}{\partial x}G(\dot{x})P(x, t|x_0)$$

$$+ \int_0^t \mathrm{d}t' \int \mathrm{d}x'\, K(x, x', t - t')P(x', t'|x_0), \tag{4.3.5}$$

which is in general nonlocal in state space and in time. A number of techniques have been developed that lead to formally exact (but intractable) expressions for the kernel operator $K(x, x', t - t')$. These techniques have generally been applied to physical processes in which the fluctuations assume a continuous range of values. In most cases the fluctuations are assumed to be Gaussian. Even if one were able to obtain an explicit form for the kernel, the connection between the evolution equation (4.3.5) and the first passage time statistics would remain obscure. In particular, the first passage time problem in this 'reduced space' can no longer be expressed as a boundary value problem when the problem is nonlocal in state space.

The double jeopardy of intractability and inability to obtain first passage time statistics has driven most authors to the implementation of approximations that make the problem tractable. The universal approximation that has been implemented in one form or another in the various techniques is the representation of the evolution equation for $P(x, t|x_0)$ as an effective Fokker–Planck equation (EFPE):

$$\frac{\partial}{\partial t} P_t = -\frac{\partial}{\partial x} [G(x)P_t] + \frac{\partial}{\partial x} g(x) \frac{\partial}{\partial x} g(x) D(x, \tau_c) P_t. \qquad (4.3.6)$$

The 'effective diffusion function' $D(x, \tau_c)$ is of $O(D)$ and goes to D as $\tau_c \to 0$. Its specific form for finite τ_c varies according to the details of the approximations made in its derivation (see below). All of the approximations rely on D being small and on various combinations of D and τ_c being small. In other words, the Langevin equation (4.1.1) with colored noise $\eta(t)$ is in each of these approaches approximated by another Langevin equation driven by Gaussian white fluctuations $\eta^w(t)$ of intensity D (Lindenberg, Shuler, Seshadri and West, 1983):

$$\dot{X}(t) = G(X) - g^2(X)[D^{1/2}(X, \tau_c)]'$$
$$+ \left[\frac{D(X, \tau_c)}{D}\right]^{1/2} g(X)\eta^w(t), \qquad (4.3.7)$$

where the prime denotes differentiation with respect to X. One reason for seeking the Fokker–Planck form (4.3.6) is that it presumably enables one to directly implement the formalism discussed in Section 4.2 for the determination of first passage time statistics.

As mentioned above, different forms for the effective diffusion function $D(x, \tau_c)$ have been proposed. This has engendered significant discussion on the relative merits of the respective techniques. It appears to us that these different forms do not arise as a result of the differences in the techniques but rather as a consequence of different approximation schemes implemented within the various techniques. It is of course true that different methods suggest different approximation at various stages; however, one should focus on the approximations and not the techniques as the source of the differences. In fact, we demonstrate that the same approximation within the different methods leads to the same EFPE. This point has already been recognized by Grigolini and his coworkers, who discussed the different EFPEs within the context of one technique (projection operators) (Faetti, Fronzoni, Grigolini and Mannella, 1986a). We instead herein work within the context of each of the different techniques and show that they all lead to the same EFPE if the same approximation is applied.

As also recognized by Grigolini and his coworkers, the entire Fokker–Planck structure (4.3.6) for any value of D and/or $\tau_c > 0$ may break down because the contributions of neglected terms may not be negligible (Faetti et al., 1986b). They show that these terms yield corrections to $D(x, \tau_c)$ of $O(D\tau_c^2)$ for

116

Gaussian noise and of $O(D\tau_c)$ for non-Gaussian noise. This breakdown often influences the asymptotic regions of low probability more than the high probability regions. In the context of our discussions, it is likely that the first passage time statistics between states in the central region of the distribution can be faithfully calculated using the EFPE. On the other hand, great caution must be exercised in using this formalism to calculate first passage times to asymptotic regions.

We reiterate a point made earlier: a big motivating factor for the pursuit of the EFPE structure (4.3.6) is that, until recently, no methods other than the Fokker–Planck equation route were available to relate a dynamical stochastic equation such as (4.1.1) to the first passage time properties. Thus, extension of the evolution equation for the probability P_t beyond the Sturm–Liouville form (4.3.6) does not in itself provide a formalism for the determination of first passage time statistics. Therefore, in the absence of the EFPE structure, radically different approaches to the first passage time problem have been developed. In this section we review the techniques and approximations that lead to the EFPE. In the following section we introduce and review some of the methods that are useful when the EFPE structure is not applicable.

4.3.1   Cumulant resummation technique

Let us begin with the solution $X_\eta(t)$ of the stochastic equation (4.1.1) for a *particular* realization of the fluctuations $\eta(t)$ and the initial condition $X_\eta(0) = x_0$. One can then define the conditional phase space distribution function

$$\rho_\eta(x, t|x_0) = \delta[x - X_\eta(t)]. \tag{4.3.8}$$

The conditional probability density $P_t \equiv P(x, t|x_0)$ that characterizes the process is defined as the average of the distribution (4.3.8) over all realizations of $\eta(t)$ in an ensemble whose members all start from the same initial value x_0:

$$P(x, t|x_0) = \langle \delta[x - X_\eta(t)] \rangle. \tag{4.3.9}$$

The evolution of P_t is determined by that of ρ_η, which is given by the Liouville equation

$$\frac{\partial}{\partial t}\rho_\eta(x, t|x_0) + \frac{\partial}{\partial x}[\dot{x}_\eta \rho_\eta(x, t|x_0)] = 0 \tag{4.3.10}$$

subject to the initial condition $\rho_\eta(x, 0|x_0) = \delta(x - x_0)$. The 'velocity' $\dot{x}_\eta$ in (4.3.10) is to be interpreted as a shorthand notation for $[G(x) + g(x)\eta(t)]$ (cf. (4.1.1)). In implementing the average (4.3.9) in (4.3.10) it is useful to define the differential deterministic and stochastic operators L_0 and $L_\eta(t)$ through their action on an arbitrary function $h(x)$:

$$L_0 h(x) = -\frac{\partial}{\partial x}[G(x)h(x)], \tag{4.3.11}$$

117

$$L_\eta(t)h(x) \equiv -\eta(t)\frac{\partial}{\partial x}[g(x)h(x)]. \tag{4.3.12}$$

It is also useful to introduce the cumulant functions $K_n(t)$, $n = 1, 2, \ldots$, the first few of which are explicitly given as follows:

$$K_1(t) = 0 \tag{4.3.13}$$

$$K_2(t) = \int_0^t d\tau_1 \int_0^{\tau_1} d\tau_2 m_2(\tau_1, \tau_2) \tag{4.3.14}$$

$$K_3(t) = \int_0^t d\tau_1 \int_0^{\tau_1} d\tau_2 \int_0^{\tau_2} d\tau_3 m_3(\tau_1, \tau_2, \tau_3) \tag{4.3.15}$$

$$K_4(t) = \int_0^t d\tau_1 \int_0^{\tau_1} d\tau_2 \int_0^{\tau_2} d\tau_3 \int_0^{\tau_3} d\tau_4 [m_4(\tau_1, \tau_2, \tau_3, \tau_4)$$
$$- m_2(\tau_1, \tau_2)m_2(\tau_3, \tau_4) - m_2(\tau_1, \tau_3)m_2(\tau_2, \tau_4)$$
$$- m_2(\tau_1, \tau_4)m_2(\tau_2, \tau_3)]. \tag{4.3.16}$$

Here

$$m_n(\tau_1 \ldots \tau_n) \equiv \langle e^{-L_0\tau_1} L_\eta(\tau_1)e^{-L_0(\tau_2 - \tau_1)}$$
$$\times L_\eta(\tau_2)\ldots e^{-L_0(\tau_n - \tau_{n-1})} L_\eta(\tau_n)e^{L_0\tau_n} \rangle, \tag{4.3.17}$$

and the expressions (4.3.13)–(4.3.17) are valid for zero-centered fluctuations of any statistics having finite moments. After a lengthy but standard procedure one obtains the *exact* evolution equation (Lindenberg et al., 1983; West and Lindenberg, 1987):

$$\frac{\partial}{\partial t}P_t = L_0 P_t + e^{L_0 t} \sum_{n=1}^{\infty} (dK_n(t)/dt)e^{-L_0 t} P_t. \tag{4.3.18}$$

The approximation that yields an EFPE is to truncate the cumulant sum in (4.3.18) at second order:

$$\frac{\partial}{\partial t}P_t = L_0 P_t + \int_0^t d\tau \langle L_\eta(t)e^{L_0\tau} L_\eta(t - \tau)e^{-L_0\tau} \rangle P_t. \tag{4.3.19}$$

The rationale for this truncation is that, in systems with white Gaussian noise, cumulants above the second vanish identically: (4.3.18) is then precisely the Fokker–Planck equation and (4.3.19) is exact. When the noise is colored the higher cumulants no longer vanish, and it is difficult to estimate the magnitude of their contributions. Thus, although the fourth-order cumulant of the fluctuations is of $O(D^2\tau_c)$, Faetti et al. (1986b) have shown that the corrections due to the fourth-order term are actually of $O(D\tau_c^2)$ for Gaussian noise and of $O(D\tau_c)$ for other statistics. We note that this is a difficulty not only with (4.3.19) but with *all* EFPEs that have been constructed to date.

Although our statements above imply it, we have not yet shown that (4.3.19) leads to an EFPE. We now show that it indeed does so with no further approximations (Lindenberg and West, 1983, 1984). To evaluate the integrand in (4.3.19) we use a commutator relation that is valid for arbitrary operators A and B:

$$e^{A}Be^{-A} = B + [A, B] + \frac{1}{2!}[A, [A, B]]$$

$$+ \frac{1}{3!}[A, [A, [A, B]]] + \cdots, \tag{4.3.20}$$

where $[\,,\,]$ denotes the commutator. We identify A with $L_0\tau$ and B with $L_\eta(t - \tau)$. One can follow a simple induction argument to show that the nth order commutator in the series (4.3.20) can be written as

$$\frac{\tau^n}{n!}[L_0, [L_0, \ldots [L_0, L_\eta(t - \tau)] \cdots]]$$

$$= -\eta(t - \tau)\frac{\tau^n}{n!}\frac{\partial}{\partial x}F_n(x), \tag{4.3.21}$$

where the functions $F_n(x)$ obey the recursion relation (Kús and Wódkiewicz, 1982)

$$F_n(x) = G'(x)F_{n-1}(x) - G(x)F'_{n-1}(x), \tag{4.3.22}$$

with $F_0(x) = g(x)$. Hence we have the explicit expression

$$e^{L_0\tau}L_\eta(t - \tau)e^{-L_0\tau} = -\eta(t - \tau)\frac{\partial}{\partial x}\sum_{n=0}^{\infty}\frac{\tau^n}{n!}F_n(x) \tag{4.3.23}$$

and consequently the integrand of (4.3.19) is given by

$$\langle L_\eta(t)e^{L_0\tau}L_\eta(t - \tau)e^{-L_0\tau}\rangle = Q(\tau)\frac{\partial}{\partial x}g(x)\frac{\partial}{\partial x}\sum_{n=0}^{\infty}\frac{\tau^n}{n!}F_n(x). \tag{4.3.24}$$

Insertion of (4.3.24) into (4.3.19) and the assumption that the correlation time is much shorter than the time of observation $(t \gg \tau_c)$ yields an EFPE of the form (4.3.6) with the diffusion function

$$D(x, \tau_c) = \int_0^{\infty}\sum_{n=0}^{\infty}\frac{\tau^n}{n!}\frac{F_n(x)}{g(x)}Q(\tau)\,d\tau. \tag{4.3.25}$$

Using the properties of the recursion relation (4.3.22) to perform the sum over n, one can write the diffusion function (4.3.25) in two equivalent forms. One is the integral expression

$$D(x, \tau_c) = \frac{D}{\tau_c}\frac{G(x)}{g(x)}\int_0^x dx'\frac{g(x')}{G^2(x)}$$

$$\times \exp\left[-\int_{x'}^x\frac{dy}{\tau_c G(y)}\right] + D(0, \tau_c). \tag{4.3.26}$$

If $G(0) = 0$ and $G'(0) = $ constant (as it is in all the examples that we consider here), then

$$D(0, \tau_c) = \frac{D}{1 - G'(0)\tau_c}.$$ (4.3.27)

The second form for the diffusion function is the formal expression

$$D(x, \tau_c) = D \frac{G(x)}{g(x)} \left[1 + \tau_c G(x) \frac{d}{dx} \right]^{-1} \frac{g(x)}{G(x)}.$$ (4.3.28)

In writing (4.3.26)–(4.3.28) we have taken the correlation function to be exponential:

$$Q(t) = \frac{D}{\tau_c} e^{-|t|/\tau_c}.$$ (4.3.29)

The EFPE with the above diffusion function has been called the 'best Fokker–Planck equation' (BFPE).

Using the techniques outlined in Section 4.2 one can now proceed to the calculation of first passage time statistics (Masoliver, West and Lindenberg, 1987). One obtains for the mean first passage time to x_p the result [cf. (4.2.11)]

$$T_1(x_p|x_0) = \int_{x_0}^{x_p} dz \frac{1}{D(z, \tau_c)g^2(z)P_{ss}(z)} \int_{x_L}^{z} dy \, P_{ss}(y),$$ (4.3.30)

where $P_{ss}(x)$ is the steady-state solution of (4.3.6):

$$P_{ss}(x) = \frac{N}{D(x, \tau_c)g(x)} \exp \left[\int^{x} dy \frac{G(y)}{D(y, \tau_c)g^2(y)} \right],$$ (4.3.31)

and where N is the normalization constant.

The procedure outlined above to derive the BFPE can easily be extended to higher dimensions (Lindenberg and West, 1984). However, as noted before, the calculation of first passage time properties in higher dimensions is problematic.

Finally, we note that the cumulant series (4.3.18) with the cumulants (4.3.14)–(4.3.16) et al. is not unique. In particular, a frequently used alternative leads to an exact evolution equation for P_t that is nonlocal in time (Mukamel, Oppenheim and Ross, 1978). When this equation is truncated at the second cumulant, one obtains an approximate equation which differs from the EFPE in various ways, one of them being that it is nonlocal in time (Sancho, Sagués and San Miguel, 1986). This equation also reduces to the exact Fokker–Planck equation in the Gaussian white noise limit, but may in fact give a better approximation to the exact evolution for noise of different statistics. In particular, it is exact for dichotomous (i.e two-valued) exponentially correlated noise whereas (4.3.6) is not. However, the route from this integro-differential evolution equation to the first passage time statistics is not clear even in one dimension.

120

4.3.2   Functional derivative technique

Another method for the treatment of processes driven by colored Gaussian noise employs functional derivative techniques. These techniques require the evaluation of correlations between the phase space distribution function $\rho_\eta(x, t|x_0)$ of (4.3.8) and the dynamical variable $X_\eta(t)$. Most of the literature using this approach is restricted to one-dimensional processes (Sancho, San Miguel, Katz and Gunton, 1982a; Sancho, San Miguel, Yamazaki and Kawakubo, 1982b).

One begins this analysis with the Liouville equation (4.3.10) and averages it over an ensemble of realizations of the fluctuations $\eta(t)$:

$$\frac{\partial}{\partial t}P_t = L_0 P_t - \frac{\partial}{\partial x}g(x)\langle \eta(t)\delta[x - X_\eta(t)]\rangle. \tag{4.3.32}$$

Novikov (1965) established a technique for evaluating the correlation function in (4.3.32), and his theorem states that

$$\langle \eta(t)\delta[x - X_\eta(t)]\rangle = \int_0^t d\tau\, Q(t - \tau)\left\langle \frac{\Delta[\delta(x - X_\eta(t))]}{\Delta\eta(\tau)} \right\rangle$$

$$= -\frac{\partial}{\partial x}\int_0^t d\tau\, Q(t - \tau)$$

$$\times \left\langle \delta[x - X_\eta(t)]\frac{\Delta X_\eta(t)}{\Delta\eta(\tau)} \right\rangle, \tag{4.3.33}$$

where $Q(t)$ is the correlation function (4.1.2) and where Δ denotes a functional derivative. Substitution of (4.3.33) into (4.3.32) gives the formally *exact* evolution equation

$$\frac{\partial}{\partial t}P_t = L_0 P_t + \frac{\partial}{\partial x}g(x)\frac{\partial}{\partial x}\int_0^t d\tau\, Q(t - \tau)$$

$$\times \left\langle \delta[x - X_\eta(t)]\frac{\Delta X_\eta(t)}{\Delta\eta(\tau)} \right\rangle. \tag{4.3.34}$$

This equation is completely equivalent to (4.3.18) and equally intractable as it stands.

Sancho *et al.* (1982a, b) expand the two-time functional derivative in (4.3.34) around the equal-time value:

$$\frac{\Delta X_\eta(t)}{\Delta\eta(\tau)} = \sum_{n=0}^\infty \frac{(-1)^n}{n!}(t - \tau)^n\left[\frac{d^n}{d\tau^n}\frac{\Delta X_\eta(t)}{\Delta\eta(\tau)}\right]_{\tau=t}. \tag{4.3.35}$$

They then retain those contributions to this series that lead to coefficients of $O(D\tau_c^n)$ and neglect terms of $O(D^m\tau_c^n)$ with $m > 1$. To carry out this evaluation they formally solve the stochastic equation (4.1.1),

$$X_\eta(t) = x_0 + \int_0^t du\{G[X_\eta(u)] + g[X_\eta(u)]\eta(u)\}, \qquad (4.3.36)$$

and then take the functional derivative,

$$\frac{\Delta X_\eta(t)}{\Delta\eta(\tau)} = g[X_\eta(\tau)] + \int_\tau^t du\{G'[X_\eta(u)] + g'[X_\eta(u)]\eta(u)\}$$

$$\times \frac{\Delta X_\eta(u)}{\Delta\eta(\tau)}, \quad t > \tau. \qquad (4.3.37)$$

To solve this integral equation for the functional derivative one differentiates (4.3.37) with respect to τ and re-integrates the resulting equation subject to the 'initial condition'

$$\left.\frac{\Delta X_\eta(t)}{\Delta\eta(\tau)}\right|_{t=\tau} = g[X_\eta(\tau)] \qquad (4.3.38)$$

to obtain

$$\frac{\Delta X_\eta(t)}{\Delta\eta(\tau)} = g[X_\eta(\tau)]\exp\int_\tau^t du\{G'[X_\eta(u)] + g'[X_\eta(u)]\eta(u)\}. \quad (4.3.39)$$

The first term in the Taylor series (4.3.35) is precisely (4.3.38). The second term in the Taylor series is obtained directly by differentiation of (4.3.39) with respect to τ:

$$\left.\frac{d}{d\tau}\frac{\Delta X_\eta(t)}{\Delta\eta(\tau)}\right|_{\tau=t} = -g^2[X_\eta(t)]\left\{\frac{G[X_\eta(t)]}{g[X_\eta(t)]}\right\}. \qquad (4.3.40)$$

Higher order terms can be similarly evaluated. When this is done and all terms of $O(D^0\tau_c^n)$ are retained in the series, substitution into (4.3.34) with the exponential correlation function (4.3.29) leads to the BFPE. Thus the cumulant resummation technique and the functional derivative technique as they have been implemented lead to the same EFPE.

We end this section by noting that insertion of the expression (4.3.39) for the functional derivative into (4.3.33) yields the alternate (still *exact*) evolution equation (Sancho *et al.*, 1982a)

$$\frac{\partial}{\partial t}P_t = L_0 P_t + \frac{\partial}{\partial x}g(x)\frac{\partial}{\partial x}\int_0^t d\tau\, Q(t-\tau)\Big\langle \delta[x - X_\eta(t)]g[X_\eta(\tau)]$$

$$\times \exp\int_\tau^t du\{G'[X_\eta(u)] + g'[X_\eta(u)]\eta(u)\}\Big\rangle. \qquad (4.3.41)$$

This particular form of the evolution equation is the starting point for the approach to be discussed next.

4.3.3 Path integral technique

Recalling the definition of the brackets $\langle\ \rangle$, the exact evolution equation (4.3.41) can be rewritten as an explicit path integral over the fluctuations (Fox, 1986a, b; R. Kariotis, 1984, private communication):

$$\frac{\partial}{\partial x}P_t = L_0 P_t + \frac{\partial}{\partial x}g(x)\frac{\partial}{\partial x}\int_0^t d\tau\, Q(t-\tau)$$

$$\times \int D\eta \rho(\eta)\delta[x - X_\eta(t)]g[X_\eta(\tau)]$$

$$\times \exp\int_\tau^t du\{G'[X_\eta(u)] + g'[X_\eta(u)]\eta(u)\}. \tag{4.3.42}$$

Here $\int D\eta$ denotes a functional path integral over the fluctuations whose distribution is specified by $\rho(\eta)$ (usually taken to be Gaussian). As observed by Fox (1986a, b), the explicit η-dependence in the exponent of (4.3.42) can be eliminated by noting that the equation of motion (4.1.1) yields the relation

$$\frac{d}{dt}g[X(t)] = g'(X)\dot{X}$$

$$= g'(X)[G(X) + g(X)\eta(t)] \tag{4.3.43}$$

so that upon formal integration

$$g[X(t)] = g[X(\tau)]\exp\left\{\int_\tau^t du\,\frac{g'[X(u)]}{g[X(u)]}\right.$$

$$\times \{G[X(u)] + g[X(u)]\eta(u)\}\bigg\}. \tag{4.3.44}$$

Substitution of (4.3.44) into (4.3.42) yields the still *exact* expression

$$\frac{\partial}{\partial t}P_t = L_0 P_t + \frac{\partial}{\partial x}g(x)\frac{\partial}{\partial x}\int_0^t d\tau\, Q(t-\tau)$$

$$\times \int D\eta\, \rho(\eta)\delta[x - X_\eta(t)]g[X_\eta(t)]$$

$$\times \exp\left[\int_\tau^t du\left\{G'[X_\eta(u)] - \frac{g'[X_\eta(u)]}{g[X_\eta(u)]}G[X_\eta(u)]\right\}\right]. \tag{4.3.45}$$

The difficulty with this formal expression is the actual implementation of the path integral due to the dependence of the trajectory $X_\eta(u)$ on the fluctuations. Various approximations can now be made to deal with this problem. One that is consistent with those used with the other techniques discussed in this section is to retain all explicit contributions of $O(D)$ to the integral term in (4.3.44) and neglect all others. There is one explicit factor D in the correlation function $Q(t-\tau)$, so that our goal is to evaluate the τ-integrand to zeroth order in D. In practice this is achieved by considering only the *deterministic* contribution to the evolution $X_\eta(u)$, i.e. $X_\eta(u)$ in the exponent is replaced by the solution $X_d(u)$ of the deterministic equation (Peacock, West and Lindenberg, 1988):

$$\dot{X}_d(t) = G[X_d(t)], \qquad (4.3.46)$$

with $X_d(0) = x_0$. Since $X_d(u)$ is independent of the fluctuations, the path integral in (4.3.45) can be carried out explicitly. First we carry out the integration in the exponent with the replacement of $X_\eta(u)$ with $X_d(u)$ and the observation that for any function $F[X(u)]$

$$F'[X_d(u)] = \frac{1}{G[X_d(u)]} \frac{d}{du} F[X_d(u)], \qquad (4.3.47)$$

where we have used (4.3.46). The integral in the exponent then becomes

$$\int_\tau^t du \frac{d}{du} \ln \left| \frac{G[X_d(u)]}{g[X_d(u)]} \right| = \ln|\hat{h}(x, t - \tau)|, \qquad (4.3.48)$$

where

$$\hat{h}(x, t - \tau) \equiv \frac{G(x)}{g(x)} \frac{g[X_d(\tau)]}{G[X_d(\tau)]} \qquad (4.3.49)$$

and where the explicit t-dependence of $\hat{h}(x, t - \tau)$ arises from the fact that the deterministic solution $X_d(\tau)$ is subject to the endpoint condition $X_d(t) = x$. We can now perform the path integration

$$P_t = P(x, t|x_0) = \int D\eta\, \rho(\eta) \delta[x - X_\eta(t)] \qquad (4.3.50)$$

to obtain

$$\frac{\partial}{\partial t} P_t = L_0 P_t + \frac{\partial}{\partial x} g(x) \frac{\partial}{\partial x} g(x) \int_0^t d\tau\, Q(t - \tau)|\hat{h}(x, t - \tau)| P_t. \qquad (4.3.51)$$

Thus one can define the time-dependent diffusion function

$$D(x, \tau_c, t) = \int_0^t d\tau\, Q(\tau) h(x, \tau), \qquad (4.3.52)$$

where

$$h(x, \tau) \equiv |\hat{h}(x, \tau)|. \qquad (4.3.53)$$

We show below that $h(x, \tau)$ is not an explicit function of t because of the endpoint condition $X_d(t) = x$. To make contact with the diffusion function obtained earlier we again consider the limit $t \gg \tau_c$ and thus replace the diffusion function by its long-time limit:

$$D(x, \tau_c) = \lim_{t \to \infty} D(x, \tau_c, t)$$

$$= \int_0^\infty d\tau\, Q(\tau) h(x, \tau). \qquad (4.3.54)$$

The functional calculus method has thus led at $O(D)$ to the EFPE (4.3.6) with the diffusion function (4.3.54).

To relate the result (4.3.54) to the BFPE diffusion function given in (4.3.25) or (4.3.26) or (4.3.28) we note that the latter function can be written as (see (4.3.25))

$$D(x, \tau_c) = \int_0^\infty d\tau \, Q(\tau) H(x, \tau), \qquad (4.3.55)$$

where

$$H(x, \tau) = \sum_{n=0}^\infty \frac{\tau^n}{n!} \frac{F_n(x)}{g(x)}. \qquad (4.3.56)$$

The recursion relation (4.3.22) can be converted to a differential equation for $H(x, \tau)$ (Masoliver, West and Lindenberg, 1987):

$$\frac{\partial}{\partial \tau} H(x, \tau) = G'(x) H(x, \tau) - G(x) \frac{\partial}{\partial x} H(x, \tau) - G(x) H(x, \tau) \frac{g'(x)}{g(x)}$$

$$\equiv O H(x, \tau). \qquad (4.3.57)$$

The initial condition for (4.3.57) is clear from (4.3.56) and is $H(x, 0) = 1$. Comparison of (4.3.55) and (4.3.56) indicates that equality of the two diffusion functions exists if $h(x, \tau) = H(x, \tau)$. The equality is trivially satisfied at $\tau = 0$. This equality can of course only be satisfied in the x-regime for which $H(x, \tau) \geq 0$. In the unphysical regime $H(x, \tau) < 0$ (discussed in Section 4.3.5) the equality cannot hold because $h(x, \tau)$ is an absolute value. In this range, equality (were it desirable) could be achieved by analytic continuation of the integral in (4.3.55), i.e. by setting $H(x, \tau) = \hat{h}(x, \tau)$. It remains to be shown that $h(x, \tau)$ satisfies the differential equation (4.3.57). Using (4.3.53) we have for $\hat{h}(x, t - \tau) > 0$

$$\frac{\partial}{\partial \tau} h(x, \tau) = \frac{G(x)}{g(x)} \left(\frac{g_d}{G_d} \right)' \frac{d}{d\tau} X_d(t - \tau)$$

$$= -\frac{G(x)}{g(x)} \left(\frac{g_d}{G_d} \right)' G_d, \qquad (4.3.58)$$

where $g_d \equiv g[X_d(t - \tau)]$, similarly for G_d, and where we have used (4.3.46) to obtain the last equality. For the right-hand side of the differential equation (4.3.57) we obtain (again noting the x-dependence of $X_d(t - \tau)$ through the endpoint condition)

$$Oh(x, \tau) = G'(x) \frac{G(x)}{g(x)} \frac{g_d}{G_d} - \frac{G^2(x)}{g(x)} \frac{g_d}{G_d} \frac{g'(x)}{g(x)}$$

$$- G(x) \left[\frac{G(x)}{g(x)} \right]' \frac{g_d}{G_d} - \frac{G^2(x)}{g(x)} \left(\frac{g_d}{G_d} \right)' \frac{dX_d(t - \tau)}{dx}. \qquad (4.3.59)$$

KATJA LINDENBERG, BRUCE J. WEST and JAUME MASOLIVER

Integration of the dynamical equation (4.3.46) gives

$$\tau = \int_{X_d(t-\tau)}^{x} dX_d\, G(X_d),$$ (4.3.60)

and taking the derivative of this solution with respect to x immediately yields

$$\frac{d}{dx}X_d(t-\tau) = \frac{G_d}{G(x)}.$$ (4.3.61)

Substitution of (4.3.61) into (4.3.59) then leads to the equality of (4.3.59) and (4.3.58), i.e.

$$\frac{\partial}{\partial \tau}h(x,\tau) = Oh(x,\tau).$$ (4.3.62)

The identity of $h(x,\tau)$ in (4.3.53) with the function $H(x,\tau)$ that appears in the BFPE (and the consequent explicit t-independence of $h(x,\tau)$) is clear. We have thus obtained the BFPE using the functional path integral technique (Peacock, West and Lindenberg, 1988).

We finally note that (4.3.52) with (4.3.53) provides an alternative but equivalent expression to (4.3.26) and (4.3.28) for the diffusion function of the BFPE.

4.3.4   Projection operator technique

Another independent approach to obtaining an EFPE was introduced by Grigolini and colleagues and is based on projection operator methods (Faetti et al., 1986a b; Grigolini, 1986). As mentioned earlier, when the correlation of the fluctuations is exponential it is possible to treat the noise itself as a dynamic process driven by white fluctuations [see (4.3.1)–(4.3.4)]. The Fokker–Planck equation for $P(x,\eta,t)$ is then given by

$$\frac{\partial}{\partial t}P(x,\eta,t) = (L_0 + L_c + L_\eta)P(x,\eta,t),$$ (4.3.63)

where L_0 and L_η have been defined in (4.3.11) and (4.3.12) but L_η is now a deterministic rather than a stochastic operator (and therefore not explicitly time dependent), and where for an arbitrary function $h(x,\eta)$

$$L_c h(x,\eta) = \left(\frac{1}{\tau_c}\frac{\partial}{\partial \eta}\eta + D\frac{\partial^2}{\partial \eta^2}\right)h(x,\eta).$$ (4.3.64)

Grigolini's approach is to project the dynamics onto the reduced phase space x, and in doing so to treat L_η as a perturbation. To do so one rewrites (4.3.63) in the interaction representation with L_η as the interaction:

$$\frac{\partial}{\partial t}\tilde{P}(x,\eta,t) = \tilde{L}_\eta(t)\tilde{P}(x,\eta,t),$$ (4.3.65)

126

where

$$\tilde{P}(x,\eta,t) = e^{-Lt}P(x,\eta,t), \tag{4.3.66}$$

$$\tilde{L}_\eta(t) = e^{-Lt}L_\eta e^{Lt}, \tag{4.3.67}$$

and

$$L = L_0 + L_c. \tag{4.3.68}$$

The reduction of (4.3.65) to an evolution equation for $P_t \equiv P(x,t) \equiv P(x,t|x_0)$ is accomplished by applying the projection operator

$$PP(x,\eta,t) = P_{eq}(\eta)\int d\eta\, P(x,\eta,t), \tag{4.3.69}$$

where $P_{eq}(\eta)$ is defined as the solution to the equation $L_c P_{eq}(\eta) = 0$. Following the familiar techniques introduced by Mori and Zwanzig, one obtains the form

$$\frac{\partial}{\partial t}P_t = L_0 P_t + \int_0^t d\tau\, K(t-\tau)P_\tau, \tag{4.3.70}$$

where

$$K(t-\tau) = P_{eq}^{-1}(\eta)PL_\eta e^{Lt}\exp_T\left[\int_\tau^t dt'(1-P)\tilde{L}_\eta(t')\right]$$
$$\times (1-P)e^{-Lt}L_\eta PP_{eq}(\eta) \tag{4.3.71}$$

and where the subscript 'T' denotes time ordering. The initial condition term that would appear in (4.3.70) vanishes if $P(x,\eta,0) = P_{eq}(\eta)P(x,0)$.

The next step of Faetti *et al.* (1986a, b) is to approximate the kernel $K(t-\tau)$ in the spirit of linear response theory, dropping all contributions of $O(L_\eta^2)$ and higher. They further expand P_τ in the integrand in a Taylor series about $\tau = t$. Finally, as in the preceding techniques, the upper limit in the integral in (4.3.70) is set to infinity. The resulting evolution equation is

$$\frac{\partial}{\partial t}P_t = L_0 P_t + \hat{W}P_t, \tag{4.3.72}$$

where the operator $\hat{W}$ is given by

$$\hat{W} \equiv \int_0^\infty d\tau\, K(\tau)e^{-L_0\tau}, \tag{4.3.73}$$

and where in this approximation (4.3.71) is replaced by

$$K(t-\tau) = P_{eq}^{-1}(\eta)PL_\eta e^{Lt}(1-P)e^{-Lt}L_\eta PP_{eq}(\eta). \tag{4.3.74}$$

Finally, Faetti *et al.* (1986a) explicitly evaluate the operator (4.3.73) and show that (4.3.51) is then precisely the BFPE. Hence, once again this equation has been obtained by retaining all the contributions to $\hat{W}$ that are of $O(D)$ and neglecting all others.

127

4.3.5 Difficulties with and alternatives to the BFPE

The diffusion function $D(x, \tau_c)$ of the BFPE has been explicitly given via three different expressions: (4.3.26), (4.3.28), and (4.3.54) with (4.3.53). There is nothing in any of these expressions that restricts the diffusion function to positive-definite values. In fact, we will present an example for which $D(x, \tau_c) \to 0$ for some finite value x_s of x. Beyond this value the sign of $D(x, \tau_c)$ is ambiguous and depends on the technique used (if $D(x, \tau_c)$ is analytic then beyond this value it can be negative). It is important to note that typically $x_s = O(\tau_c^{-\alpha})$ where $\alpha > 0$ and hence $x_s \to \infty$ as $\tau_c \to 0$. If one blindly applies the theory to values of $x > x_s$ then one often finds an unphysical buildup of probability that is an artifact of the BFPE. To circumvent this artifact one can augment the BFPE with a reflecting boundary condition near x_s. It should be noted that the effective 'restoring force' $G - g^2(D^{1/2})'$ (cf. (4.3.7)) is negative provided x is below x_s by an amount of $O(D\tau_c^\beta)$ with $\beta > 0$. Thus one can set the reflecting boundary at a distance within this order of x_s, e.g. at the point where the restoring force vanishes. By construction there is no flux of probability across this point, which then defines one boundary of the region of support.

The appearance of a finite limit to the region of support when $D(x, \tau_c) \to 0$ for a finite value of x has elicited a certain amount of criticism of the BFPE. This has prompted a number of investigators to seek EFPEs that do not exhibit this behavior, i.e. whose diffusion functions are positive definite for all x (see below). We emphasize that the quality of the BFPE depends on the physical property of interest. Those properties that do not specifically involve P_t for $x \geqslant x_s$ when $x_0 < x_s$ may be well represented by the BFPE. For instance, in Figure 4.1 we show three steady state distributions $P_{ss}(x)$ for an example considered subsequently (Section 4.3.6). The solid curve is the steady-state solution of the BFPE and has support in the region $(0, 2)$, while the other two curves are solutions of related EFPEs that have infinite support. Any property dependent on $P_{ss}(x)$ for $0 \leqslant x \leqslant 2$ will not vary significantly with the choice of distribution. Even first passage times within this interval would be insensitive to this choice. On the other hand, the first passage time to a point $x > 2$ would be quite different for the different choices, and would in fact be infinite based on the BFPE. The EFPE solutions would yield finite first passage times, but there again the validity of the result is completely unknown. Thus one must exercise caution in the determination of physical properties (particularly in asymptotic regimes) using any EFPE. It should be stressed that the BFPE diffusion function need not vanish for finite x (see Section 4.3.6), but even then caution must be exercised with its use in asymptotic regimes.

Three alternatives to the BFPE are frequently mentioned in the literature. These alternatives arise from the implementation of different approximations within the techniques presented earlier. The first was introduced by Sancho et al. (1982a) to avoid passing through $\underline{D}(x, \tau_c) = 0$. When Sancho et al. (1982a, b) first used the functional derivative approach, they in fact calculated

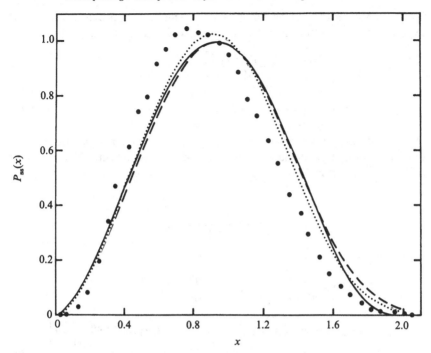

Figure 4.1. Steady-state distributions for the system $\dot{X} = a(X - X^3) + X\eta(t)$ with $a\tau_c = 0.167$ and $D\tau_c = 0.0635$. ——, BFPE; -----, exponentiation of Sancho *et al.* (1982a); ......, EFPE of Fox (1986b); ●, analog simulation of Sancho *et al.* (1982b).

the diffusion function only to lowest order in τ_c, i.e. (cf. (4.3.28))

$$D(x, \tau_c) \approx D \left\{ 1 - \tau_c \left[G(x) \frac{g'(x)}{g(x)} - G'(x) \right] \right\}. \tag{4.3.75}$$

This approximation leads to a vanishing (and subsequently negative) diffusion function even in cases where the full BFPE does not. To solve this problem they suggested an (admittedly *ad hoc*) exponentiation procedure in the calculation of the probability density. Thus they replaced (4.3.31) with (Sancho *et al.*, 1982a)

$$P_{ss}(x) = \frac{N}{g(x)} \exp \left\{ - \tau_c \left[G'(x) - G(x) \frac{g'(x)}{g(x)} \right] \right. $$
$$\left. + \int^x dy \frac{G(y)}{Dg^2(y)} \left(1 - \tau_c \left[G'(y) - G(y) \frac{g'(y)}{g(y)} \right] \right) \right\}. \tag{4.3.76}$$

This density has support in the entire range of values of x.

Fox has recently addressed the support problem from the point of view of the path integral technique and arrives at a result closely related to (4.3.76)

129

(Fox, 1986b). In (4.3.45) he implements a 'small τ_c' approximation in the exponent by expanding the integral to first order in $(t - \tau)$ around t:

$$\int_\tau^t du \left[G'[X_\eta(u)] - \frac{g'[X_\eta(u)]}{g[X_\eta(u)]} G[X_\eta(u)] \right]$$

$$= (t - \tau) \left[G'(x) - \frac{g'(x)}{g(x)} G(x) \right]. \tag{4.3.77}$$

(Note that this *ad hoc* assumption makes the integral independent of the dynamics and, in particular, independent of the fluctuations.) The resulting diffusion function is

$$D_F(x, \tau_c) = \frac{D}{1 - \tau_c \left[G'(x) - \frac{g'(x)}{g(x)} G(x) \right]} \tag{4.3.78}$$

giving rise to

$$P_{ss}(x) = \frac{N}{g(x)} \left\{ 1 - \tau_c \left[G'(x) - \frac{g'(x)}{g(x)} G(x) \right] \right\}$$

$$\times \exp \left\{ \int^x dy \frac{G(y)}{Dg^2(y)} \left[1 - \tau_c \left(G'(y) - G(y) \frac{g'(y)}{g(y)} \right) \right] \right\}. \tag{4.3.79}$$

This is equivalent to (4.3.76) with the first exponential expanded to lowest order in τ_c. The relation of Fox's result to the BFPE has been discussed extensively by Faetti *et al.* (1986a).

The third procedure that has been implemented by some is the 'decoupling theory' of Hänggi, Mroczkowski, Moss and McClintock (1985) for additive colored fluctuations $(g(x) = 1)$. In this approach the state-dependent diffusion function is replaced by the constant

$$D_H = \frac{D}{1 - \tau_c \langle G'(x) \rangle}. \tag{4.3.80}$$

Since the average $\langle G'(x) \rangle$ depends on the unknown probability density P_t, Hänggi *et al.* replace the average with its (also unknown) steady-state value $\langle G'(x) \rangle_{ss}$. This value has then either been calculated self consistently or it has been treated as a phenomenological parameter.

Each of the techniques presented in Sections 4.3.1–4.3.4 can in principle be extended beyond their present regions of applicability. However, although the necessary extensions for the calculation of P_t may be clear (again in principle), the connection with the first passage time statistics is obscure. The only systematic analysis of these extensions has been made by Faetti *et al.* (1986a). One important result of this analysis is the conclusion that 'higher order' derivative corrections contribute terms which would alter the diffusion function at $O(D\tau_c^2)$ for Gaussian noise and $O(D\tau_c)$ for noise of other statistics (Faetti *et al.*, 1986b). One must therefore conclude that EFPEs are at best reliable for small D and small τ_c.

To overcome these limitations requires an entirely different approach to the problem. As mentioned in the introduction to this chapter, one way to deal with colored noise is to simplify its statistics. This is the approach reviewed in Section 4.4. A new theory valid for Gaussian colored noise has recently been developed by Doering, Hagan and Levermore (1987). They use a singular perturbation approach to solve the full two-dimensional problem (4.3.63) (cf. (4.3.1) and (4.3.2)) for small τ_c. This is in contrast to the other methods which rely on eliminating the explicit dependence on the fluctuations $\eta(t)$, thereby reducing the equation to one dimension. Doering *et al.* then connect the probability density $P(x, \eta, t)$ to the first passage time statistics using a novel idea first introduced by Büttiker and Landauer (1982). In the EFPE approach it is always assumed that the effective Fokker–Planck form entitles one to use white noise boundary conditions to obtain first passage time statistics. Doering *et al.* show that this assumption may not be valid. The true boundary conditions are in general τ_c-dependent and may introduce corrections not contained in P_t even if P_t could be calculated exactly (see Section 4.4 for an example). They finally obtain an expansion of the mean first passage time in terms of τ_c. This expansion is of the form

$$T_1(x_p|x_0) = T_1^0[1 + O(\gamma\tau_c)^{1/2} + O(\gamma\tau_c) + \cdots]. \tag{4.3.81}$$

Here $T_1^0 \equiv T_1(x_p|x)$ when $\tau_c = 0$, i.e. it is the white noise limit of the mean first passage time. The correction of $O(\gamma\tau_c)^{1/2}$ arises from the modified boundary conditions and does not occur in any of the EFPE theories considered earlier (see Section 4.4). The parameter γ is a measure of the deterministic dynamics of the system in the absence of boundaries. This is essentially the first term that appears in a small τ_c expansion of the mean first passage time based on an EFPE theory. The relative importance of these two terms is clearly system-specific. In Section 4.4 we consider a free diffusion process in which only the $\tau_c^{1/2}$ term appears. In Section 4.3.6, on the other hand, we consider an example in which the $\tau_c^{1/2}$ term is at most apparent only for extremely small τ_c while the linear contribution dominates over a broad range of values of τ_c.

4.3.6   Nonlinear systems driven by Gaussian colored noise

Example of bistability

The validity and relevance of the theories discussed in this section can ultimately only be judged on the basis of experimental results. Few controlled experiments in one-dimensional systems are available, and thus the work of Moss *et al.*, which provides such results, takes on special significance (Hänggi *et al.*, 1985). The analog computer experiments of Moss have been used as test cases by most workers in this field.

The experimental circuit that is appropriate for the first passage time discussion is a bistable system described by the dynamic equation

$$\dot{X} = aX - bX^3 + \eta(t), \tag{4.3.82}$$

which is of the form (4.1.1) with $g(X) = 1$. The noise $\eta(t)$ is Gaussian and exponentially correlated, as in (4.3.4). The 'potential'

$$V(X) = \tfrac{1}{4}bX^4 - \tfrac{1}{2}aX^2 \tag{4.3.83}$$

has minima at $\pm (a/b)^{1/2}$ and a maximum at $X = 0$. If the noise $\eta(t)$ were white, then the steady-state distribution $P_{ss}(x) = \lim_{t \to \infty} P(x, t|x_0)$ associated with (4.3.82) would have maxima at $x = \pm (a/b)^{1/2}$ and a minimum at $x = 0$. Because the noise is colored, the precise location of the maxima are shifted from the above values (see below); the bimodal structure may even disappear entirely if τ_c is sufficiently large. The experiments of Moss *et al.* were carried out in the regime where $P_{ss}(x)$ is indeed bimodal: they measured the mean sojourn time T_s near one peak of the distribution.

The experimental results for T_s as a function of the correlation time τ_c are shown in Figure 4.2 for various values of the noise intensity D. It should be noted that the measured values D_{expt} may be ambiguous due to large experimental fluctuations and that the mean value of D itself drifts downwards as τ_c decreases (F. Moss, 1986, private communication). The 'white noise' $(\tau_c \simeq 0)$ values of T_s therefore correspond in each case to a significantly lower value of D than the reported value D_{expt}. The experimental mean sojourn time is written in the form

$$T_s = \frac{\pi}{a2^{1/2}} \exp[\Delta\phi_{expt}(\tau_c)/D_{expt}], \tag{4.3.84}$$

where the measured function $\Delta\phi_{expt}(\tau_c)$ is usually assumed to be of the linear form

$$\Delta\phi_{expt}(\tau_c) = \frac{a^2}{4b}(1 + C_{expt}\tau_c) \tag{4.3.85}$$

for small τ_c. Here $a^2/4b$ is the white noise limit of $\Delta\phi_{expt}$ and C_{expt} is the experimental slope of the data (see Figure 4.3). Note that the slope in Figure 4.3 is D_{expt}-dependent and that it increases with increasing D_{expt}.

Let us next consider the theoretical predictions for T_s on the basis of the BFPE. We identify the mean sojourn time T_s with the mean first passage time $T_1(0|x_m)$ from one of the two maxima at $\pm x_m$ of the steady-state distribution to its minimum at $x = 0$. To evaluate T_1 we first need to calculate the diffusion function $D(x, \tau_c)$ and the steady-state distribution so that (4.3.30) can be applied. The diffusion function is obtained using any of the techniques discussed earlier in this section and is given in terms of the hypergeometric

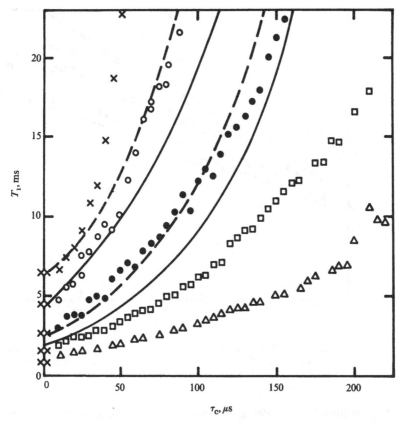

Figure 4.2. Mean first passage time $T_1(0|x_m)$ against correlation time τ_c with $a = b = 1$ for several values of D. $\times$, $D_{\text{expt}} = 0.073$; $\bigcirc$, 0.083; $\bullet$, 0.114; $\square$, 0.153; $\triangle$, 0.212. The data is taken from Moss *et al.* (Hänggi *et al.*, 1985) with the permission of the authors. ———, (4.3.90) with $D = D_{\text{expt}} = 0.114$ (lower) and $D = D_{\text{expt}} = 0.083$ (upper); ----, $D = 0.10$ (lower) and $D = 0.075$ (upper).

function as (Masoliver, West and Lindenberg, 1987)

$$D(x, \tau_c) = \frac{D}{1 - a\tau_c} F\left[-\frac{3}{2}; 1; \frac{1}{2}\left(\frac{1}{a\tau_c} + 1 \right); \frac{bx^2}{a} \right]. \tag{4.3.86}$$

This diffusion function is positive definite if $a\tau_c < 1$, and therefore the BFPE in this case yields a nonzero distribution for all x. It can be seen by straight-forward analysis that to evaluate $\ln T_1$ to linear order in τ_c (see (4.3.84) and (4.3.85)) it is sufficient to expand $D^{-1}(x, \tau_c)$ to linear order in τ_c. To this degree of accuracy one can thus replace (4.3.86) with

$$D(x, \tau_c) \simeq \frac{D}{1 - \tau_c(a - 3bx^2)}, \tag{4.3.87}$$

133

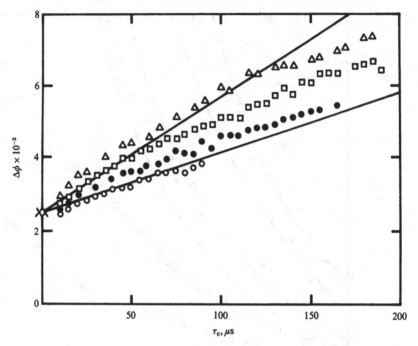

Figure 4.3. The data of Figure 4.2 replotted to show $\Delta\phi_{expt}$ against τ_c (taken from Moss *et al.* (Hänggi *et al.*, 1985), with the permission of the authors). The solid lines are calculated from (4.3.91) with $D = 0.212$ (upper line) and $D = 0.114$ (lower line).

which is also positive definite provided $a\tau_c < 1$. Note that (4.3.87) is the diffusion function directly obtained from Fox's small τ_c approximation to the diffusion function using the functional calculus approach (Fox, 1986b). The steady-state distribution is obtained from (4.3.31) and is bimodal for $a\tau_c < 1$ with a minimum at $x = 0$ and maxima at $x = \pm x_m$, where

$$x_m = \left(\frac{a}{b}\right)^{1/2}\left[1 + \frac{3bD\tau_c}{a} - O(bD\tau_c^2)\right].$$ (4.3.88)

Equation (4.3.28) cannot be evaluated in closed form for this example. However, if we write the mean first passage time in the form (4.3.84), viz.

$$T_1(0|x_m) = \frac{\pi}{a2^{1/2}}\exp[\Delta\phi(D, \tau_c)/D],$$ (4.3.89)

then we can evaluate $\Delta\phi$ analytically to leading order in D and τ_c. This level of approximation is consistent with the empirical fit (4.3.85). One obtains the result (Masoliver, West and Lindenberg, 1987)

$$T_1(0|x_m) \simeq \frac{\pi}{a2^{1/2}}\exp\left[\frac{a^2}{4bD} + \frac{3}{2}a\tau_c\right]$$ (4.3.90)

First passage time problems for non-Markovian processes

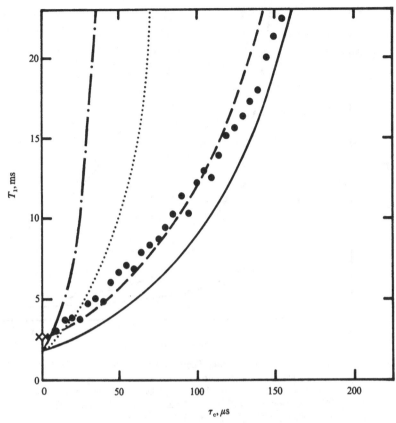

Figure 4.4. Experimental and theoretical results for $T_1(0|x_m)$ against τ_c with $a = b = 1$. ●, experiments of Moss *et al.* (Hänggi *et al.*, 1985) with $D_{expt} = 0.114$; ——, (4.3.90) with $D = D_{expt} = 0.114$; ----, (4.3.90) with $D = 0.10$. ......, theory of Hänggi *et al.* (1985), equation (9b) with $D = D_{expt}$ and $\langle x^2 \rangle = 0.86$, ·-·-·-·, extrapolated Fox's calculation (see equation (115) of Fox, (1987a)).

so that

$$\Delta\phi = \frac{a^2}{4b} + \frac{3}{2}aD\tau_c. \tag{4.3.91}$$

The leading corrections to (4.3.91) are of $O(D^2\tau_c)$ and $O(D\tau_c^2)$.

The comparison of the theoretical results with the experiments of Moss *et al.* are shown in Figures 4.2 and 4.3. Although the restriction $\tau_c < 1/a$ implies that validity of the theoretical results should be restricted to $\tau_c < 100 \, \mu s$, the quality of the fits remains good even beyond this region. In Figure 4.2 the two solid curves represent the results of (4.3.90) with $a = b = 1$. The lower solid curve is for $D = 0.114$, while the lower dashed curve is for $D = 0.10$. Thus within a 10% range (well within experimental variability) the theory brackets most of the

135

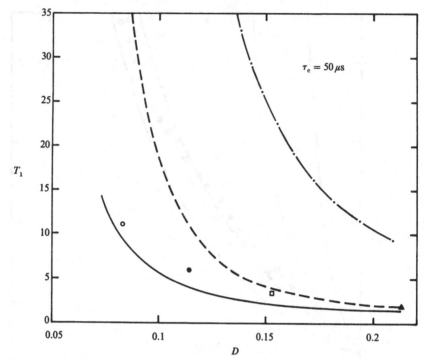

Figure 4.5. Experimental and theoretical results for $T_1(0|x_m)$ against D with $a = b = 1$ when $\tau_c = 50\,\mu s$ (corresponding to $\tau_c = 0.5$ in dimensionless form). $\circ$, $\bullet$, $\square$, $\blacktriangle$, experimental results; ———, (4.3.90);----, theory of Hänggi *et al.* (1985, equation (9b)) with $\langle x^2 \rangle = 0.85$; $-\cdots-$, extrapolated Fox's calculations (see equation (115) of Fox, (1986a)).

data. The upper solid and dashed curves, respectively, correspond to the choices $D = 0.083$ and $D = 0.075$. Note that the systematic downward shift of the experimental value of D with decreasing τ_c if included in the theoretical predictions would increase the predicted mean first passage time at small τ_c, thereby further improving the theoretical fit to the experiments.

We note that to date the above theory provides the best fit to the experimental results of Moss *et al.* Thus, in Figure 4.4 we exhibit the experimental results for $D_{expt} = 0.114$, our theoretical results for $D = 0.114$ (solid curve) and $D = 0.10$ (dashed curve) as in Figure 4.2, the result obtained from Hänggi's 'decoupling approximation' (dotted curve), and the curve obtained by an extrapolation of Fox's calculated mean first passage time (Fox, 1986a) (dot–dashed curve). In Figure 4.5 the results for the D-dependence of the mean first passage time for a particular value of τ_c are shown for the various theories.

On the basis of these comparisons we can arrive at a number of conclusions. An EFPE (and, in particular, the BFPE) is sufficient to *quantitatively* describe

136

the mean first passage time of a bistable process driven by weak colored Gaussian fluctuations with a short but finite correlation time τ_c. In the experimental range covered by Moss *et al.* the logarithm of the sojourn time is essentially linear in τ_c. Any EFPE with a diffusion function whose inverse is correct to $O(\tau_c)$ reproduces the measured results. Therefore the BFPE or use of Fox's EFPE in our equations leads to agreement with experiments whereas Fox's calculated T_1 (Fox, 1986a) and Hänggi's result (Hänggi *et al.*, 1985) do not. The quantitative agreement with the BFPE extends to the D-dependence of T_1.

Let us now consider the deviations of $\ln T_1$ from the linear τ_c-dependence discussed above. The large τ_c deviations presumably reflect higher order contributions that have not been taken into account in the present calculations but that are contained (at least in part) in a less approximate description of the evolution of P_t. At the other extreme (small τ_c), the analog results of Moss *et al.* are not reliable because of the experimental drift in the intensity D_{expt} of the fluctuations. To explore this regime a digital simulation of the system (4.3.82) to obtain the mean first passage time $T_1(0|x_m)$ was carried out by Grigolini *et al.* (Hänggi, Marchesoni and Grigolini, 1984). Recently R. Mannella (1986, private communication) has provided us with the updated set of data points shown in Figure 4.6. The deviation from linear behavior is clear. The various EFPEs are indicated in the figure, and clearly cannot duplicate the observed behavior over the entire τ_c range. The BFPE is best for larger τ_c, while Hänggi *et al.* best reproduce the very small τ_c behavior. The result of Fox cannot be extrapolated to $\tau_c \geqslant 1$, and does not reproduce the observations at small τ_c.

Zero-dimensional Landau–Ginzburg equation

In Section 4.3.5 we provided a general discussion of the limitations of the BFPE. In particular, we addressed the question of the finite region of support necessitated in problems where $D(x, \tau_c) \to 0$ for finite x. It is instructive to briefly consider a simple example of a system in which this occurs. The system is similar to (4.3.82) but with multiplicative fluctuations:

$$\dot{X} = aX - bX^3 + X\eta(t). \qquad (4.3.92)$$

This is the zero-dimensional Landau–Ginzburg equation, and is of the form (4.1.1) with $g(X) = X$. Although the 'potential' (4.3.83) has a bimodal structure, the distribution function $P_{\text{ss}}(x)$ here has a unimodal structure for white noise $\eta(t)$. For colored noise there is some evidence of a bimodal structure for some parameter ranges (Sancho *et al.*, 1982a).

The BFPE diffusion function that one obtains for Gaussian colored fluctuations $\eta(t)$ is (Lindenberg and West, 1983, West and Lindenberg, 1987)

$$D(x, \tau_c) = D\left[1 - \frac{2b\tau_c x^2}{1 + 2a\tau_c}\right], \qquad (4.3.93)$$

137

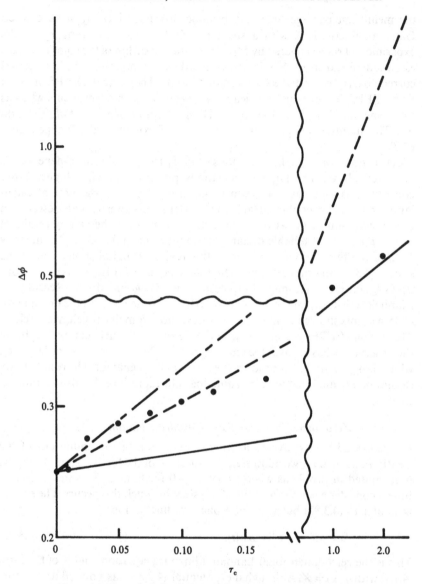

Figure 4.6. Numerical simulation results and theoretical results for $\Delta\phi$ against τ_c with $a = b = 1$. ●, digital simulation of R. Mannella (1986, private communication); ———, BFPE result (4.3.91); ----, theory of Hänggi *et al.* (1985, their equation (9b)); –·–·–, Fox's calculation (equation (115) of Fox, 1986a).

which vanishes when $x^2 = (1 + 2a\tau_c)/2b\tau_c \equiv x_s^2$. In Figure 4.1 we showed the steady-state distributions arising from the BFPE, the EFPE of Fox (1986b), and the exponential 'small-τ_c' solution of Sancho *et al.* (1982a) for one set of parameter values. The results of the digital simulations of Sancho *et al.* (1982b)

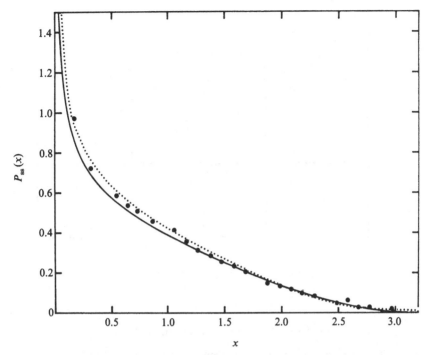

Figure 4.7. Steady-state distributions for the same system as in Figure 4.1 [$\dot{X} = a(X - X^3) + X\eta(t)$] but with parameter values $a\tau_c = 1/18$ and $D\tau_c = 1/12$. ——, BFPE; ......, exponention of Sancho *et al.* (1982a) and Fox's EFPE (Fox, 1986b) (indistinguishable); ●, digital simulations of Sancho *et al.* 1982a).

are also shown. The differences among the theories are no larger than those between any one theory and the data. In Figure 4.7 the steady-state distributions for a different set of parameter values are shown, as are the results of the analog simulation of Sancho *et al.* (1982a). One would be hard put to quantify the difference between either theoretical curve and the data.

4.4 Discontinuous colored (mostly dichotomous and shot) fluctuations

In Section 4.3 we discussed methods for dealing with the first passage time problem that centers on the retention of the Fokker–Planck structure. This approach in effect replaces the process (4.1.1) driven by colored noise $\eta(t)$ with another process (4.3.7) driven by white noise $\eta^w(t)$. The argument is that one can then bring the well-established first passage time theory for white noise to bear. (As mentioned before, this latter argument has been brought into question by the recent work of Doering.)

139

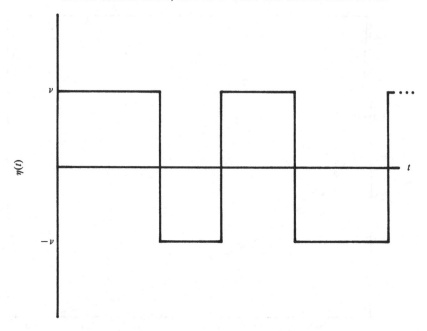

Figure 4.8. A typical realization of a symmetric dichotomous process.

An alternative approach that has been taken by a number of authors is to simplify the statistical properties of the (colored) noise so drastically that one can deal with the first passage time problem exactly for arbitrary noise intensity and correlation time. Dichotomous fluctuations and a particular form of shot noise are some examples that we consider herein.

As mentioned earlier in this chapter, the exact structure of the equations describing the evolution of probabilities for the colored-noise-driven process will in general be integral rather than differential (unless the noise is included as a variable as described in Section 4.3). We thus set out to derive such integral equations. There are a number of ways of carrying out such a program, the most transparent of which is perhaps that of explicitly following the trajectories of the process. This procedure will be reviewed below. More recently we have developed a formulation that allows us to incorporate the trajectories in a way that avoids some of the unnecessary intermediate steps in their construction (Weiss, Masoliver, Lindenberg and West, 1988).

4.4.1 Stochastic trajectory analysis technique (STAT)

Colored dichotomous noise

Consider a one-dimensional dynamical system described by (4.1.1) and let us take $\eta(t)$ to be a dichotomous process. Thus $\eta(t)$ alternately takes on the values v and $-v$ with $v > 0$ (see Figure 4.8). The times Δt that $\eta(t)$ retains a given value are governed by a distribution $\Psi(\Delta t)$, i.e. the time intervals form a renewal process (Masoliver, 1987; Masoliver, Lindenberg and West, 1986a–c), and the switching rate λ_s from one value of $\eta(t)$ to the other is given by

$$\lambda_s^{-1} = \int_0^\infty d(\Delta t) \, \Delta t \, \Psi(\Delta t). \tag{4.4.1}$$

If $\eta(t)$ is a dichotomous Markov process, then this switching time distribution is exponential:

$$\Psi(\Delta t) = \lambda e^{-\lambda \Delta t}, \tag{4.4.2}$$

where λ^{-1} is the average residence time in either state. Thus $\lambda^{-1} = \lambda_s^{-1}$ is the average time between switches. A simple calculation shows that the correlation function of the fluctuations is related to $\Psi(\Delta t)$ as follows:

$$\langle \eta(t)\eta(\tau) \rangle = v^2 [\Phi_{\text{even}}(t - \tau) - \Phi_{\text{odd}}(t - \tau)], \tag{4.4.3}$$

where

$$\Phi_{\text{even}}(t) \equiv \sum_{n=0}^\infty \Phi_{2n}(t), \tag{4.4.4}$$

$$\Phi_{\text{odd}}(t) \equiv \sum_{n=0}^\infty \Phi_{2n+1}(t), \tag{4.4.5}$$

and

$$\Phi_n(t) \equiv \lambda_s \int_t^\infty dt'' \int_0^t dt_n \int_0^{t_n} dt_{n-1} \dots \int_0^{t_2} dt_1$$

$$\times \int_{-\infty}^0 dt' \, \Psi(t'' - t_n)\Psi(t_n - t_{n-1})\dots\Psi(t_2 - t_1)\Psi(t_1 - t'). \tag{4.4.6}$$

The function $\Phi_n(t - \tau)$ is simply the probability per unit time that exactly n switches have occurred in the time interval $(t - \tau)$. If the switching time distribution has the exponential form (4.4.2), then (4.4.3) becomes

$$\langle \eta(t)\eta(\tau) \rangle = \frac{D}{\tau_c} e^{-|t - \tau|/\tau_c}, \tag{4.4.7}$$

where

$$D \equiv v^2 \tau_c \tag{4.4.8}$$

and

$$\tau_c \equiv \frac{1}{2\lambda}. \tag{4.4.9}$$

The theory developed below is not restricted to the form (4.4.2). We assume that $f(X) \pm vg(X)$ is smooth and such that the solution $X(t)$ of (4.1.1) does not become infinite in a finite time.

Let $X_\pm(t)$ be the solution of (4.1.1) when $\eta(t) = \pm v$. Let us for the sake of discussion restrict our analysis to functions $G(X)$ and $g(X)$ for which $F_+(X) \equiv G(X) + vg(X) > G(X) - vg(X) \equiv F_-(X)$ for all X within the region of

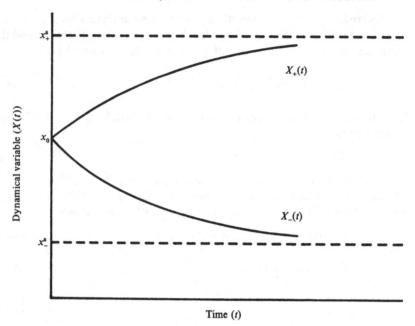

Figure 4.9. Dynamical variable as a function of time for two values of $\eta(t)$. $X_+(t)$ is the trajectory when $\eta(t) = v$, and $X_-(t)$ is the trajectory when $\eta(t) = -v$. The trajectories approach their respective asymptotic fixed points x^s_+ and x^s_-.

interest (see below) and for which $F_+(X)$ and $F_-(X)$ are monotonic. Within this region it then follows that $X_+(t) \geqslant X_-(t)$. We are interested in the first passage time for $X(t)$ to reach certain values, say $-L$ and L. If $F_+(X)$ and/or $F_-(X)$ have asymptotically stable fixed points x^s_+ and x^s_-, respectively at which $F_j(X)(j = +, -)$ vanish, then

$$\lim_{t \to \infty} X_\pm(t) = x^s_\pm \tag{4.4.10}$$

with $x^s_+ > x^s_-$. The system trajectory cannot leave the interval $[x^s_-, x^s_+]$ if it starts within this interval (see Figure 4.9). The first passage time problem is then meaningful only if $L < x^s_+$ and/or $-L > x^s_-$.

The trajectory of the process $X(t)$ for a particular realization of the fluctuations can be followed in detail. Thus, let us define the trajectory functions $\phi_\pm(t)$ via the inverse relation

$$\phi_\pm^{-1}(X) \equiv \int^X \frac{dX'}{F_\pm(X')}. \tag{4.4.11}$$

If, for instance, $X(0) = x_0$, $\eta(0) = v$ and η retains this value for a time interval of duration $\geqslant t$, then

$$X(t) = \phi_+[t + \phi_+^{-1}(x_0)]. \tag{4.4.12}$$

142

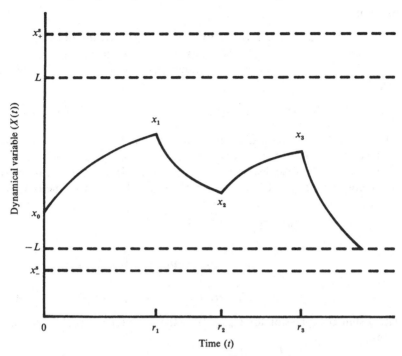

Figure 4.10. A typical trajectory for a dynamical process $X(t)$ starting at x_0. This trajectory is driven by the dichotomous process of Figure 4.8. The time increment labels are related to those in the text by $t_i = r_i - r_{i-1}$.

If, e.g., $G(X) = 0$ and $g(X) = 1$, (4.4.12) is the simple ballistic trajectory $X_+(t) = x_0 + vt$. A typical trajectory associated with Figure 4.8 is shown in Figure 4.10 and is given by the incremental values

$$X(t) = \phi_+(t + \phi_+^{-1}(x_0)), \quad 0 \leqslant t \leqslant t_1$$

$$X(t) = \phi_-(t + \phi_-^{-1}(\phi_+(t_1 + \phi_+^{-1}(x_0)))), \quad t_1 \leqslant t \leqslant t_1 + t_2 \quad (4.4.13)$$

$$X(t) = \phi_+(t + \phi_+^{-1}(\phi_-(t_2 + \phi_-^{-1}(\phi_+(t_1 + \phi_+^{-1}(x_0)))))),$$

$$t_1 + t_2 \leqslant t \leqslant t_1 + t_2 + t_3,$$

and so on. The distribution $\Psi(\Delta t)$ governs the time intervals $t_1, t_2, t_3, \cdots$. We introduce the notation $x_n \equiv X(t_n)$ so that for $\eta(0) = \pm v$ we have

$$x_{2n-1} = \phi_\pm(t_{2n-1} + \phi_\pm^{-1}(x_{2n-2})) \quad (4.4.14)$$

and

$$x_{2n} = \phi_\mp(t_{2n} + \phi_\mp^{-1}(x_{2n-1})). \quad (4.4.15)$$

Our goal is to calculate the conditional first passage time probability density $f(t, L|x_0)$ (see (4.2.5)). To calculate $f(t, L|x_0)$ it is useful to define the auxiliary

143

probability

$$f_n(t, L|x_0)\, dt = \text{probability that the first crossing of } L$$
$$\text{or } -L \text{ occurs during the } n\text{th time interval}$$
$$\text{in the time range } (t, t + dt). \tag{4.4.16}$$

Clearly

$$f(t, L|x_0) = \sum_{n=1}^{\infty} f_n(t, L|x_0). \tag{4.4.17}$$

Let us for the sake of the discussion here assume that a switch to v occurred at exactly $t = 0$. The probability density $f_n(t, L|x_0)$ can be constructed in terms of a sequence of switches leading to a trajectory that is constrained not to cross $\pm L$ before time t and to cross within the interval $(t, t + dt)$. One obtains

$$f_{2n}(t, L|x_0) = \int_0^{\tau_1} dt_1\, \Psi(t_1) \int_0^{\tau_2} dt_2\, \Psi(t_2) \dots$$
$$\times \int_0^{\tau_{2n-1}} dt_{2n-1}\, \Psi(t_{2n-1}) \int_{\tau_{2n}}^{\infty} dt_{2n}\, \Psi(t_{2n})$$
$$\times \delta[t - (t_1 + t_2 + \dots + t_{2n-1} + \tau_{2n})] \tag{4.4.18}$$

with a similar expression for f_{2n-1}. Here the τ_i are functions of the switching times:

$$\tau_{2m-1} \equiv \int_{x_{2m-2}}^{L} \frac{dX}{F_+(X)}, \tag{4.4.19}$$

$$\tau_{2m} \equiv \int_{x_{2m-1}}^{-L} \frac{dX}{F_-(X)}, \tag{4.4.20}$$

and the τ-limits in (4.4.18) insure that no crossing takes place before time t. The delta function in (4.4.18) insures a crossing at time t. Substitution of (4.4.18) into (4.4.17) yields the desired probability density.

A more tractable formulation is obtained by Laplace transforming (4.4.18) with the definition

$$\tilde{f}(s) \equiv \int_0^{\infty} dt\, e^{-st} f(t). \tag{4.4.21}$$

The Laplace transform of (4.4.18) makes evident an integral recursion relation connecting the nth and $(n + 2)$nd auxiliary probabilities. When this recursion relation is summed over all n the result is an integral equation for the first passage time probability density:

$$\tilde{f}(s, L|x_0) = \tilde{f}_1(s, L|x_0) + \tilde{f}_2(s, L|x_0) + \int_0^{\tau_1} dt_1\, e^{-st_1} \Psi(t_1)$$
$$\times \int_0^{\tau_2} dt_2\, e^{-st_2} \Psi(t_2) \tilde{f}(s, L|x_2), \tag{4.4.22}$$

144

where we recall that x_2 is a function of t_2 and t_1 (see (4.4.14)). Equation (4.4.22) is the result that replaces the EFPE formulation of the first passage time statistics.

Equation (4.4.22) has been obtained with the conditions that $\eta(0^+) = v$ and that a switch to this value occurred exactly at $t = 0$. Here 0^+ indicates that $t \rightarrow 0$ from above. Both of these conditions are easily relaxed (Masoliver, Lindenberg and West, 1986a–c). We again note that (4.4.22) is valid for general $\Psi(t)$, i.e. the dichotomous process $\eta(t)$ need not be Markovian.

For some forms of $\Psi(t)$ the integral equation (4.4.22) can be transformed to a differential equation. For instance, if $\eta(t)$ is a dichotomous Markov process then $\Psi(t)$ takes the exponential form (4.4.2) and (4.4.22) can be converted to (Masoliver, Lindenberg and West, 1986c)

$$\frac{\partial^2}{\partial x_0^2}\tilde{f}(s,L|x_0) + \left[\frac{F'_+(x_0)}{F_+(x_0)} - \frac{s+\lambda}{F_-(x_0)} - \frac{s+\lambda}{F_+(x_0)}\right]\frac{\partial}{\partial x_0}\tilde{f}(s,L|x_0)$$

$$+ \frac{s(s+2\lambda)}{F_+(x_0)F_-(x_0)}\tilde{f}(s,L|x_0) = 0 \qquad (4.4.23)$$

together with the boundary conditions (which follow directly from the integral equation)

(i) $\tilde{f}(s,L|L) = 1,$ \qquad (4.4.24)

(ii) $\left.\frac{\partial}{\partial x_0}\tilde{f}(s,L|x_0)\right|_{x_0=-L} = \frac{[-\lambda + (s+\lambda)\tilde{f}(s,L|-L)]}{F_+(-L)}.$ \qquad (4.4.25)

The first boundary condition insures that a process initiated at the upper boundary with positive slope is immediately trapped with certainty. The second boundary condition is not of the usual form for a Fokker–Planck process (cf. discussion in Section 4.3.4) in which trapping at the lowest boundary would also be guaranteed if the process starts there, i.e. $\tilde{f}(s,L|-L) = 1$. The physical interpretation of (4.4.25) is not straightforward.

A differential equation for the mean first passage time $T_1(L|x_0)$ (and also for higher moments) can be obtained from the relation (4.2.6):

$$T_1(L|x_0) = -\left.\frac{\partial}{\partial s}\tilde{f}(s,L|x_0)\right|_{s=0}. \qquad (4.4.26)$$

The derivative of (4.4.23) yields (cf. (4.2.8))

$$\frac{\partial^2}{\partial x_0^2}T_1(L|x_0) + \left[\frac{F'_+(x_0)}{F_+(x_0)} - \frac{\lambda}{F_-(x_0)} - \frac{\lambda}{F_+(x_0)}\right]\frac{\partial}{\partial x_0}T_1(L|x_0)$$

$$= \frac{2\lambda}{F_+(x_0)F_-(x_0)}. \qquad (4.4.27)$$

The boundary conditions obtained from (4.4.24) and (4.4.25) are (cf. (4.2.12))

(i) $T_1(L|L) = 0,$ \qquad (4.4.28)

$$\text{(ii)} \quad \frac{\partial}{\partial x_0} T_1(L|x_0)\bigg|_{x_0 = -L} = \frac{\lambda T_1(L| - L) - 1}{F_+(-L)}. \tag{4.4.29}$$

Equation (4.4.29) has an unusual structure and again reflects the fact that the colored nature of the fluctuations affects not only the differential equation for T_1 but also its boundary conditions.

The solution of the problem (4.4.27)–(4.4.29) is given by (cf. (4.2.13))

$$T_1(L|x_0) = \int_L^{x_0} dx\, V(x) e^{-M(x)} + C \int_L^{x_0} dx\, e^{-M(x)}, \tag{4.4.30}$$

where

$$M(x) \equiv \int^x dy \left[\frac{F'_+(y)}{F_+(y)} - \frac{\lambda}{F_-(y)} - \frac{\lambda}{F_+(y)} \right], \tag{4.4.31}$$

$$V(x) \equiv \int^x dy \frac{2\lambda e^{M(y)}}{F_+(y) F_-(y)}, \tag{4.4.32}$$

$$C \equiv \frac{-V(-L) e^{-M(-L)} + \frac{1}{F_+(-L)} \left[\lambda \int_L^{-L} dx\, V(x) e^{-M(x)} - 1 \right]}{e^{-M(-L)} - \frac{\lambda}{F_+(-L)} \int_L^{-L} dx\, e^{-M(x)}}. \tag{4.4.33}$$

Colored shot noise

Let us now return to (4.1.1) and consider a process $X(t)$ driven by a random process $\eta(t)$ defined as follows (Masoliver, 1987):

$$\eta(t) = \sum_i \gamma_i \delta(t - r_i). \tag{4.4.34}$$

The noise thus consists of random impulses of strength γ_i at the times r_i, i.e. $\eta(t)$ is a type of shot noise. We restrict our analysis to positive γ_i and assume the γ_i and r_i to be mutually uncorrelated random variables with distribution functions $\Lambda(\gamma_i)$ and $\Psi(r_i - r_{i-1})$. Note that an exponential $\Psi(\Delta t)$ leads to delta-correlated (i.e. white) shot noise while other forms of $\Psi(\Delta t)$ lead to colored shot noise. The effect of shot noise is to cause the trajectory $X(t)$ to jump discontinuously, the jumps being of random height and occurring at random times (see Figure 4.11). Again we are interested in the statistics of first passage across L or $-L$. If $G(X)$ leads to a monotonic deterministic trajectory which has a stable fixed point at x_s, then the details of the analysis will depend on the relative location of x_0, x_s, L and $-L$. For concreteness we take the configuration shown in Figure 4.11 (other configurations have also been considered by Masoliver (1987)). For ease of analysis we take $g(X) = 1$.

Let us take $\gamma_0 = 0$ and let us assume that this 'zero-strength' impulse occurs exactly at $t = r_0 = 0$ (these conditions can easily be relaxed). The trajectory $X(t)$ for a particular realization of $\eta(t)$ consists of intervals of deterministic

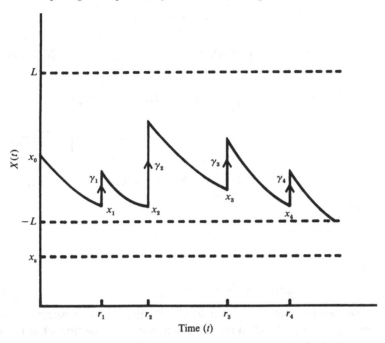

Figure 4.11. Trajectory for a process driven by shot noise. The fixed point x_s lies below the boundaries at L and $-L$.

evolution governed by $G(X)$ and discontinuous jumps governed by the shot noise. This trajectory is now given by the incremental values

$$
\left.
\begin{aligned}
X(t) &= \phi(t + \phi^{-1}(x_0)), \quad 0 \leqslant t \leqslant t_1 \\
X(t) &= \phi(t + \phi^{-1}(\gamma_1 + \phi(t_1 + \phi^{-1}(x_0)))), \\
&\qquad t_1 \leqslant t \leqslant t_1 + t_2 \\
X(t) &= \phi(t + \phi^{-1}(\gamma_2 + \phi(t_2 + t_1 \\
&\qquad + \phi^{-1}(\gamma_1 + \phi(t_1 + \phi^{-1}(x_0)))))), \, t_1 + t_2 \leqslant t \leqslant t_1 + t_2 + t_3
\end{aligned}
\right\}
$$

$$(4.4.35)$$

and so on. Here we have defined the trajectory function $\phi(t)$ via the inverse relation

$$
\phi^{-1}(X) \equiv \int^X \frac{\mathrm{d}X'}{G(X')}. \tag{4.4.36}
$$

It is again convenient to introduce the auxiliary probabilities $f_n(t, L|x_0)$ defined in (4.4.16). As before, one can construct as expression for these probability densities by explicitly following the constrained trajectories that do not cross $\pm L$ before time t but cross one or the other at time t. For the configuration shown in Figure 4.11 one obtains

$$f_n(t, L|x_0) = \int_0^{\tau_1} dt_1 \, \Psi(t_1) \int_0^{L-x_1} d\gamma_1 \Lambda(\gamma_1) \dots \int_0^{\tau_{n-1}} dt_{n-1} \Psi(t_{n-1})$$

$$\times \int_0^{L-x_{n-1}} d\gamma_{n-1} \Lambda(\gamma_{n-1}) \left\{ \int_{\tau_n}^{\infty} dt_n \Psi(t_n) \right.$$

$$\times \delta[t - (t_1 + \dots + t_{n-1} + \tau_n)] + \int_0^{\tau_n} dt_n \Psi(t_n) \quad (4.4.37)$$

$$\left. \times \int_{L-x_n}^{\infty} d\gamma_n \Lambda(\gamma_n) \delta[t - (t_1 + \dots + t_{n-1} + t_n)] \right\},$$

where

$$\tau_m \equiv \int_{x_{m-1}+\gamma_{m-1}}^{-L} \frac{dx}{G(x)} \quad (4.4.38)$$

and

$$x_m = \phi(t_m + \phi^{-1}(\gamma_{m-1} + x_{m-1})). \quad (4.4.39)$$

Within the curly brackets { } in (4.4.37) the first term represents a crossing of L or $-L$ at time t during a period of deterministic evolution while the second term represents a crossing of L via a jump. The Laplace transform of (4.4.37) with respect to time leads to a recursion relation connecting $\tilde{f}_{n+1}$ with $\tilde{f}_n$. When summed over n, the result is an integral equation for $\tilde{f}(s, L|x_0)$ (cf. (4.4.22)):

$$\tilde{f}(s, L|x_0) = \tilde{f}_1(s, L|x_0)$$

$$+ \int_0^{\tau_1} dt_1 \, e^{-st_1} \Psi(t_1) \int_0^{L-x_1} d\gamma_1 \, \Lambda(\gamma_1) \tilde{f}(s, L|x_1 + \gamma_1).$$

$$(4.4.40)$$

The relation (4.4.26) immediately leads to an integral equation for the mean first passage time:

$$T_1(L|x_0) = \tau_1 \int_{\tau_1}^{\infty} dt_1 \, \Psi(t_1) + \int_0^{\tau_1} dt_1 \, t_1 \Psi(t_1)$$

$$+ \int_0^{\tau_1} dt_1 \, \Psi(t_1) \int_0^{L-x_1} d\gamma_1 \, \Lambda(\gamma_1) T_1(L|x_1 + \gamma_1). \quad (4.4.41)$$

Similar integral equations are obtained for other relative configurations of x_0, x_s, L and $-L$.

For some forms of the time interval distribution $\Psi(\Delta t)$ and of the jump size distribution $\Lambda(\gamma_i)$ it is again possible to convert the integral equation (4.4.41) to a differential equation. This can be done, for instance, if the set of random jump times is a Poisson sequence so that $\Psi(\Delta t)$ takes the exponential form (4.4.2) and if the jump sizes are exponentially distributed,

$$\Lambda(\gamma_i) = \frac{1}{\gamma} e^{-\gamma_i/\gamma}, \qquad (4.4.42)$$

where γ is the average jump size. Equation (4.4.41) can then be converted to the differential equation (cf. (4.2.8) and (4.4.27))

$$\frac{\partial^2}{\partial x_0^2} T_1(L|x_0) + \left[\frac{G'(x_0)}{G(x_0)} - \frac{\lambda}{G(x_0)} - \frac{1}{\gamma} \right] \frac{\partial}{\partial x_0} T_1(L|x_0)$$

$$= \frac{1}{\gamma G(x_0)} \qquad (4.4.43)$$

together with the boundary conditions

(i) $\quad T_1(L|-L) = 0,$ \qquad (4.4.44)

(ii) $\quad \dfrac{\partial}{\partial x_0} T_1(L|x_0)\bigg|_{x_0=L} = \dfrac{1}{G(L)}[\lambda T_1(L|L) - 1].$ \qquad (4.4.45)

The interpretation of the structure of the boundary condition (4.4.45) is again not straightforward. The solution of the problem (4.4.43)–(4.4.45) is (cf. (4.2.13) and (4.4.30))

$$T_1(L|x_0) = \int_{-L}^{x_0} dx \frac{e^{M(x)}}{G(x)} \left[\frac{1}{\gamma} \int^x dx' e^{-M(x')} + C \right], \qquad (4.4.46)$$

where

$$M(x) \equiv \frac{x}{\gamma} + \gamma \int^x \frac{dx'}{G(x')} \qquad (4.4.47)$$

and

$$C = \frac{-1 + \dfrac{\lambda}{\gamma} \int_{-L}^{L} dx \dfrac{e^{M(x)}}{G(x)} \int^x dy\, e^{-M(y)} - \dfrac{1}{\gamma} e^{M(L)} \int^L dx\, e^{-M(x)}}{e^{M(L)} - \lambda \int_{-L}^{L} dx \dfrac{e^{M(x)}}{G(x)}}. \qquad (4.4.48)$$

The exact enumeration of trajectories which forms the basis of the previous method can become difficult and even intractable as the complexity of the system and/or the noise increases. It is therefore desirable to reformulate the theory in such a way as to minimize the combinational difficulties that accompany the exact enumeration of trajectories. A beginning in that direction has been made with a certain amount of success (Weiss *et al.*, 1987).

4.4.2 Examples

Colored dichotomous noise

Let us first consider the simplest dynamical system driven by colored dichotomous fluctuations, that being (4.3.1) with $G(X) = 0$ and $g(X) = 1$, i.e, the free process

$$\dot{X}(t) = \eta(t). \tag{4.4.49}$$

Implementation of the solution (4.4.30) for the mean first passage time with (4.4.31)–(4.4.33) is straightforward, as is the corresponding solution when the initial value of $\eta(t)$ is $-v$. An average of these two solutions equally weighted leads to the mean first passage time (Masoliver, Lindenberg and West, 1986a–c, Masoliver, West and Lindenberg, 1986)

$$T_1(L|x_0) = \frac{L^2 - x_0^2}{2D} + L(\tau_c/D)^{1/2}, \tag{4.4.50}$$

where again $D \equiv v^2 \tau_c$ (cf. (4.4.8) and (4.4.9)). When (4.4.50) is averaged over a uniform initial distribution between $-L$ and L, one obtains

$$T_1(L) \equiv \frac{1}{2L} \int_{-L}^{L} dx_0 \, T_1(L|x_0)$$

$$= \frac{L^2}{3D} + L(\tau_c/D)^{1/2}. \tag{4.4.51}$$

The corresponding results for a (diffusive) process (4.4.49) driven by white Gaussian fluctuations of strength D are the first terms on the right side of (4.4.50) and (4.4.51). These results are shown in Figures 4.12 and 4.13. Note that the mean first passage time for the system driven by the dichotomous Markov process is longer than that of the diffusive process for all parameter values.

We can write (4.4.51) in the form (4.3.81),

$$T_1(L) = T_1^0 \left[1 + 3^{1/2} \left(\frac{\tau_c}{T_1^0} \right)^{1/2} \right], \tag{4.4.52}$$

where $T_1^0 \equiv L^2/3D$ is the white noise result. Thus the only correction to T_1^0 is of the form found by Doering, Hagan and Levermore (1987) for Gaussian colored noise (cf. Section 4.3.5); since the dynamical process (4.4.49) is otherwise a free process, there are no $O(\tau_c)$ corrections in (4.4.52). We stress that the result (4.4.52) is exact.

It is instructive to compare these results to those obtained for an example in which $\Psi(t)$ is not simply exponential. Since most other techniques have been restricted to exponential correlation functions for the fluctuations, this comparison is particularly interesting. One of the several examples that we have worked out corresponds to a situation in which two successive switches must be separated by at least a time t_0, i.e. t_0 is a 'recovery time' or 'refractory period' (Masoliver, West and Lindenberg, 1986). One such switching time distribution is

$$\Psi(\Delta t) = 0, \quad \Delta t < t_0$$

$$= \lambda e^{-\lambda(\Delta t - t_0)}, \quad \Delta t > t_0. \tag{4.4.53}$$

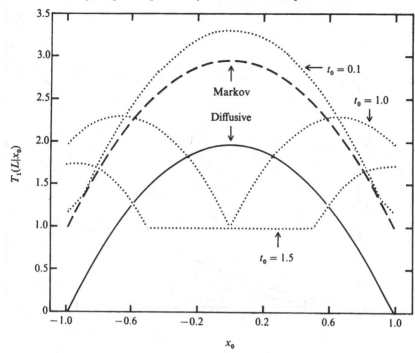

Figure 4.12. Mean first passage time $T_1(L|x_0)$ as a function of the initial location x_0 with parameter values $\lambda = 2$, $v = 1$ and $L = 1$. The free processes shown are the diffusive one, the one driven by dichotomous Markov fluctuations, and the inertial one for three values of t_0.

Once again it is possible to convert the integral equation (4.4.21) to a differential equation for $\tilde{f}(s, L|x_0)$ and thence to a differential equation for the mean first passage time to $\pm L$ together with appropriate boundary conditions. The differential equation and the boundary conditions are similar to (4.4.26)–(4.4.28) provided $x_0 < L - vt_0$ if $\eta(0) = v$ and provided $x_0 > vt_0 - L$ if $\eta(0) = -v$. If x_0 does not satisfy these conditions then the first interval takes the process to a boundary in a time $T_1(L|x_0) = (L - x_0)/v$ if $\eta(0) = v$ and $(L + x_0)/v$ if $\eta(0) = -v$. If x_0 does satisfy the above conditions then at least one switch may occur before the process reaches a boundary; in this case one finds

$$T_1^{\pm}(L|x_0) = \frac{\lambda}{v^2}(1 + \lambda t_0)(L^2 - x_0^2) - \lambda t_0^2$$

$$+ \frac{L}{v}(1 - \lambda^2 t_0^2) \mp \frac{x_0}{v}(1 + \lambda t_0), \qquad (4.4.54)$$

where the superscript on T_1 denotes the sign of the initial value of $\eta(t)$. The mean first passage time $T_1(L|x_0)$ is a properly weighted average over these

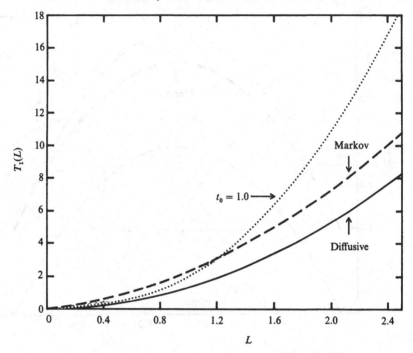

Figure 4.13. Averaged mean first passage time $T_1(L)$ as a function of L with parameter values $\lambda = 2$ and $v = 1$. The free processes correspond to those depicted in Figure 4.12.

results. The mean first passage time averaged over a uniform initial distribution is

$$T_1(l) = L/v, \quad L < vt_0/2$$

$$= \frac{2\lambda}{3v^2}(1 + \lambda t_0)L^2 + \frac{1}{v}(1 - \lambda^2 t_0^2)L$$

$$- \frac{\lambda t_0^2}{2}(1 - \lambda t_0) + \frac{v\lambda t_0^3}{6L}(1 - \tfrac{1}{2}\lambda t_0), \quad L > vt_0/2. \qquad (4.4.55)$$

These results are compared with those of the previous example in Figures 4.12 and 4.13. For small t_0 the behavior of $T_1(L|x_0)$ in Figure 4.12 essentially coincides with that of the dichotomous Markov process. As t_0 increases, the boundaries at L and $-L$ are reached more rapidly when the process begins near the center of the interval because of those realizations that arrive at the boundary during the first interval. The effect becomes more pronounced as t_0 increases. The flat portion of the curve for $t_0 = 1.5$ represents realizations that reach a boundary during the first interval regardless of the initial direction of

152

the motion. This occurs in the interval $L - vt_0 < x_0 < vt_0 - L$. In this range

$$T_1(L|x_0) = \frac{1}{2v}[(L + x_0) + (L - x_0)] = \frac{L}{v}. \tag{4.4.56}$$

These results emphasize the fact that the most rapid arrival at a boundary does not always occur when a process begins its migration near a boundary. However, the inertial mean first passage time may exceed the Markov one for all values of x_0 if t_0 is sufficiently small compared to L/v and λ^{-1}. The averaged mean first passage time $T_1(L)$ is shown in Figure 4.13 for $t_0 = 1$. For small L this average reflects the behavior just discussed, resulting in an inertial $T_1(L)$ that lies between the diffusive and Markov values. For large L (t_0 held fixed) the inertial result crosses above the Markov one, in analogy with the $t_0 = 0.1$ result in Figure 4.12. In this range the ratio of the inertial to the Markov and diffusive results approaches the value $(1 + \lambda t_0) > 1$.

Finally, it is instructive to examine the traditional parameters D and τ_c that result from the 'inertial' form (4.4.53). Although it is difficult to obtain an analytic expression for the correlation function (4.4.3), one can calculate its first two moments exactly. One obtains

$$D = \frac{v^2}{2\lambda(1 + \lambda t_0)} \tag{4.4.57}$$

and

$$\tau_c = \frac{1}{2\lambda} - \frac{t_0}{2}(1 + t_0\lambda + \tfrac{1}{3}t_0^2\lambda^2). \tag{4.4.58}$$

These expressions reduce to (4.4.8) and (4.4.9) when $t_0 \to 0$ (as they should). The 'correlation time' τ_c in (4.4.58) is, however, not positive definite, becoming negative for sufficiently large t_0. This behavior together with the decrease of D with increasing t_0 indicates that the correlation function oscillates around zero. Such a correlation function is of course not well characterized by its low order moments.

Let us next consider a somewhat more complex process than (4.4.49), i.e. one with a linear deterministic evolution (Masoliver, Lindenberg and West, 1986c):

$$\dot{X} = -\mu X + \eta(t), \tag{4.4.59}$$

with $\mu > 0$. The process can only reach L or $-L$ if these boundaries lie within the interval $(-v/\mu, v/\mu)$ defined by the asymptotically stable fixed points of the trajectories $X_-(t)$ and $X_+(t)$. The solution (4.4.46) can only be evaluated in closed form for certain relative values of the parameters λ, v and μ. For $v = 1$, $\lambda = 2$, $\mu = 1$ and $L < 1$ one obtains

$$T_1(L|x_0) = \frac{1}{2}\ln\left(\frac{1 - x_0^2}{1 - L^2}\right) + \frac{2}{3}\frac{(x_0^2 - 2)}{(1 - x_0^2)^2}$$

$$-\frac{(L^2 + L - 4)}{2(1+L)(1-L)^2}. \tag{4.4.60}$$

The corresponding result for a diffusive process is

$$T_1(L|x_0) = 8 \int_{x_0}^{L} du\, e^{2u^2} \int_0^u dy\, e^{-2y^2}. \tag{4.4.61}$$

These results are shown in Figure 4.14, where they are compared to those for a free process. When averaged over a uniform initial distribution in the interval $(-L, L)$, (4.4.60) leads to

$$T_1(L) = \frac{-3L^3 + 2L^2 + 3L}{3(1+L)(1-L)^2}, \tag{4.4.62}$$

while for the diffusive process

$$T_1(L) = 2\left[\frac{e^{2L^2}}{L} \int_0^L dy\, e^{-2y^2} - 1\right]. \tag{4.4.63}$$

These results are shown in Figure 4.15 together with those for the corresponding free process. Note that the mean first passage times for the bound process (4.4.59) are longer than those of the free process (4.4.49), and that the first arrival at L or $-L$ occurs more rapidly for the diffusive process than for the system driven by dichotomous colored noise.

Shot noise

Let us begin again with the simple 'free process' described by (4.4.49), but with $\eta(t)$ being given by (4.4.34). In this case it is particularly straightforward to solve the integral equations (4.4.40) and (4.4.41) directly. Thus, (4.4.41) reduces to (Masoliver, 1987)

$$T_1(L|x_0) = \frac{1}{\lambda} + \int_0^{L-x_0} d\gamma_1\, \Lambda(\gamma_1)\, T_1(L|x_0 + \gamma_1) \tag{4.4.64}$$

whose solution is

$$T_1(L|x_0) = \frac{1}{\lambda}\underline{L}^{-1}\left\{\frac{1}{s[1 - L[\Lambda(L-x_0)]]}\right\}, \tag{4.4.65}$$

where $\underline{L}$ denotes the Laplace transform and $\underline{L}^{-1}$ its inverse. Explicit expressions can be given for specific forms of the distribution $\Lambda(\gamma_i)$. For the exponential distribution (4.4.42) one obtains

$$T_1(L|x_0) = \frac{1}{\lambda}\left(1 + \frac{L-x_0}{\gamma}\right). \tag{4.4.66}$$

If, on the other hand, all the jumps are of the same size, i.e.

$$\Lambda(\gamma_i) = \delta(\gamma - \gamma_i), \tag{4.4.67}$$

154

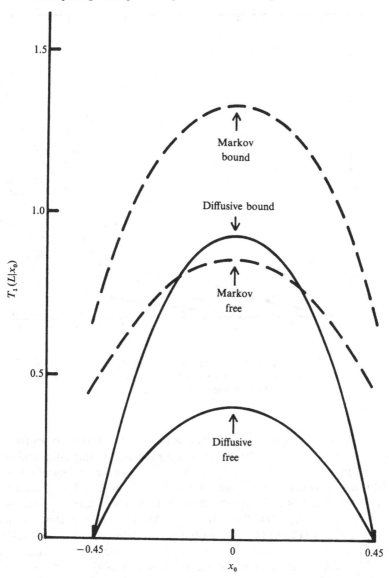

Figure 4.14. Mean first passage time $T_1(L|x_0)$ as a function of the initial location x_0 for bound ($\mu = 1$) and free processes with $\lambda = 2$, $\nu = 1$ and $L = 0.45$. The processes shown are diffusive (solid curves) and driven by dichotomous Markov fluctuations (dashed).

then

$$T_1(L|x_0) = \frac{1}{\lambda}\left[1 + \sum_{n=1}^{\infty} \theta(L - x_0 - n\gamma) \right], \qquad (4.4.68)$$

where $\theta(x)$ is the Heaviside function.

155

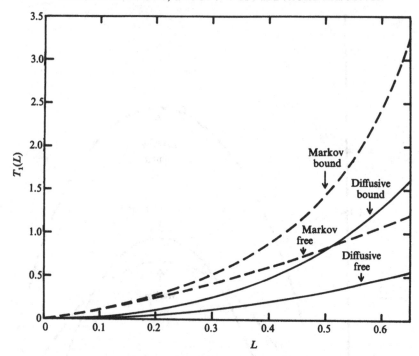

Figure 4.15. Averaged mean first passage time $T_1(L)$ as a function of L for bound ($\mu = 1$) and free processes with $\lambda = 2$ and $\nu = 1$. The processes shown correspond to those depicted in Figure 4.14.

Let us next consider the bound process (4.4.59). For this problem the fixed point occurs at $x_s = 0$ and hence the relative location of this point and of the boundaries at $\pm L$ is not as shown in Figure 4.11. For the configuration $-L < x_s < L$ the process can only reach the boundaries at L and $-L$ via a jump and not by deterministic dynamical evolution. The integral equations for $\tilde{f}(s, L|x_0)$ and $T_1(L|x_0)$ are therefore somewhat different from those shown in (4.4.40) and (4.4.41). For the exponential distribution of jump sizes (4.4.42) the differential equation, boundary conditions, and solution for $T_1(L|x_0)$ are also suitably modified from those shown in (4.4.46)–(4.4.48). To obtain analytic results it is again necessary to choose particular values of the parameters. For $\mu = \lambda = 1$ one finds (Masoliver, 1987)

$$T_1(L|x_0) = -\ln L + Ei(L/\gamma) + \ln|x_0|$$

$$+ \frac{\gamma}{x_0}(e^{x_0/\gamma} - 1) - Ei(x_0/\gamma), \qquad (4.4.69)$$

where $Ei(u)$ is the exponential-integral function. If the jumps are all of the

156

same size (cf. (4.4.67)) then one finds a simple analytic solution when $\mu = \lambda = 1$ and $L = \gamma$:

$$T_1(L|x_0) = \begin{cases} 1, & 0 < x_0 \leqslant L \\ 2, & -L \leqslant x_0 < 0. \end{cases} \qquad (4.4.70)$$

This extremely simple result is a consequence of the particular choice of parameters.

4.4.3   Other methods

The first passage time statistics for the dynamical system (4.1.1) driven by exponentially correlated dichotomous fluctuations has been considered by other methods. These methods have been developed specifically for Markov fluctuations and are not generalizable to other correlation properties.

The usual procedure starts with the idea of extending the phase space to include the dichotomous fluctuations as a dynamical variable, in the spirit of (4.3.1) and (4.3.2). The resulting system is Markovian and the evolution of the probability density $P(x, \eta, t|x_0, \eta_0)$ can be expressed in terms of the master equation (Kitahara, Horsthemke, Lefever and Inaba, 1980)

$$\frac{\partial}{\partial t} P(x, v, t|x_0, \eta_0) = -\frac{\partial}{\partial x}[G(x) + g(x)v]P(x, v, t|x_0, \eta_0)$$

$$+ 2\lambda[P(x, -v, t|x_0, \eta_0)$$

$$- P(x, v, t|x_0, \eta_0)] \qquad (4.4.71)$$

$$\frac{\partial}{\partial t} P(x, -v, t|x_0, \eta_0) = -\frac{\partial}{\partial x}[G(x) - g(x)v]P(x, -v, t|x_0, \eta_0)$$

$$+ 2\lambda[P(x, v, t|x_0, \eta_0)$$

$$- P(x, -v, t|x_0, \eta_0)], \qquad (4.4.72)$$

where $(2\lambda)^{-1}$ is the correlation of the dichotomous noise (cf. (4.4.2) and (4.4.7)–(4.4.9)). The procedure is akin in spirit to that of Section 4.3.2 where the first passage time moments are obtained from the backward Kolmogorov equation (4.2.7) with appropriate boundary conditions. Here once again one can construct the backward equations arising from (4.4.71) and (4.4.72), but it is difficult to formulate the appropriate boundary conditions. The problem lies in insuring that in the reduced system (i.e. when the values of $\eta(t)$ are summed over) there be no backflow of probability into the interval $(-L, L)$ (Hänggi and Riseborough, 1983; Rodriguez and Pesquera, 1986; Sancho, Sagués and San Miguel, 1986; Van den Broeck and Hänggi, 1984).

Let $W(x, \eta, t|x_0, \eta_0)$ be the probability density for $X(t)$ to be in the interval $(x, x + dx)$ and for $\eta(t) = \eta$ subject to the constraint that $X(\tau)$ has not crossed

$\pm L$ for $0 \leqslant \tau \leqslant t$ (cf. Section 4.2). The probability density $W_t(x_0, v_0) \equiv W(x, \eta, t | x_0, v_0)$ satisfies the backward equation (Rodríguez and Pesquera, 1986)

$$\frac{\partial}{\partial t} W_t(x_0, v) = [G(x_0) + vg(x_0)] \frac{\partial}{\partial x_0} W_t(x_0, v)$$

$$+ 2\lambda[W_t(x_0, -v) - W_t(x_0, v)], \tag{4.4.73}$$

$$\frac{\partial}{\partial t} W_t(x_0, -v) = [G(x_0) - vg(x_0)] \frac{\partial}{\partial x_0} W_t(x_0, -v)$$

$$+ 2\lambda[W_t(x_0, v) - W_t(x_0, -v)] \tag{4.4.74}$$

together with the initial and boundary conditions

$$W_0(x_0, \eta_0) = \theta(x_0 + L)\theta(L - x_0) \tag{4.4.75}$$

$$W_t(x_0, \eta_0) = 0 \quad \text{for} \quad |x_0| > L \tag{4.4.76}$$

$$W_t(-L^+, -v) = W_t(L^-, v) = 0, \tag{4.4.77}$$

where $u^+(u^-)$ indicates that u is approached from above (below). The condition (4.4.77) insures that a process starting near a boundary with initial velocity towards the boundary is guaranteed to exit the interval (cf. (4.4.24)). The relation between the boundary condition (4.4.76) and our earlier condition (4.4.25) is not at all clear. However, the boundary conditions obtained from (4.4.75) and (4.4.76) for the mean first passage time are identical to (4.4.28) and (4.4.29).

Rodríguez and Pesquera (1986) have solved (4.4.73) and (4.4.74) with the conditions (4.4.75)–(4.4.77). Their analysis subsumes earlier ones carried out for specific examples. As they point out, their results agree with those obtained from our trajectory analysis technique.

4.5   Summary

We end with a brief summary of the highlights of this review.

For processes driven by continuous (usually Gaussian) colored noise we discussed the rationale for the search of an effective Fokker–Planck equation (EFPE). Once such an equation is constructed the assumption is made that the usual white noise construction of first passage time measures can be directly adopted. We showed that all available techniques when implemented to lowest order in the noise intensity lead to the so-called 'best Fokker–Planck equation' (BFPE). We then discussed the relation of other EFPEs to the BFPE. The BFPE was used to calculate mean first passage times for a bistable system for which analog experimental results are available; one finds remarkably close agreement over large ranges of parameter values.

We addressed in some detail a number of theoretical limitations of the BFPE and other EFPEs and noted that the sources of these limitations (but not their resolution) are well understood. Two directions have been followed in efforts to overcome these difficulties. One is the solution of the higher dimensional Markovian problem onto which the colored noise problem can be mapped. The other is the analysis of systems in which the statistics of the colored fluctuations have been simplified.

This second avenue has led to the analysis of first passage time statistic of systems driven by dichotomous and shot colored noise. We discussed a trajectory analysis method that allows the exact calculation of these statistics for simple model systems. This analysis shows clearly the ways in which the EFPE approach may fail.

A few general observations can be made about the effects of coloring the noise. Increasing the correlation time of the fluctuations tends to narrow the distribution of the process and therefore to increase the mean first passage time. These effects can often be dramatic, as can those due to changes in the noise statistics. In this review we have only explicitly calculated mean first passage times, although higher moments (and in fact the entire distribution) can also be calculated from these theories. Such calculations might be interesting since the higher moments are probably even more sensitive to the effects of color and detailed noise statistics.

Acknowledgements

This research was supported in part by Grant/Contract DE-FG03-86ER13606 from the Department of Energy and in part by NSF grants ATM-8507820 and ATM-8509353.

References

Büttiker, M. and Landauer, R. 1982. In *Nonlinear Phenomena at Phase Transitions and Instabilities* (T. Riste, ed.), pp. 111–43. NATO ASI Series. New York: Plenum.
Doering, C. R. 1986. *Phys. Rev. A* **34**, 2564.
Doering, C. R., Hagan, S. and Levermore, C. D. 1987. *Phys. Rev. Lett.* **59**, 2129.
Faetti, S., Fronzoni, L., Grigolini, P. and Mannella R. 1986a. Preprint.
Faetti, S., Fronzoni, L., Grigolini, P., Palleschi, V. and Tropiano, G. 1986b. *J. Stat. Phys.* (in press).
Fox, R. F. 1986a. *Phys. Rev. A* **33**, 467.
Fox, R. F. 1986b. *Phys. Rev. A* **34**, 4525.
Grigolini, P. 1986. *Phys. Lett.* **119A**, 157.
Hänggi, P., Marchesoni, F. and Grigolini, P. 1984. *Z. Phys. B* **56**, 333.
Hänggi, P., Mroczkowski, T. J., Moss, F. and McClintock, P. V. E. 1985. *Phys. Rev. A* **32**, 695.
Hänggi, P. and Riseborough, P. 1983. *Phys. Rev. A* **27**, 3379.
Kitahara, K., Horsthemke, W., Lefever, R. and Inaba, Y. 1980. *Prog. Theor. Phys.* **64**, 1233.

Kuś, M. and Wódkiewicz, K. 1982. *Phys. Lett.* **90A**, 23.

Lindenberg, K., Shuler, K. E., Seshadri, V. and West, B. J. 1983. In *Probabilistic Analysis and Related Topics* (A. T. Bharucha-Reid, ed.), vol. 3, pp. 81–125. New York: Academic Press.

Lindenberg, K. and West, B. J. 1983. *Physica* **119A**, 485.

Lindenberg, K. and West, B. J. 1984. *Physica* **128A**, 25.

Lindenberg, K. and West, B. J. 1986. *J. Stat. Phys.* **42**, 201.

Masoliver, J. 1987. *Phys. Rev. A* **35**, 3918.

Masoliver, J., Lindenberg, K. and West, B. J. 1986a. *Phys. Rev. A* **33**, 2177.

Masoliver, J., Lindenberg, K. and West, B. J. 1986b. *Phys. Rev. A* **34**, 1481.

Masoliver, J., Lindenberg, K. and West, B. J. 1986c. *Phys. Rev. A* **34**, 2351.

Masoliver, J., West, B. J. and Lindenberg, K. 1986. In *Transport and Relaxation in Random Materials* (J. Klafter, R. J. Rubin and M. F. Shlesinger, eds.), pp. 357–75. Singapore: World Scientific Publishing.

Masoliver, J., West, B. J. and Lindenberg, K. 1987. *Phys. Rev. A* **35**, 3086.

Mukamel, S., Oppenheim, I. and Ross, J. 1978. *Phys. Rev. A* **17**, 1978.

Novikov, E. A. 1965. *Sov. Phys.-JETP* **20**, 1290.

Peacock, E., West, B. J. and Lindenberg, K. 1988. *Phys. Rev. A* **37**, 3530.

Pontryagin, L., Andronov, A. and Vitt, A. 1933. *Zh. Eksp. Theor. Fiz.* **3**, 165–80.

Rodriguez, M. A. and Pesquera, L. 1986. *Phys. Rev. A* **34**, 4532.

Sancho, J. M., Saugés, F. and San Miguel, M. 1986. *Phys. Rev. A* **33**, 3399.

Sancho, J. M., San Miguel, M., Katz, S. L. and Gunton, J. D. 1982a. *Phys. Rev. A* **26**, 1589.

Sancho, J. M., San Miguel, M., Yamazaki, H. and Kawakubo, T. 1982b. *Physica* **116A**, 560.

Schuss, Z. 1980. *Theory and Applications of Stochastic Differential Equations*. New York: Wiley.

Van den Broeck, C. and Hänggi, P. 1984. *Phys. Rev. A* **30**, 2730.

Wang, M. C. and Uhlenbeck, G. E. 1945. *Rev. Mod. Phys.* **17**, 323.

Weiss, G. H. 1966. *Adv. Chem. Phys.* **13**, 1.

Weiss, G. H., Masoliver, J., Lindenberg, K. and West, B. J. 1987. *Phys. Rev. A* **36**, 1435.

West, B. J. and Lindenberg, K. 1987. In *Studies in Statistical Mechanics* (J. Lebowitz, ed.), vol. 13, pp. 159–266. Amsterdam: North-Holland.

Note added in proof

Since this review was completed, a number of investigators have made important contributions to the subject. Particularly noteworthy is the work of J. F. Luciani and A. D. Verga (*J. Stat. Phys.* **50**, 567, 1988), M. M. Kłosek-Dygas, B. J. Matkowsky and Z. Schuss (*Phys. Rev. A*, in press, 1988), and G. P. Tsironis and P. Grigolini (*Phys. Rev. A*, in press, 1988).

5   The projection approach to the Fokker–Planck equation: applications to phenomenological stochastic equations with colored noises

PAOLO GRIGOLINI

5.1   Introduction

The Fokker–Planck equation plays an important role within many fields of investigation due to the fact that the complex aspect of a many-body interaction is simulated by a simple diffusion term, thereby offering the chance of getting analytical solutions to describe, for instance, the steady-state statistical distribution of a variable of interest. It is therefore a problem of paramount importance to establish under which restrictive conditions the picture of the dynamical and statistical behavior of a physical system can rely on an equation of motion (of the probability distribution of the variables of interest) with a Fokker–Planck structure.

To discuss this basic problem let us consider the paradigmatic case of Brownian motion. This is usually pictured (Uhlenbeck and Ornstein, 1930; Wang and Uhlenbeck, 1945) via the Langevin equation

$$\dot{a} = -\gamma a(t) + f_a(t). \tag{5.1.1}$$

a is the velocity of a Brownian particle moving, for simplicity, in a one-dimensional space and $f_a(t)$ is a white Gaussian noise defined by

$$\langle f_a(0) f_a(t) \rangle = 2D_a \delta(t), \tag{5.1.2}$$

with

$$D_a = \gamma k_{\mathrm{B}} T / m, \tag{5.1.3}$$

where k_{B} is the Brownian constant, T is the absolute temperature and m is the mass of the Brownian particle. Equation (5.1.3) relates the intensity of the stochastic force $f(t)$ to the mean value of the variable a,

$$\langle a^2 \rangle_{\mathrm{eq}} = k_{\mathrm{B}} T / m, \tag{5.1.4}$$

through the friction γ, and it is usually referred to as the fluctuation-dissipation relation. It is widely accepted that (5.1.1) is associated with the following equation of motion for the time evolution of the probability distribution of the variable a, $\sigma(a; t)$,

$$\frac{\partial}{\partial t} \sigma(a; t) = \gamma \frac{\partial}{\partial a} \left(a + \langle a^2 \rangle_{\mathrm{eq}} \frac{\partial}{\partial a} \right) \sigma(a; t). \tag{5.1.5}$$

PAOLO GRIGOLINI

This is immediately recognized as being a special case of the more general form

$$\frac{\partial}{\partial t}\sigma(a;t) = \left\{\frac{\partial}{\partial a}\varphi(a) + \frac{\partial}{\partial a}(a)\frac{\partial}{\partial a}\Phi(a)\right\}\sigma(a;t),$$ (5.1.6)

which, in turn, is the one-dimensional version of the multidimensional structure

$$\frac{\partial}{\partial t}\sigma(\mathbf{a};t) = \left\{\frac{\partial}{\partial a_i}\varphi(\mathbf{a}) + \sum_{ij}\frac{\partial}{\partial a_i}\psi_i(\mathbf{a})\frac{\partial}{\partial a_i}\Phi_j(\mathbf{a})\right\}\sigma(\mathbf{a};t).$$ (5.1.7)

Now the question arises of developing a method to associate the stochastic motion of a generic (in general multidimensional) variable **a**, not necessarily Markovian, with the Fokker–Planck structure of (5.1.7). This is a challenging problem due to the fact that (5.1.7) has a clear Markovian character.

A first treatment of this problem can be found in the pioneer work of Stratonovich (1963) (see also Chapter 2 of this volume). A review of many more attempts was recently given by Grigolini and Marchesoni (1985). Another review paper on the subject of elimination of fast variables is that of Van Kampen (1985). Van Kampen and Oppenheim (1986) have recently used a technique of elimination of fast variables to build up the Fokker–Planck equation for the Brownian motion of a particle with a large mass in a bath of light particles. The resulting Fokker–Planck equation is formally derived from the rigorous microscopic Liouville equation describing the dynamics of the whole system. Examples of the same kind (concerning one-dimensional systems) are provided by Ferrario et al. (1985) and Grigolini and Marchesoni (1985). Our purpose in this chapter is much less ambitious than that of deriving the Fokker–Planck equation from a rigorous Liouville equation. We must point out, however, that when dealing with nonlinear systems our task is no less arduous than that of deriving the Fokker–Planck equation from a rigorous Liouville equation. In the latter case the additional difficulties stemming from the presence of virtually infinite freedom degrees do not allow us to properly appraise the subtle aspects of the interplay of nonlinear interactions and stochastic forces (the stochastic forces aim at mimicking the infinite freedom degrees). We must also take into account the ever increasing importance of analog and digital simulation (see Chapters 7–9, Volume 3). These simulation techniques have by now reached a so-advanced technical level that they can be used as reliable tests for the predictions of theoretical investigations. Nonlinear electric circuits, furthermore, are to be conceived, by themselves, as real physical systems whose dynamical behavior represents a challenge for theoretical investigations. The theoretical predictions of this chapter (and many others in this volume) can be easily supported or 'falsified' (Popper, 1959) by comparison with the results of simulation 'experiments'. This is so precisely because simple nonlinear systems are the subject of our study.

162

The approach to the Fokker–Planck equation described in this chapter stems from a theoretical background characterized by two major achievements in the field of modern statistical physics. The former is the projection method of Zwanzig (1960). This allows us to write the exact expression of the diffusion equation driving the motion of the stochastic variables *a* via a contraction over the set of the 'irrelevant' variables. Throughout the present chapter this latter set is also referred to as the 'thermal bath' of the system of interest, and it is assumed to be characterized by a *finite* time scale τ (this is why the influence of the 'thermal bath' on the system is simulated by colored noises). The resulting diffusion equation will exhibit a structure more general than that of (5.1.7) and it will be necessary to make additional assumptions for this important structure to be recovered. In addition to that of exploring a time region much longer than τ (this condition has already been stressed in the pioneer work of Stratonovich, 1963) we shall also rely on the linear response approximation which is the basic assumption behind the Kubo linear response theory (Kubo, 1957, 1966). This theory is precisely the latter of the two significant achievements characterizing the theoretical background from which the theoretical approach of this chapter develops.

The spirit of our treatment is as follows. Let us assume that many variables $q_1, q_2, \ldots$ interact among themselves via deterministic couplings. Let us also assume that in the absence of this mutual interaction the motion of each variable would be described by the standard Langevin equation of (5.1.1) (this includes also the case where γ, $f(t)$, or both, vanish). We can therefore write without ambiguity the equation of motion for the time evolution of the probability distribution $\rho(\mathbf{q}; t)$,

$$\frac{\partial}{\partial t}\rho(\mathbf{q}; t) = \mathscr{L}\rho(\mathbf{q}; t). \tag{5.1.8}$$

If the variables $\mathbf{q}$ represent all the possible variables of the system (and therefore neither frictions nor stochastic forces are present) $\mathscr{L}$ coincides with the rigorous Liouvillian operator of the system. The description can be widely simplified if suitable fluctuation–dissipation processes are introduced to simulate the influence of infinite other variables. In this case $\mathbf{q}$ stands for a set of a few variables, some of which in the absence of the mutual interaction would be driven by Langevin equations of the same kind as (5.1.1). The operator $\mathscr{L}$ consists then of a sum of operators of the same form as that of (5.1.5) plus interaction terms with the character of rigorous microscopic Liouville operators. The set $\mathbf{q}$ is then divided in a relevant (**a**) and irrelevant (**b**) part. Correspondingly, the operator $\mathscr{L}$ is divided into $\mathscr{L}_a, \mathscr{L}_b$ and the interaction term $\mathscr{L}_1$. The equation of motion for the distribution of **a** has to be built up via contraction over the variables **b**.

It must be pointed out that if the mutual interaction (and especially that between **a** and **b**) is linear, the problem reduces to a trivial matrix diagonalization which, when the dimension of the whole system exceeds three, requires the use of a computational approach without involving, however, any conceptual subtlety. We shall focus therefore our attention on the case of nonlinear interactions.

To make clear our way of proceeding let us consider the paradigmatic case of the stochastic oscillator ($x \equiv q_1$, $v \equiv q_2$, $m = 1$)

$$\dot{x} = v$$

$$\dot{v} = -\frac{\partial V}{\partial x} - \gamma v + f_v(t), \tag{5.1.9}$$

where, according to (5.1.2) and (5.1.3), $f_v(t)$ is a white Gaussian noise defined by

$$\langle f_v(0) f_v(t) \rangle = 2D\delta(t) = 2\gamma \langle v^2 \rangle \delta(t). \tag{5.1.10}$$

The operator $\mathscr{L}$ in this case reads

$$\mathscr{L} = -v\frac{\partial}{\partial x} + V'\frac{\partial}{\partial x} + \gamma\left[\frac{\partial}{\partial v}v + \langle v^2 \rangle \frac{\partial^2}{\partial v^2}\right]. \tag{5.1.11}$$

If we perform the identification $\mathbf{a} \equiv x$, $\mathbf{b} = v$, then our goal coincides precisely with that of (for instance) San Miguel and Sancho (1980). Of course, as already pointed out above, this problem becomes relevant only when dV/dx is not a linear function of x.

A problem more significant for its implications on the field of rate processes (especially those activated via radiation fields) comes into play when the stochastic oscillator of (5.1.9) is assumed to interact with an 'irrelevant' oscillator. The system is then described by ($m = 1$)

$$\left.\begin{aligned}
\dot{x} &= v \\[4pt]
\dot{v} &= -\frac{\partial V}{\partial x} - \gamma v + f_v(t) - q'(x)y \\[4pt]
\dot{y} &= w \\[4pt]
\dot{w} &= -\omega_R^2 y - \alpha q(x) - \lambda w + f_w(t),
\end{aligned}\right\} \tag{5.1.12}$$

where $f_w(t)$ is the usual white Gaussian noise defined by

$$\langle f_w(0) f_w(t) \rangle = 2\lambda \langle w^2 \rangle \delta(t) \tag{5.1.13a}$$

and

$$q'(x) \equiv dq/dx. \tag{5.1.13b}$$

We shall focus on the following two cases:

(i) $\alpha = 0$. This means that the dynamics of the 'irrelevant' oscillator does not depend on that of the system of interest. y mimics the influence of a radiation field with frequency ω_R and coherence lifetime $\tau_c \equiv 1/\lambda$. Let us consider the parameter

$$r_c \equiv 2\lambda/\omega_R. \tag{5.1.14}$$

When $r_c \ll 1$ the correlation function $\langle y(0)y(t) \rangle$ is virtually an oscillatory function of time with frequency ω_R and this will be referred to as the coherent condition. When $r_c \gg 1$ the correlation function $\langle y(0)y(t) \rangle$ decays with the exponential law $\exp(-\Gamma t)$, where $\Gamma = \omega_R^2/\lambda$. This will be referred to as the incoherent regime.

The study of the coherent regime is especially illuminating, since it will allow us to establish a contact with the celebrated linear response theory of Kubo (1966) (see Sections 5.2 and 5.3). This will serve also the purpose of establishing what really are the physical properties described in a virtually exact way by keeping the intensity of noise very weak, i.e. the intensity of the radiation field in the example under discussion.

(ii) $\alpha = 1$. This means that the interaction between the oscillator of interest and the 'irrelevant' one has a conservative Hamiltonian nature. The detailed balance condition holds, the equilibrium condition of the whole system is well known and the Fokker–Planck equation resulting from the contraction over the 'irrelevant' oscillator must be compatible with this important information.

In the last few years many investigators have focused on the study of stochastic differential equations with the general structure ($\xi_m(t)$ and $\xi_a(t)$ are two generic noises, usually assumed to be white)

$$\dot{x} = \varphi(x) + g(x)\xi_m(t) + \xi_a(t). \tag{5.1.15}$$

For a wide discussion of this stochastic differential equation, especially in the case $\mathrm{d}g/\mathrm{d}x = 0$, see, for instance, Schenzle and Brand (1979). A noise depending on the state of the variable of interest like $\xi_m(t)$ of (5.1.15) is usually termed 'multiplicative' (Fox, 1972, 1978). For historical examples of noises dependent on the state of system see also Büttiker and Landauer (1982) and Landauer (1962), as well as Chapter 1 of this volume. In the present chapter we shall devote ourselves to the problem of how to derive (5.1.15) from the more complex system of (5.1.12). This can be done via repeated application of a naïve method as follows. Let us assume that the variable w is so fast as to be always found in its equilibrium state $\dot{w} = 0$. This assumption allows us to derive the explicit expression of w from the last equation of the set of (5.1.12). Thus (5.1.12) can be replaced by

$$\left. \begin{aligned} \dot{x} &= v \\ \dot{v} &= \frac{\partial V}{\partial x} - \gamma v + f_v(t) - q'(x)y \\ \dot{y} &= -\Gamma y - \frac{\alpha}{\lambda}q(x) + F_w(t), \end{aligned} \right\} \tag{5.1.16a}$$

where

$$\Gamma \equiv \omega_R^2/\lambda$$
$$F_w \equiv f_w/\lambda.$$

165

PAOLO GRIGOLINI

By applying the same method to v we obtain from (5.1.16a)

$$\left.\begin{aligned}\dot{x} &= -\frac{1}{\gamma}\frac{\partial V}{\partial x}+\frac{f(t)}{\gamma}-\frac{q'(x)}{\gamma}y\\\dot{y} &= -\Gamma y-\frac{\alpha}{\lambda}q(x)+F_w(t).\end{aligned}\right\}$$ (5.1.16b)

Now let us assume that Γ is large enough as to render y a fast variable. The naïve method applied to y leads us to

$$\dot{x} = -\frac{1}{\gamma}\frac{\partial V}{\partial x}+\frac{\alpha q(x)q'(x)}{\gamma\omega_R^2}+\frac{f(t)}{\gamma}-\frac{q'(x)}{\gamma}\frac{F_w(t)}{\Gamma},$$ (5.1.17a)

which has indeed the same structure as (5.1.15). In Sections 5.4 and 5.5 we shall deal with the problem of deriving from (5.1.12) (or from (5.1.16a)) an equation of motion for the variable x alone with a more rigorous method. Note that the Stratonovich and Itô prescriptions for the integration of (5.1.17a) associate it with the diffusion equations

$$\frac{\partial}{\partial t}\sigma(x;t) = \left\{\frac{\partial}{\partial x}\left[\left(\frac{\partial V}{\partial x}\right)\frac{1}{\gamma}-\frac{\alpha q(x)q'(x)}{\gamma\omega_R^2}\right]\right.$$
$$\left. +Q\frac{\partial}{\partial x}q'(x)\frac{\partial}{\partial x}q'(x)\right\}\sigma(x;t)$$ (5.1.17b)

and

$$\frac{\partial}{\partial t}\sigma(x;t) = \left\{\frac{\partial}{\partial x}\left[\left(\frac{\partial V}{\partial x}\right)\frac{1}{\gamma}-\frac{\alpha q(x)q'(x)}{\gamma\omega_R^2}\right]\right.$$
$$\left. +Q\frac{\partial^2}{\partial x^2}q'^2(x)\right\}\sigma(x;t),$$ (5.1.17c)

respectively, the intensity of noise Q being defined by

$$Q \equiv \frac{\langle y^2\rangle}{\gamma^2\Gamma}.$$ (5.1.17d)

The former equation can also be written as follows

$$\frac{\partial}{\partial t}\sigma(x;t) = \left\{\frac{\partial}{\partial x}\left(\left[\left(\frac{\partial V}{\partial x}\right)\frac{1}{\gamma}-\frac{\alpha q(x)q'(x)}{\gamma\omega_R^2}\right]-Qq(x)q'(x)\right)\right.$$
$$\left. +Q\frac{\partial^2}{\partial x^2}q'^2(x)\right\}\sigma(x;t),$$ (5.1.17e)

which shows that the two proposals generate different drift terms. According to Van Kampen (1981) this disconcerting discrepancy stems from the fact that when $dg/dx \neq 0$ the tacit assumption that $\varphi(x)$ (we are referring to (5.1.15)) is to be identified with the macroscopic law might turn out to be invalid. We believe

166

that the naïve method of replacing (5.1.12) with (5.1.15) wipes off relevant pieces of information such as the ratio of the inertia time scale, $1/\gamma$, to the noise correlation time, $\tau \equiv 1/\Gamma$,

$$R \equiv \Gamma/\gamma. \tag{5.1.17f}$$

The projection approach illustrated in the next section allows us to write a diffusion equation for the variable x alone without losing any relevant piece of information on the system of (5.1.12). This will allow us to establish that for values of R intermediate between $R = \infty$ and $R = 0$ neither the Itô nor the Stratonovich prescription applied to (5.1.17a) can lead to the right form of diffusion equation. When studying the case $\alpha = 0$, (5.1.17a) supplemented by the Stratonovich and Itô prescriptions is shown to produce at $R = 0$ and $R = \infty$, respectively, the correct form of diffusion equation (see Section 5.4). However, if $\alpha = 1$ the failure of (5.1.17a) to produce a correct diffusion equation is total: whatever value the parameter R is given, neither the Itô nor the Stratonovich prescription can associate (5.1.17a) with a reliable form of diffusion equation. This is so because the naïve method of elimination of fast variables does not retain the important information on the constraint imposed by the canonical character of the system (see Section 5.5).

This chapter is organized as follows. Section 5.2 is devoted to illustrating the main rules of the projection calculus, which is basically a perturbation expansion in the interaction term $\mathscr{L}_1$. The physical meaning of the truncation at the order $\mathscr{L}_1^2$ is illustrated in Section 5.3. This section is of special importance in that it shows that the predictions of the linear response theory concern the change by unit time induced by external excitation on the second moment of the variable of interest rather than the steady-state distribution itself (the latter is a property well distinct from the former one). Section 5.4 is devoted to studying the case where the excitation radiation field is totally incoherent ($r_c \ll 1$, $\alpha = 0$). The corresponding case with $\alpha = 1$ is studied in Section 5.5. Section 5.6 is devoted to discussing the system of (5.1.16b) with $\alpha = 0$, i.e. the system object of many recent investigations (Fox, 1986; Hänggi, Marchesoni and Grigolini, 1984; Lindenberg and West, 1983; Sancho, San Miguel, Katz and Gunton, 1982).

5.2   General theory

Our departure point is given by the equation of motion

$$\frac{\partial}{\partial t}\rho(\mathbf{a},\mathbf{b};t) = \mathscr{L}(\mathbf{a},\mathbf{b};t), \tag{5.2.1}$$

where $\mathbf{a}$ and $\mathbf{b}$ denote sets of 'relevant' and 'irrelevant' variables, respectively. The operator $\mathscr{L}$, in principle, could stand for the classical Poisson brackets, and $\mathbf{a}$ and $\mathbf{b}$ could denote coordinates and momenta of the 'relevant' (or Brownian) and 'irrelevant' particles of a physical system of Hamiltonian

nature. In other words, (5.2.1) could stand for the classical Liouville equation
and this could be a rigorous starting point to study the motion of Brownian
particles (Grigolini and Marchesoni, 1985; Van Kampen and Oppenheim,
1986). It must also be pointed out that the projection method described in this
chapter could even be applied to a quantum mechanical Liouville equation. It
has indeed been originally envisaged as a perturbation technique to study the
decay of metastable quantum mechanical systems (Grigolini, 1974, 1975).
However, in this chapter we have mainly in mind the already 'reduced' systems
illustrated in the preceding section. Thus, for the sake of generality, it would be
better to regard (5.2.1) as being a generalized (classical) Liouville equation.
This means that the operator $\mathscr{L}$ also contains second-order derivatives,
mimicking the diffusion effects of stochastic forces, and the corresponding
dissipation terms (a rigorous classical Liouville equation would contain only
first-order derivatives).

This perturbation technique is based on a repartition of the operator $\mathscr{L}$
of (5.2.1) into an unperturbed part, $\mathscr{L}_0$, and an interaction part, $\mathscr{L}_1$. The unperturbed part $\mathscr{L}_0$, in turn, can be expressed as the sum of a part involving
only the a-system, $\mathscr{L}_a$, and a part involving only the b-system, $\mathscr{L}_b$. Thus
we have

$$\mathscr{L} = \mathscr{L}_0 + \mathscr{L}_1, \tag{5.2.2}$$

where

$$\mathscr{L}_0 \equiv \mathscr{L}_a + \mathscr{L}_b. \tag{5.2.3}$$

We shall show in Section 5.5 that the repartition of $\mathscr{L}$ into a perturbed and
unperturbed part is not unambiguously defined and must be done whilst
bearing in mind the nature of the physical system under study.

A further significant ingredient of our projection technique is, of course, the
projection operator P. Throughout this chapter we shall rely on the following
projection operator

$$P\rho(\mathbf{a}, \mathbf{b}; t) \equiv \rho_{eq}(\mathbf{b}) \int d\mathbf{b}\, \rho(\mathbf{a}, \mathbf{b}; t), \tag{5.2.4}$$

where $\rho_{eq}(\mathbf{b})$ denotes the equilibrium distribution of the 'irrelevant' system,
thereby satisfying

$$\mathscr{L}_b \rho_{eq}(\mathbf{b}) = 0. \tag{5.2.5}$$

After application of P the actual probability distribution $\rho(\mathbf{a}, \mathbf{b}; t)$ is replaced
by the product of $\rho_{eq}(\mathbf{b})$ by

$$\sigma(\mathbf{a}; t) \equiv \int d\mathbf{b}\, \rho(\mathbf{a}, \mathbf{b}; t). \tag{5.2.6}$$

(Throughout the present chapter we shall denote by σ the reduced probability
distributions, i.e. those obtained via contraction from the complete equation of
motion of (5.2.1). In some cases we shall apply again the contraction procedure

to already contracted probability distributions. The probability distributions obtained by repeated application of the contraction procedure will be denoted by the symbol ψ.)

As a basic prescription dictated by the need for the limitations of the standard time-dependent perturbation techniques to be overcome (Grigolini, 1975, 1976) the generalized Liouville equation (5.2.1) must be rewritten in the interaction picture corresponding to the repartition of (5.2.2). This means

$$\frac{\partial}{\partial t}\tilde{\rho} = \mathscr{L}_1(t)\tilde{\rho}(t), \tag{5.2.7}$$

where

$$\tilde{\rho}(t) \equiv \exp\{-\mathscr{L}_0 t\}\rho(t) \tag{5.2.8}$$

and

$$\mathscr{L}_1(t) \equiv \exp\{-\mathscr{L}_0 t\}\mathscr{L}_1 \exp\{\mathscr{L}_0 t\}. \tag{5.2.9}$$

Equation (5.2.7) is still of the same form as (5.2.1), the new 'generalized' Liouvillian operator being dependent of time. This is the price we must pay to point out the leading role of the interaction term $\mathscr{L}_1$. By applying the well known Zwanzig projection method (Zwanzig, 1960) to (5.2.7) we shall be in a position to relate the relaxation properties of the system of interest to the correlation functions associated with the interaction term (in this chapter this basically means the autocorrelation function of the (coloured) external noise y of (5.1.12)). It has been pointed out (Grigolini, 1985) that this projection method is the counterpart in the 'Schrödinger picture' of the continued fraction methods used in the 'Heisenberg picture'. Within this representation the relation between relaxation properties of the system of interest and correlation function of the interaction term is made immediately evident by the Green function method of Pastori-Parravicini and coworkers (Giannozzi, Grosso, Moroni and Pastori-Parravicini, 1988). For a further significant example of application of the projection method in the 'Heisenberg picture' we would like to mention the investigation work of Lee (1982).

By applying the Zwanzig projection method to (5.2.7) we obtain, after a straightforward algebra (note that we assume $P\mathscr{L}_1 P = 0$)

$$\frac{\partial}{\partial t}\sigma(\mathbf{a}; t) = \mathscr{L}_a\sigma(\mathbf{a}; t) + \int_0^t K(t-s)\sigma(\mathbf{a}; s)\,ds, \tag{5.2.10}$$

where (the arrow denotes time ordering)

$$K(t-s) = \rho_{eq}(b)^{-1}\exp(\mathscr{L}_a t)P\mathscr{L}_1(t)\overleftarrow{\exp}\left[\int_s^t dt'(1-P)\mathscr{L}_1(t')\right]$$

$$\cdot \mathscr{L}_1(s)P\rho_{eq}(b). \tag{5.2.11}$$

The interested reader can find details on the (straightforward) algebra leading to (5.2.10) from (5.2.7) in the papers of Nordholm and Zwanzig (1975). Details

PAOLO GRIGOLINI

on the algebra of time-ordered exponentials can be found, for instance, in a paper by Muus (1972).

The author would like to stress that, as an effect of applying our contraction procedure to the 'irrelevant' freedom degrees, the Markovian nature of the complete system (no memory of its past story) is lost and the reduced system of (5.2.10) exhibits a non-Markovian character. From the time convolution of the second term on the right-hand side of (5.2.10) we see indeed that the time evolution of the reduced system depends on its state at earlier times. The relevant aspect of a non-Markovian (generalized) master equation obtained via contraction over 'irrelevant' freedom degrees has been discussed by Nordholm and Zwanzig (1975) and Zwanzig (1960).

It must be remarked that an additional preparation term should appear on the right-hand side of (5.2.10) if the initial condition

$$\rho(\mathbf{a}, \mathbf{b}; 0) = \rho_{eq}(\mathbf{b}) \cdot \sigma(\mathbf{a}, 0) \tag{5.2.12}$$

is not satisfied. A review on the effects of this preparation term can be found in Grigolini (1981). In this chapter these effects will be disregarded as they should not play any relevant role in determining the steady states (which are the major objectives of our investigation).

Before proceeding we must point out the following relevant mathematical properties:

$$\mathscr{L}_b P(\ldots) = 0, \tag{5.2.13}$$

$$P\mathscr{L}_b(\ldots) = 0. \tag{5.2.14}$$

The former is the consequence of the fact that, due to (5.2.4), we have $\mathscr{L}_b P(\cdots) = \mathscr{L}_b \rho_{eq}(\mathbf{b}) \int d\mathbf{b}(\cdots)$, from which, by using (5.2.5), we derive precisely (5.2.13). The latter is derived by applying $\mathscr{L}_b$, via integration by parts, on the left. Throughout this chapter $\mathscr{L}_b$ is a differential operator which starts acting with $\partial/\partial b_i$ (b_i denotes a generic component of the vector $\mathbf{b}$). Thus (5.2.14) can be immediately derived on condition that the population of the system vanishes at the boundaries (or on condition that the system is periodic).

By using the properties of (5.2.13) and (5.2.14) we rewrite (5.2.11) as follows:

$$K(t - s) = \rho_{eq}(\mathbf{b})^{-1} P\mathscr{L}_1 \exp(\mathscr{L}_0 t)$$

$$\times \overleftarrow{\exp}\left[\int_s^t dt'(1 - P)\mathscr{L}_1(t')\right](1 - P)$$

$$\times \exp(-\mathscr{L}_0 s)\mathscr{L}_1 P\rho_{eq}(\mathbf{b}). \tag{5.2.15}$$

Equation (5.2.10) with $K(t - s)$ given by (5.2.11) or the equivalent expression of (5.2.15) is equivalent to the complete description of (5.2.1) (if the preparation effects can be neglected). To get more workable equations we must (a) replace the time-ordered exponential with its (more convenient) expansion in a Taylor like series (Section 5.2.1), and (b) take into account the time convoluted nature of (5.2.10) (Section 5.2.2).

170

5.2.1 Expansion of the time-ordered exponential in a Taylor like series

By exploiting the invariance of the multitime stationary correlation functions by time translation, $K(s_0)$ (via expansion in a Taylor like series) can be written as follows:

$$K(s_0) = \sum_{n=2}^{\infty} K_n(s_0), \tag{5.2.16a}$$

where

$$K_2(s_0) = \rho_{eq}^{-1}(\mathbf{b})P\mathcal{L}_1 \exp(\mathcal{L}_0 s_0)\mathcal{L}_1 P\rho_{eq}(\mathbf{b}) \tag{5.2.16b}$$

$$K_{n+2}(s_0) = \rho_{eq}^{-1}(\mathbf{b}) \int_0^{s_0} ds_1 \int_0^{s_1} ds_2 \ldots \int_0^{s_{n-1}} ds_n P\mathcal{L}_1$$

$$\times \exp(\mathcal{L}_0 s_0)(1-P)\mathcal{L}_1(s_1)\ldots(1-P)\mathcal{L}_1(s_{n-1})$$

$$\times (1-P)\mathcal{L}_1(s_n)\mathcal{L}_1 P\rho_{eq}(\mathbf{b}), \quad n = 1, 2, \ldots. \tag{5.2.16c}$$

In the remainder of this chapter we shall principally illustrate results stemming from $K_2(s_0)$. Note that the disentanglement of the time convolution of (5.2.10) can, in principle, lead to terms of order higher than $\mathcal{L}_1^2$ even when $K(s_0)$ is replaced by its second-order approximation. We shall adopt a disentanglement prescription eventually resulting in a purely second-order expression.

5.2.2 Replacement of the time convolution of (5.2.10) with a time-dependent diffusion term

Equation (5.2.10) can also be written as follows:

$$\frac{\partial}{\partial t}\sigma(\mathbf{a}; t) = \mathcal{L}_a\sigma(\mathbf{a}; t) + \int_0^t ds\, K(s)\sigma(\mathbf{a}; t-s). \tag{5.2.17}$$

By developing $\sigma(\mathbf{a}; t-s)$ into a Taylor power series around $s = 0$ we obtain from (5.2.17)

$$\frac{\partial}{\partial t}\sigma(\mathbf{a}; t) = \mathcal{L}_a\sigma(\mathbf{a}; t) + \sum_{r=0}^{\infty}\frac{1}{r!}\int_0^t ds\, K(s)(-s)^r$$

$$\cdot \frac{\partial^r}{\partial t^r}\sigma(\mathbf{a}; t). \tag{5.2.18}$$

To get rid of the time derivatives appearing on the right-hand side of (5.2.18) without introducing extra terms of order higher than the second in $\mathcal{L}_1$ we shall adopt the following approximation:

$$\frac{\partial^r}{\partial t^r}\sigma(\mathbf{a}; t) = \mathcal{L}_a^r\sigma(\mathbf{a}; t). \tag{5.2.19}$$

By replacing (5.2.19) into the second term on the right-hand side of (5.2.18) we obtain

$$\frac{\partial}{\partial t}\sigma(\mathbf{a};t) = \mathscr{L}_a\sigma(\mathbf{a};t) + \sum_{r=0}^{\infty}\frac{1}{r!}\int_0^t ds\, K(s)(-s)^r\mathscr{L}_a^r\sigma(\mathbf{a};t), \qquad (5.2.20)$$

which can also be written in the equivalent form

$$\frac{\partial}{\partial t}\sigma(\mathbf{a};t) = \mathscr{L}_a\sigma(\mathbf{a};t) + \int_0^t ds\, K(s)e^{-\mathscr{L}_a s}\sigma(\mathbf{a};t). \qquad (5.2.21)$$

Note that when $K(s)$ in (5.2.21) is replaced by $K_2(s)$, the second term on the right-hand side of (5.2.22) turns out to be of the order $\mathscr{L}_1^2$.

As to the problem of evaluating the corrections to this approximated result, it has been shown (Faetti, Fronzoni, Grigolini and Mannella, 1988) that a promising way of solving it consists in replacing (5.2.10) with

$$\frac{\partial}{\partial t}\sigma(\mathbf{a};t) = \mathscr{L}_a\sigma(\mathbf{a};t) + \int_0^t ds\, W(s)e^{-\mathbb{Q}s}\sigma(\mathbf{a};t), \qquad (5.2.22)$$

where

$$W(s) \equiv K(s)e^{-\mathscr{L}_a s} \qquad (5.2.23)$$

and the operator $\mathbb{Q}$ is defined by

$$e^{-\mathbb{Q}s} \equiv e^{\mathscr{L}_a s}e^{(\mathscr{L}_a + \mathbb{D})s}. \qquad (5.2.24)$$

$\mathbb{D}$ denotes a diffusion operator to be determined via the following perturbation scheme (note that $\mathbb{D} = \mathbb{D}_2 + \mathbb{D}_3 + \cdots$ and $W_n(s) \equiv K_n(s)e^{-\mathscr{L}_a s}$)

$$\left.\begin{aligned}
\frac{\partial}{\partial t}\sigma(\mathbf{a};t) &= \mathscr{L}_a\sigma(\mathbf{a};t) + \int_0^t ds\, W_2(s)\sigma(\mathbf{a};t) \\
\frac{\partial}{\partial t}\sigma(\mathbf{a};t\to\infty) &= \mathscr{L}_a\sigma(\mathbf{a};t) + \mathbb{D}_2\sigma(\mathbf{a};t), \\
\mathbb{D}_2 &\equiv \int_0^{\infty} W_2(s)\, ds
\end{aligned}\right\} \qquad (5.2.25)$$

$$\left.\begin{aligned}
\frac{\partial}{\partial t}\sigma(\mathbf{a};t) &= \mathscr{L}_a\sigma(\mathbf{a};t) + \int_0^t ds(W_2(s) + W_3(s))\sigma(\mathbf{a};t) \\
\frac{\partial}{\partial t}\sigma(\mathbf{a};t\to\infty) &= \mathscr{L}_a\sigma(\mathbf{a};t) + (\mathbb{D}_2 + \mathbb{D}_3)\sigma(\mathbf{a};t), \\
\mathbb{D}_3 &\equiv \int_0^{\infty} W_3(s)\, ds
\end{aligned}\right\} \qquad (5.2.26)$$

The exponential of (5.2.24) starts exerting its influence at the fourth order in

$\mathscr{L}_1$ as it is clearly illustrated by

$$\frac{\partial}{\partial t}\sigma(\mathbf{a};t) = \mathscr{L}_a\sigma(\mathbf{a};t)$$

$$+ \int_0^t ds[W_2(s)e^{Q_2 s} + W_3(s) + W_4(s)]\sigma(\mathbf{a};t), \qquad (5.2.27)$$

where

$$e^{Q_2 t} \equiv e^{\mathscr{L}_a t}e^{(\mathscr{L}_a + D_2)t}. \qquad (5.2.28)$$

This equation can be used to determine the fourth-order contribution to $\mathbb{D}$. In this chapter we shall not deal with the problem of determining corrections higher than $\mathscr{L}_1^2$, but in Sections 5.5 and 5.6 we shall illustrate a (simple) calculation up to the order $\mathscr{L}_1^3$ and the results of former investigations at the order $\mathscr{L}_1^4$.

The approximation at the order $\mathscr{L}_1^2$ (from the next section it will become clear why this can be termed linear response approximation) reads

$$\frac{\partial}{\partial t}\sigma(\mathbf{a};t) = \mathscr{L}_a\sigma(\mathbf{a};t) + \int_0^t ds\langle\tilde{\mathscr{L}}_1(s)e^{\mathscr{L}_a s}\mathscr{L}_1 e^{-\mathscr{L}_a s}\sigma(\mathbf{a};t), \qquad (5.2.29a)$$

where

$$\tilde{\mathscr{L}}_1(t) \equiv e^{-\mathscr{L}_b t}\cdot\mathscr{L}_1 e^{\mathscr{L}_b t}. \qquad (5.2.29b)$$

This result is obtained from (5.2.21) with $K(s)$ replaced by $K_2(s)$ of (5.2.16b). Equation (5.2.29a), by taking the invariance of the stationary correlation functions by time translation into account, is immediately recognized as coincident with both the result of the partial ordering prescription of Mukamel, Oppenheim and Ross (1978) and the fundamental result on which most of the applications illustrated in the 1976 report of Van Kampen (Van Kampen, 1976) are based. The best Fokker–Planck approximation proposed by Lindenberg and West (1983) and Sancho et al. (1982) stem precisely from an explicit treatment of (5.2.29a) in the one-dimensional case (see Section 5.6). In later sections we shall illustrate some applications of this choice, (5.2.21) and (5.2.29a) to many-dimensional cases. We would like to point out that the projection method affords a way of discussing the limits of validity of this family of Fokker–Planck equations and of determining the corresponding corrections (Faetti et al., 1988).

In the related literature (see Mukamel, Oppenheim and Ross, 1978) we find also the prescription

$$\frac{\partial}{\partial t}\sigma(\mathbf{a};t) = \mathscr{L}_a\sigma(\mathbf{a};t) + \int_0^t ds\langle\mathscr{L}_1(t-s)$$

$$\times \exp\{\mathscr{L}_a(t-s)\}\mathscr{L}_1\rangle\sigma(\mathbf{a};s). \qquad (5.2.29c)$$

This is obtained from (5.2.10) by replacing $K(t-s)$ with $K_2(t-s)$ of (5.2.16b). Equation (5.2.29c) is introduced by Mukamel, Oppenheim and Ross (1978) as resulting from the adoption of the chronological ordering prescription. As correctly pointed out by Sancho, Sagués and San Miguel (1986), (5.2.29c) is a rigorously exact result when the statistics of the 'thermal bath' are driven by a two-state operator (see also Horsthemke and Lefevèr, 1984). Due to its non-Markovian structure (the time convoluted form of the second term on the right-hand side of (5.2.29c)) this equation also involves terms of order higher than $\mathscr{L}_1^2$. It is not possible, therefore, to interpret it (see next section) in terms of a linear response to the excitation produced by its 'thermal bath'. (Note that our definition of 'thermal bath' applies also to the excitation radiation field.) An appealing aspect of the choice of (5.2.29a) (this beautiful property is not completely shared by all the many-dimensional cases studied in this chapter) is that in the limit $t \to \infty$ it leads to a standard Fokker–Planck structure at any order in τ, whereas at higher orders in $\mathscr{L}_1$ also nonstandard diffusion terms can appear and the standard terms can be disregarded by keeping weak the strength of the interaction $\mathscr{L}_1$ (Faetti et al., 1988; Grigolini, 1986; Lindenberg and West, 1983; Sancho et al., 1982). This is the reason why in the one-dimensional case the diffusion structure corresponding to the choice of (5.2.29a) is usually termed the best Fokker–Planck approximation (Masoliver, Lindenberg and West, 1987; Sancho et al., 1982).

Before concluding this section, we would like to remark that the Zwanzig projection approach to the Fokker–Planck equation illustrated in this chapter must not be confused with the well known Fokker–Planck approach based on Zwanzig (1961), recently reviewed and extended by Grabert (1982). This approach rests on projecting the Liouville equation on the space spanned by the linear and nonlinear functions of the macroscopic variable of interest. The projection operator is then directly applied to the Liouville equation rather than to the interaction picture, as done in this section. Unfortunately, we are not aware of applications of this method to simple systems of the same kind as that of (5.1.12) but of the case of the anharmonic oscillator mentioned in Section 5.3. This prevents us from making an exhaustive comparison between the two methods.

5.3 The second-order truncation as a linear response approximation

This section is devoted to pointing out the relations between the truncation at the order $\mathscr{L}_1^2$ of the perturbation expansion illustrated in the preceding section and the linear response theory (Kubo, 1957, 1966). This discussion is rendered more transparent by using the model of the anharmonic oscillator excited via a radiation field mentioned in Section 5.1, (5.1.12). Equation (5.1.12) has precisely this physical meaning when $\alpha = 0$. We shall assume $q'(x) = -\omega_1^2$, i.e. to be independent of x. The system (x, v) of (5.1.12) interacts with both the *bona fide* 'thermal bath' at the temperature T and the system (y, w). The latter is a special kind of 'thermal bath' as, due to the absence of the back reaction term

($\alpha = 0$), energy can only flow from (y, w) into (x, v) whereas the reverse process is forbidden. The energy entering the system from (x, w) does not accumulate, being in part dissipated through the *bona fide* 'thermal bath'. If V is a multiple-well potential, the fluctuating force $f_v(t)$ would allow the Brownian particle (with space coordinate x and mass $m = 1$) to jump from one into another potential well (Kramers, 1940). This process is accelerated by the process of energy pumping from the system (y, w) (Faetti, Grigolini and Marchesoni, 1982; Marchesoni and Grigolini, 1983). As done by Fonseca and Grigolini (1986), we shall focus on the case $r_c \ll 1$, where

$$\langle y(0)y(t)\rangle = \langle y^2 \rangle \cos \omega_R t\, e^{-\lambda t/2}, \tag{5.3.1}$$

thereby making a process of resonant activation possible (see also Carmeli and Nitzan, 1985).

We identify **a** with (x, v) and **b** with (y, w). Thus we have

$$\mathscr{L}_a = -v\frac{\partial}{\partial x} + \left(\frac{\partial V}{\partial x}\right)\frac{\partial}{\partial v} + \gamma\left(\frac{\partial}{\partial v}v + k_B T\frac{\partial^2}{\partial v^2}\right),$$

$$\mathscr{L}_b = -w\frac{\partial}{\partial y} + \omega_R^2 y\frac{\partial}{\partial w} + \lambda\left(\frac{\partial}{\partial w}w + \langle w^2\rangle_{eq}\frac{\partial^2}{\partial w^2}\right),$$

$$\mathscr{L}_1 = -\omega_I^2 y\frac{\partial}{\partial v}. \tag{5.3.2}$$

Then (5.2.29a) reads (for $t \to \infty$)

$$\frac{\partial}{\partial t}\sigma(x, v; t) = \left\{\mathscr{L}_a + \frac{\langle y^2\rangle_{eq}}{2}\omega_I^4\frac{\partial}{\partial v}\sum_\mu\int_0^\infty dt\right.$$

$$\left.\times \exp\left[\left(\left(\mathscr{L}_a^X + i\mu\omega_R - \frac{\lambda}{2}\right)t\right)\frac{\partial}{\partial v}\right]\right\}\sigma(x, v; t). \tag{5.3.3}$$

Note that $\mathscr{L}_a^X$ is a superoperator acting on the operator $\partial/\partial v$ and defined, in general, by

$$\mathscr{L}_a^X A \equiv \mathscr{L}_a A - A\mathscr{L}_a \tag{5.3.4}$$

(A is a generic operator). Note also that to obtain (5.3.3) use was made of the important relation

$$Ae^{\mathscr{L}_a^X t} = e^{\mathscr{L}_a t} \cdot Ae^{-\mathscr{L}_a t} \tag{5.3.5}$$

(Muus, 1972).

It has been shown (Fronzoni, Grigolini, Mannella and Zambon, 1986a) that (5.3.4) can also be derived from a completely deterministic process of

175

excitation, i.e. when (5.1.12) (with $\alpha = 0$) is replaced by

$$\left.\begin{array}{l} \dot{x} = v \\[6pt] \dot{v} = -\dfrac{\partial V}{\partial x} - \gamma v + f_v(t) + A\cos(\omega_R t + \varphi), \end{array}\right\} \tag{5.3.6}$$

with

$$A = \omega_1^2 (2\langle y^2 \rangle_{eq})^{1/2}, \tag{5.3.7}$$

provided that the projection operator of the preceding section is replaced by unity, reminiscent of that recently used by Janssen (1986), taking the average of periodic function of time over a period. Both the result of Fronzoni *et al.* (1986a) and Janssen (1986) would seem to imply that a diffusion process might stem directly from the contraction over the fast variables of the system without involving any further condition. From the remainder of this section, however, it will become evident that a Fokker–Planck type of equation such as (5.3.3) can be trusted only when the contributions of order higher than $\mathscr{L}_1^2$ can be neglected. In the case of a purely deterministic system this poses further conditions which are probably equivalent to those behind the transition from deterministic to stochastic behavior (Marchesoni, Sparpaglione, Ruffo and Grigolini, 1982).

The balance between the energy absorbed from the radiation field and that dissipated makes it possible to reach a steady state. From (5.3.3) we obtain that the energy absorbed per unit time in this steady state is

$$P(\omega_R) \equiv \left[\frac{d}{dt}\frac{\langle v^2 \rangle}{2}\right]_{rad.} = \sum_{\mu=\pm 1}\frac{\langle y^2 \rangle}{4}\omega_1^4 \int_0^{+\infty} dt \int_{-\infty}^{+\infty} dx \int_{-\infty}^{+\infty} dv\, v^2 \frac{\partial}{\partial v}$$
$$\times e^{\mathscr{L}_a t} e^{(i\mu\omega_R - \lambda/2)t}\frac{\partial}{\partial v}e^{-\mathscr{L}_a t}\sigma_{ss}(x,v), \tag{5.3.8}$$

where $\sigma_{ss}(x,v)$ denotes the distribution of the variables (x,v) corresponding to this steady state, and 'rad' denotes 'from the radiation field'. In the spirit of the linear response theory let us make the assumption that $\sigma_{ss}(x,v)$ on the right-hand side of (5.3.8) can be replaced by

$$\sigma_{eq}(x,v) \propto e^{-V/k_B T}e^{-v^2/2k_B T}. \tag{5.3.9}$$

This assumption is justified by remarking that the corrections to it are of the order $\langle y^2 \rangle$ or higher. Their influence on $P(\omega_R)$ in (5.3.8) would be of the order $\langle y^2 \rangle^n$ with $n > 2$ thereby becoming negligible compared to the order $\langle y^2 \rangle$ for $\langle y^2 \rangle \to 0$. Let us focus on the right-hand side of (5.3.8). By substitution into it of the derivative with respect to v of $\sigma_{eq}(x,v)$ and applying to the left the first $\partial/\partial v$ via integration by parts, we obtain

$$P(\omega_R) = \frac{\langle y^2 \rangle}{2}\omega_1^4 \sum_{\mu=\pm}\int_{-\infty}^{+\infty} dx \int_{-\infty}^{+\infty} dv\, v$$

$$\times \int_0^\infty \exp\left[\left(\mathscr{L}_a + i\mu\omega_R - \frac{\lambda}{2}\right)t\right]$$

$$\times (v/k_B T)\exp(-\mathscr{L}_a t)\sigma_{eq}(x,v). \tag{5.3.10}$$

As $\mathscr{L}_a\sigma_{eq}(x,v) = 0$, we can replace (5.3.10) with

$$P(\omega_R) = \langle y^2\rangle \sum_{\mu=\pm1}\int_0^\infty \frac{\langle v(0)v(t)\rangle}{2k_B T}e^{i\mu\omega_R t}\,dt, \tag{5.3.11}$$

where $v(t) = ve^{\mathscr{L}_a^+ t}$ and $\mathscr{L}_a^+$ is the operator adjoint to that of (5.3.2). This is precisely the prediction of the linear response theory (Kubo, 1957, 1966). The beautiful aspect of the Kubo theory is that (in the spirit of perturbation techniques) an excitation dependent property $P(\omega_R)$ is determined in terms of the behavior of the unperturbed system ($\langle v(0)v(t)\rangle$ is the velocity autocorrelation function in the absence of excitation).

By using the same method as that of the next section (for the sake of brevity we do not report the corresponding demonstration and we omit writing the explicit expression of the two-dimensional differential equation obeyed by $\Phi_{1\mu}(x,y)$ and $\Phi_{2\mu}(x,y)$) (5.3.3) can be written (in the limiting case $\lambda \to 0$) as follows:

$$\frac{\partial}{\partial t}\sigma(x,v;t) = \left\{\mathscr{L}_a + \frac{\langle y^2\rangle_{eq}}{2}\omega_I^4\sum_{\mu=\pm1}\frac{1}{i\mu\omega_R}\right.$$

$$\times \frac{\partial^2}{\partial v^2}\Phi_{1\mu}(x,v) + \langle y^2\rangle_{eq}\tfrac{1}{2}\omega_I^4\sum_{\mu=\pm1}\frac{1}{i\mu\omega_R}$$

$$\left.\times \frac{\partial}{\partial v}\frac{\partial}{\partial x}\Phi_{2\mu}(x,v)\right\}\sigma(x,v;t). \tag{5.3.12}$$

The function $\Phi_{1\mu}(x,v)$ is related to the linear response expression of (5.3.11) via (as it is easily seen by evaluation of the excitation contribution to $d\langle v^2\rangle/dt$)

$$P(\omega_R) = \langle y^2\rangle_{eq}\frac{\omega_I^4}{2}\text{Re}\{\langle \Phi_{1\mu}\rangle_{ss}/i\mu\omega_R\}. \tag{5.3.13}$$

The corrections to the predictions of the linear response theory resulting from a more accurate average over the excited steady-state distribution have been more recently discussed by Fronzoni et al. (1986a). The result of the investigation of these authors indicates that (5.3.11) affords a reliable prediction of the mean value $P(\omega_R)$, as defined by (5.3.8). This kind of prediction, however, has not to be confused with that on the whole steady-state distribution. The prediction of (5.3.11) can indeed turn out to be correct even when that on $\sigma_{eq}(x,v)$, as provided by (5.3.3), is completely invalidated.

To shed light on this relevant aspect let us consider the case where the potential V is given the form

$$V = \tfrac{1}{2}\omega_0^2 x^2 + \tfrac{1}{4}\beta x^4. \tag{5.3.14}$$

177

The effective anharmonic strength of this potential is not independent of the temperature T. Upon increase of T the population density of the regions of higher potential energy becomes larger and larger, thereby making the Brownian particle 'feel' the influence of the anharmonic term, the second term on the right-hand side of (5.3.14), to a larger and larger extent. According to the method of statistical linearization (see West, Rovner and Lindenberg, 1983, and references therein) the effective oscillation frequency is

$$\Omega_{\text{eff}}^2 = \omega_0^2 + 3\beta\langle x^2 \rangle, \tag{5.3.15}$$

which, if β is weak enough, leads to

$$\Omega_{\text{eff}} = \left(\omega_0^2 + 3\beta\frac{k_B T}{\omega_0^2} \right)^{1/2} \sim \omega_0 + 2\alpha, \tag{5.3.16}$$

where (we adopt the notation of Fronzoni, Grigolini, Mannella and Zambon, 1985)

$$\alpha \equiv 3\beta\frac{k_B T}{4\omega_0^3}. \tag{5.3.17}$$

Let us consider the case where $\omega_0 \gg \alpha$. This means the anharmonic contribution to the energy $E = v^2/2 + V$ being negligible. It makes sense, therefore, to rewrite (5.3.3) in terms of the new variables

$$\left. \begin{aligned} \tilde{x} &= x \\ E &= \tfrac{1}{2}v^2 + \tfrac{1}{2}\omega_0^2 x^2 \end{aligned} \right\}. \tag{5.3.18}$$

This results in

$$\begin{aligned} \frac{\partial}{\partial x} &= \frac{\partial}{\partial \tilde{x}} + V'(x)\frac{\partial}{\partial E} \\ \frac{\partial}{\partial v} &= \frac{\partial}{\partial E}[2(E - V(x))]^{1/2}. \end{aligned} \tag{5.3.19}$$

The differential operator of (5.3.3) must be changed by using the transformation rules of (5.3.18) and (5.3.19). In the resulting Fokker–Planck equation a contraction over x is then made according to the Stratonovich method. Wide use of this method has been made by Lindenberg and West, and the reader can consult, for instance, Lindenberg and Seshadri (1981) and references therein. We supplement the use of the Stratonovich method with the assumption that $\Phi_{1\mu}(x,v)$ and $\Phi_{2\mu}(x,v)$, appearing on the right-hand side of (5.3.12), can be replaced by their mean values. We then obtain

$$\frac{\partial}{\partial t}\sigma(E;t) = \left[\gamma\frac{\partial}{\partial E}E + (k_B T + P(\omega_R))\frac{\partial}{\partial E}E\frac{\partial}{\partial E} \right]\sigma(E;t), \tag{5.3.20}$$

which, in turn, leads to the Boltzmann-like steady state

$$\sigma_{ss}(E) \propto \exp\left[\frac{-E}{k_B T + P(\omega_R)/\gamma}\right].$$ (5.3.21)

Equation (5.3.21) is not explicitly dependent on the anharmonic strength α. Nevertheless, the anharmonic nature of the potential V exerts its influence on $P(\omega_R)$ and, therefore, on the effective temperature

$$T_{eff} = T + P(\omega_R)/(\gamma k_B).$$ (5.3.22)

The range of validity of (5.3.20) and (5.3.21) can be straightforwardly established in the linear case ($\alpha = 0$), where an exact general expression for $\sigma_{ss}(E)$ has been obtained by Ben-Jacob, Bergman, Carmeli and Nitzan (1981). For (5.3.21) to be recovered from the exact result of Ben-Jacob *et al.* the radiation field must be weak enough to satisfy the condition

$$S \ll \gamma,$$ (5.3.23)

where

$$S \equiv A/[2(2k_B T)^{1/2}]$$ (5.3.24)

(note that A is related to $\langle y^2 \rangle_{eq}$ via (5.3.7)). On the other hand, in the linear case, the absorption spectrum $P(\omega_R)$ is proven (Fronzoni *et al.*, 1986a) to have a Lorentzian shape completely independent of the intensity of the radiation field. On an intuitive physical ground this can be immediately understood by remarking that in the ideal linear case the Brownian particle always 'feels' the same frequency ω_0 regardless of whether it is led by the external disturbance to explore the small or the large x region. (In the linear case no reference time scale to define the concept itself of small and large x region is available!)

This last statement seems to contradict the fact that when the radiation field intensity does not satisfy (5.3.23) the picture provided by (5.3.18) is no longer reliable. This seemingly contradictory aspect can be accounted for by noticing that the absorption spectrum $P(\omega_R)$ (the latter prediction) is nothing but the amount of energy absorbed per unit time. In the linear case this latter prediction, as expressed via (5.3.11), turns out to be valid even in the presence of radiation fields so strong as to completely invalidate the predictions on $\sigma_{ss}(E)$ ((5.3.21), which indirectly reflects (5.3.3)).

It is remarkable that when $\alpha > \gamma$ the predictions of (5.3.18) on the steady-state distribution seem to be valid (Fonseca, Fronzoni and Grigolini, 1985; Fronzoni *et al.*, 1986a) within the range of radiation field intensity defined by

$$S^2 \ll \alpha\gamma,$$ (5.3.25)

a condition less restricted than that of (5.3.23). This is so because when the condition $\gamma \sim \alpha$ applies an additional broadening mechanism appears which renders it legitimate to truncate the perturbation expansion of Section 5.2 at the second order without having recourse to the more severe condition $S \ll \gamma$.

The calculation of the absorption spectrum can be made by using both

PAOLO GRIGOLINI

(5.3.11) and (5.3.13). Fonseca and Grigolini (1986), Fronzoni *et al.* (1985, 1986a), used the former method. They pointed out that when $\gamma \gg \alpha$ the absorption spectrum is a Lorentzian curve with linewidth $\gamma/2$ (half intensity half width) centered at the frequency $\omega = \omega_0 + 2\alpha$. When $\gamma \sim \alpha$ a transition takes place to a new regime characterized by a maximum at $\omega = \omega_0 + \alpha$ (see also Dykman and Krivoglaz, 1984). When $\gamma \ll \alpha$ the linewidth is basically α itself. This means that the method of statistical linearization leading to the effective frequency of (5.3.16) is completely invalidated in the new regime. If the set of two coupled differential equations (the counterpart of (5.4.2), see below) leading to the explicit form of $\Phi_{1\mu}$ and $\Phi_{2\mu}$ is solved without having recourse to the mean-field approximation this critical region does not pass unnoticed.

We can conclude this section by saying that from relating the second-order approximation of the approach of Section 5.2 to the linear response theory we do not gain direct evidence on the validity of this approximation (recall the Van Kampen criticism on the linear response theory – Van Kampen, 1971). Rather we can shed light on the interplay of fluctuation and dissipation necessary for this almost universally accepted theory to be valid. The linear response theory, on the other hand, is widely (and successfully) used to account for the results of real spectroscopical measurements. Rather than contradicting this strengthens our conviction that the predictions on the change by unit time produced by the external excitation on the second moment of the variable of interest (as resulting from the 'best Fokker–Planck approximation' of Section 5.2) are correct at the low intensities of external disturbances.

5.4 The case of a totally incoherent radiation field

Let us focus our attention on the system of (5.1.16a) with $\alpha = 0$, $f_v = 0$ and $q'(x)$ denoted by the symbol $-p(x)$. In this case the linear-response approximation of Section 5.2 leads us to

$$\frac{\partial}{\partial t}\sigma(x,v;t) = \mathscr{L}\,\sigma(x,v;t) \equiv \mathscr{L}_a\sigma(x,v;t)$$

$$+ \left\{\frac{1}{\Gamma}\langle y^2 \rangle_{\text{eq}}\frac{\partial}{\partial v}p(x)\frac{1}{(1 - \mathscr{L}_a^x/\Gamma)}\frac{\partial}{\partial v}p(x)\right\}\sigma(x,v;t)$$

$$\tag{5.4.1a}$$

with

$$\mathscr{L}_a = -v\frac{\partial}{\partial x} + \left(\frac{\partial V}{\partial x}\right)\frac{\partial}{\partial v} + \gamma\frac{\partial}{\partial v}v \tag{5.4.1b}$$

(for the definition of $\mathscr{L}_a^x$ see (5.3.4)). The diffusion term on the right-hand side of (5.4.1a) can also be written in the following form:

$$\frac{\partial}{\partial v}p(x)\frac{1}{(1 - \mathscr{L}_a^x/\Gamma)}\frac{\partial}{\partial v}p(x) = \frac{\partial}{\partial v}p(x)\frac{\partial}{\partial v}\Phi_1(x,v)$$

180

$$+ \frac{\partial}{\partial v} p(x) \frac{\partial}{\partial x} \Phi_2(x, v), \qquad (5.4.2)$$

where $\Phi_1(x, v)$ and $\Phi_2(x, v)$ are suitable functions of x and v to be defined so as to satisfy

$$\frac{\partial}{\partial v} p(x) = \left(1 - \frac{\mathscr{L}_a^X}{\Gamma} \right) \left\{ \frac{\partial}{\partial v} \Phi_1(x, v) + \frac{\partial}{\partial x} \Phi_2(x, v) \right\}. \qquad (5.4.3)$$

By making explicit the effect of the commutator superoperator $\mathscr{L}_a^X$ on the right-hand side of (5.4.3) we see that this equation is fulfilled provided that

$$p(x) = \Phi_1 + \tau \left[v \left(\frac{\partial}{\partial x} \Phi_1 \right) - V' \left(\frac{\partial}{\partial v} \Phi_2 \right) \right.$$

$$\left. + V'' \Phi_2 + \gamma \Phi_1 - \left(\frac{\partial}{\partial v} \Phi_1 \right) v \right] \qquad (5.4.4)$$

$$0 = \Phi_2 + \tau \left[-\Phi_1 + v \left(\frac{\partial}{\partial x} \Phi_2 \right) - V' \left(\frac{\partial}{\partial v} \Phi_2 \right) - \gamma v \left(\frac{\partial}{\partial v} \Phi_2 \right) \right].$$

Let us assume that all derivatives of both Φ_1 and Φ_2 vanish (quasi-harmonic approximation). Then we immediately get from (5.4.4) the following approximated solution:

$$\Phi_2 = \tau \Phi_1 \qquad (5.4.5)$$

$$\Phi_1 = \frac{p(x)}{(1 + \tau^2 V''(x) + \gamma \tau)}.$$

In the special case where $p(x)$ is independent of x and the potential $V(x)$ is harmonic, (5.4.5) turns out to be the exact solution of (5.4.4). This explains why we adopt the definition of quasi-harmonic approximation. This kind of approximated solution has many advantages over that adopted in a recent paper (Fronzoni *et al.*, 1986b). The approximation adopted by these authors is equivalent to assuming that V can be replaced by an effective potential V_{eff} defined by $V''_{\text{eff}} = \langle V'' \rangle$ (this is nothing but the statistical linearization mentioned in the preceding section). In the case $p(x) = 1$ this assumption leads to

$$\left. \begin{array}{l} \Phi_2 = \tau \Phi_1 \\[2ex] \Phi_1 = \dfrac{1}{(1 + \tau^2 \langle V'' \rangle + \gamma \tau)}. \end{array} \right\} \qquad (5.4.6)$$

The limits of this approximation have been already pointed out in the preceding section. In the case under discussion here the mean-field approximation would not make it possible to reproduce the statistical correlation between x and v (the counterpart of that between x and $\dot{x}$ recently predicted by the theory of Jung and Risken, 1985, and checked via analog simulation by Moss and McClintock, 1985) which, on the contrary, is correctly described when using (5.4.5) (see Riccardo Mannella, Chapter 7, Volume 3).

Note that (5.4.6) has been proven (Fronzoni et al., 1986b) to lead to the correct dependence of both x- and v-probability distribution on the correlation time τ. This advantageous aspect is shared by (5.4.5).

It is also worth noting that the expansion up to the order τ of (5.4.5) would produce a result coincident with that of Schenzle and Tél (1985), who predict that the kinetic energy part is renormalized so that the corresponding distribution becomes narrower upon increase of τ. On the contrary, both mean-field approximation and the quasi-harmonic approximation of (5.4.5) predict that even the x-probability distribution becomes narrower upon increase of τ. This prediction is supported by the results of analog simulation (Fronzoni et al., 1986b).

Let us now discuss the result of a further contraction (over the variable v). To reach this result we use the following repartition of the operator $\mathscr{L}$ of (5.4.1a) (supplemented by (5.4.2) and (5.4.4)):

$$\mathscr{L}_a = 0 \tag{5.4.7a}$$

$$\mathscr{L}_b = \gamma \frac{\partial}{\partial v} v \tag{5.4.7b}$$

$$\mathscr{L}_1 = -v\frac{\partial}{\partial x} + \left(\frac{\partial V}{\partial x}\right)\frac{\partial}{\partial v} + \frac{\langle y^2 \rangle}{\Gamma}\left[\frac{\partial}{\partial v}p(x)\frac{\partial}{\partial v}\Phi_1(x,v)\right.$$
$$\left. + \frac{\partial}{\partial v}p(x)\frac{\partial}{\partial x}\Phi_2(x,v)\right]. \tag{5.4.7c}$$

Note that the second term on the right-hand side of (5.4.7c) being characterized by a differential operator of the second order in $\partial/\partial v$ will make the first nonvanishing contribution appear at the order $\mathscr{L}_1^3$. Therefore, by using (5.2.22) (with $t \to \infty$) at the order $\mathscr{L}_1^3$, we immediately get for the probability distribution of the variable x alone, $\psi(x;t) \equiv \int dv\, \rho(x,v;t)$, the following equation of motion:

$$\frac{\partial}{\partial t}\psi(x;t) = \left\{\frac{1}{\gamma}\frac{\partial}{\partial x}\left(\frac{\partial V}{\partial x}\right) + \frac{\langle y^2 \rangle}{\Gamma\gamma}\frac{\partial^2}{\partial x^2}p(x)\langle\Phi_2(x,v)\rangle\right.$$
$$\left. + \frac{\langle y^2 \rangle}{\Gamma\gamma^2}\frac{\partial}{\partial x}p(x)\frac{\partial}{\partial x}\langle\Phi_1(x,v)\rangle\right\}\psi(x,t). \tag{5.4.8}$$

From (5.4.5) we see that the ratio of the intensity of the second term to that of the third term on the right-hand side of (5.4.8) is γ/Γ. When $\gamma \sim \Gamma$ these terms therefore produce effects of comparable intensity.

If the terms of order τ^2 in (5.4.5) are disregarded the explicit form of (5.4.8) reads (with R defined by (5.1.17f))

$$\frac{\partial}{\partial t}\psi(x;t) = \left\{ \frac{\partial}{\partial x}\frac{V'}{\gamma} + \frac{Q}{(1+R)}R\frac{\partial}{\partial x}p(x)\frac{\partial}{\partial x}p(x) \right.$$

$$\left. + \frac{Q}{(1+R)}\frac{\partial^2}{\partial x^2}p^2(x) \right\}\psi(x;t), \qquad (5.4.9)$$

thereby recovering a well known result (Faetti *et al.*, 1984; Graham and Schenzle, 1982).

In the case $\gamma \gg \Gamma$, (5.4.8) supplemented by (5.4.6) becomes

$$\frac{\partial}{\partial t}\psi(x;t) = \left\{ \frac{\partial}{\partial x}\frac{V'}{\gamma} + \frac{\langle y^2 \rangle}{\gamma\Gamma^2}\frac{\partial}{\partial x}p(x)\frac{\partial}{\partial x}\frac{p(x)}{(1+\tau\gamma+\tau^2 V'')} \right\}\psi(x;t),$$

$$(5.4.10)$$

from which we derive (by neglecting 1 compared to $\tau\gamma$ and $\tau^2 V'$)

$$\frac{\partial}{\partial t}\psi(x;t) = \left\{ -\frac{\partial}{\partial x}\varphi(x) + \frac{\langle y^2 \rangle}{\Gamma}\frac{\partial}{\partial x}\rho(x)\frac{\partial}{\partial x}\frac{g(x)}{(1-\tau\varphi'(x))} \right\}\psi(x;t),$$

$$(5.4.11)$$

where $\varphi \equiv -V'/\gamma$ and $g(x) \equiv p(x)/\gamma$. Equation (5.4.11) coincides with the result recently found by Fox (1986) when no inertia is present ($\gamma \to \infty$). We thus see that to recover this limiting case (no inertia) we cannot neglect the contribution of the order τ^2, $\tau^2 V''$, of (5.4.10).

We have, in part, answered a basic question posed in Section 5.1, that concerning whether (5.1.17b) or (5.1.17c) must be used. The Stratonovich form, (5.1.17b), seems to be the correct form when inertia is not allowed at any extent and noise is not totally white, i.e. when we must set $R = 0$ in (5.4.9). The results of analog simulation carried out by McClintock and Moss (1985) and Smythe *et al.* (Smythe, Moss and McClintock, 1983; Smythe, Moss, McClintock and Clarkson, 1983) fit this conclusion. It must be pointed out, however, that if the system under study is not characterized by the complete absence of inertia, then the problem is raised of assessing whether or not the time scale of the noise is longer or shorter than the inertia time scale (i.e. whether $R < 1$ or $R > 1$). By inertia time scale we in general mean the relaxation time of those fast variables, contraction over which results in precisely the equation of motion of the variable of interest introduced with phenomenological arguments. When the noise time scale is shorter than that of inertia we would be tempted to say that the Itô form must be preferred to the Stratonovich one. However, this is far away from being a general rule. This is precisely as the system of (5.1.16a) with $\alpha = 0$ behaves. In the next section we shall study another illuminating, and more dramatic, case documenting the complete failure of the naïve method of elimination of fast variables, where no special condition can be found to guarantee the validity of either the Stratonovich or the Itô form.

As to the 'experimental' check of (5.4.9) by means of analog simulation, the recent work of Faetti *et al.* (1984) shows that the predictions of (5.4.9) on the mean value of x^2 in the steady state are correct in the weak Q range. After the work of Mannella *et al.* (1986) it became clear that the deviations from the predictions of (5.4.9) in the large Q range must be ascribed to the influence of a weak additive stochastic force which is always present in some measure in electronic simulators (for theoretical discussions of this issue see also Fedchenia and Usova (1983) and Sancho *et al.* (1982)).

5.5 On the problem of dividing $\mathscr{L}$ into a perturbation and an unperturbed part

Let us come back to the system of (5.1.12). Let us assume now that $\alpha = 1$. The stochastic forces $f_v(t)$ and $f_w(t)$, and their respective friction terms $-\gamma v$ and $-\lambda w$, are assumed to mimic the influence on the system x, v, y, w of a 'thermal bath' at the temperature T_1 and one at the temperature T_2, respectively. We set $T_2 \geqslant T_1$.

When $T_1 = T_2 \equiv T$, due to the presence of the feedback term $-\alpha q(x)$ in the last equation of the set of (5.1.12), the system is proven to be characterized by the canonical equilibrium distribution

$$\rho_{eq}(x, v, y, w) \propto \exp\left[-(2V + v^2 + \omega_R^2 y^2 + q(x)y)/2k_B T\right]. \quad (5.5.1)$$

If we are interested only in the equilibrium distribution of the variable x, via contraction on v, y and w we obtain

$$\sigma_{eq}(x) \propto \exp(-\Phi_{eff}(x)/k_B T)$$
$$\Phi_{eff}(x) = V(x) - q^2(x)/2\omega_R^2. \quad (5.5.2)$$

This means that, due to its coupling with the 'irrelevant' set of variables v, y and w, the bare potential driving the motion of x is renormalized so as to assume the form of (5.5.2). If we are also interested in the dynamical properties of the variable x it is convenient to apply to this problem the theory of Section 5.2 so as to build up a reduced Fokker–Planck equation only concerning the variable x. This must be done, however, with this important caution in mind. The constraint of (5.5.2) has to be satisfied. If the repartition of the operator $\mathscr{L}$ associated to (5.1.12) with $\alpha = 1$ is not properly done and, as a consequence of that, this constraint is not fulfilled, the resulting Fokker–Planck equation might exhibit the same phenomenon of noise-induced phase transition (Faetti *et al.*, 1982) as the reduced Fokker–Planck equation stemming from (5.1.12) with $\alpha = 0$ (i.e. (5.4.9)). As clearly pointed out by Faetti, Grigolini and Marchesoni (1982), the noise-induced transition phenomena can be interpreted as being the consequences of an energy pumping process associated with the interaction between the x-system and a radiation field.

To render our discussion more transparent let us study the reduced version of (5.1.12) as it is provided by (5.1.16a)($\alpha = 1$). Let us assume that $\mathbf{a} \equiv (x, v)$ and

$\mathbf{b} \equiv y$. It has been proven (Faetti, Fronzoni and Grigolini, 1985) that the important constraint of (5.5.2) is fulfilled provided that $\mathscr{L}_b$ is defined as follows:

$$\mathscr{L}_b \equiv \frac{q(x)}{\lambda}\frac{\partial}{\partial y} + \Gamma\frac{\partial}{\partial y}y + \frac{k_B T_2}{\lambda}\frac{\partial^2}{\partial y^2}, \tag{5.5.3}$$

thereby resulting in

$$\mathscr{L}_1 + \mathscr{L}_a = -v\frac{\partial}{\partial x} + \frac{\partial V}{\partial x}\frac{\partial}{\partial v} + \gamma\left\{\frac{\partial}{\partial v}v + \langle v^2\rangle\frac{\partial^2}{\partial v^2}\right\} + yq'(x)\frac{\partial}{\partial v}. \tag{5.5.4}$$

It is then convenient to operate the change of variables $\tilde{x} = x, \tilde{y} = y + q(x)/\omega$ and, similarly to Section 5.4 (see (5.3.13) and (5.3.14)), the corresponding change of derivatives. We thus obtain

$$\mathscr{L}_b = \Gamma\left\{\frac{\partial}{\partial\tilde{y}}\tilde{y} + \frac{k_B T_2}{\lambda}\frac{\partial^2}{\partial\tilde{y}^2}\right\}, \tag{5.5.5}$$

$$\mathscr{L}_a = -v\frac{\partial}{\partial x} + \Phi'_{\text{eff}}\frac{\partial}{\partial v} + \gamma\left\{\frac{\partial}{\partial v}v + k_B T_1\frac{\partial^2}{\partial v^2}\right\}, \tag{5.5.6}$$

$$\mathscr{L}_1 = -\frac{q'(\tilde{x})v}{\omega_R^2}\frac{\partial}{\partial v} + q'(x)\tilde{y}\frac{\partial}{\partial v}, \tag{5.5.7}$$

(from now on we shall drop the tilde over x and y). By using the significant property $\mathscr{L}_1 P = q'(x)y(\partial/\partial v + v/k_B T_2)P$ (the advantageous consequence of the repartition of $\mathscr{L}$ above illustrated) we immediately obtain from the second-order approximation of Section 5.2

$$\frac{\partial}{\partial t}\sigma(x,v;t) = \left\{-v\frac{\partial}{\partial x} + \Phi_{\text{eff}}\frac{\partial}{\partial v} + \gamma\left[\frac{\partial}{\partial v}v + k_B T_1\frac{\partial^2}{\partial v^2}\right]\right.$$
$$\left. + \frac{k_B T_2}{\Gamma\omega_0^2}q'\frac{\partial}{\partial v}\frac{1}{(1-\tau\mathscr{L}_a^X)}\left(\frac{\partial}{\partial v} + \frac{v}{kT_2}\right)\right\}\sigma(x,v;t), \tag{5.5.8}$$

the symbols having the usual meaning. The result of Fronzoni and Grigolini (1984) is recovered from (5.5.8) by making the assumption $(1-\tau\mathscr{L}_a^X)^{-1} \simeq 1$. If the more accurate approximation $(1-\tau\mathscr{L}_a^X)^{-1} \simeq 1 + \tau\mathscr{L}_a^X$ is done, (5.5.8) becomes

$$\frac{\partial}{\partial t}\sigma(x,v;t) = \left\{-v\frac{\partial}{\partial x} + \Phi'_{\text{eff}}(x)\frac{\partial}{\partial v} + \frac{\partial}{\partial v}\left[\gamma + \frac{q'(x)^2(1+\tau\gamma)}{\Gamma\omega_R^2}\right]\right.$$
$$\times\left[\frac{\partial}{\partial v}v + k_B T_1\frac{\partial^2}{\partial v^2}\right] + \frac{\tau k_B T_2}{\Gamma\omega_R^2}q'(x)^2\frac{\partial}{\partial v}\left(\frac{\partial}{\partial x} + \frac{\Phi'_{\text{eff}}(x)}{k_B T_2}\right)$$
$$\left. + \frac{k_B T_2\tau}{\Gamma\omega_R^2}q'(x)q''(x)\frac{\partial}{\partial v}v\left[\frac{\partial}{\partial v} + \frac{v}{k_B T_2}\right]\right.$$

185

$$+ k_B(T_2 - T_1)\frac{q'(x)^2(1 + \tau\gamma)}{\Gamma\omega_R^2}\frac{\partial^2}{\partial v^2}\Big\}\sigma(x, v; t). \qquad (5.5.9)$$

This is the generalization to the case of finite τ of the result of Faetti, Fronzoni and Grigolini (1985) which, in turn, produces the same kind of fluctuation–dissipation process as that discussed by Lindenberg and Seshadri (1981). By setting in (5.5.9) $\tau = 0$, $T_1 = T_2 = T$ and keeping $k_B T/\Gamma\omega_R^2$ and $k_B T/\Gamma\omega_R^2$ constant, we obtain indeed the effective damping $\gamma_{\text{eff}} = \gamma + q'(x)^2/\Gamma\omega_R^2$, which is the dissipation counterpart of the stochastic force $f_v(t) - q'(x(t))y''(t)$. Papers of remarkable interest on this subject are those by Lindenberg and Seshadri (1981), Ramshaw (1985) and Ramshaw and Lindenberg (1986).

Let us operate a further contraction over v (on (5.5.9)) at $T_1 = T_2 = T$. It is convenient to identify the third term on the right-hand side of (5.5.9) with $\mathscr{L}_b$. $\mathscr{L}_1$ is provided by the remaining terms and $\mathscr{L}_a = 0$. At the order $\mathscr{L}_1^2$ (at this order the fifth term on the right-hand side of (5.5.9) does not provide any contribution) we obtain the following equation of motion for $\psi(x; t) \equiv \int dv\, \sigma(x, v; t)$

$$\frac{\partial}{\partial t}\psi(x; t) = k_B T \left\{ \frac{\partial}{\partial x}\left[\frac{1}{\gamma + \dfrac{q'(x)^2(1 + \tau\gamma)}{\Gamma\omega_R^2}} \right] \right.$$

$$\left. + \frac{\tau}{\Gamma\omega_R^2}\frac{\partial}{\partial x}\left[\frac{q'(x)}{\gamma + \dfrac{q'(x)^2(1 + \tau\gamma)}{\Gamma\omega_R^2}} \right] \right\}$$

$$\times \left(\frac{\partial}{\partial x} + \frac{\Phi'_{\text{eff}}(x)}{k_B T} \right)\psi(x; t). \qquad (5.5.10)$$

At the first order in $q'(x)^2$, (5.5.10) does not conflict with the result of a former calculation (Fonseca, Grigolini and Pareo, 1985) carried out in the case $\gamma \to \infty$ and resulting in

$$\frac{\partial}{\partial t}\psi(x; t) = \left\{ \frac{\partial}{\partial x}\left(\frac{k_B T}{\gamma} - \frac{k_B T}{\omega_0^2\gamma^2\Gamma} \right)\left(\frac{\partial}{\partial x} + \frac{\Phi'_{\text{eff}}}{k_B T} \right) \right\}\psi(x; t). \qquad (5.5.11)$$

In the general case $T_2 \geqslant T_1$ the direct contraction over both v and y at the fourth order of the perturbation technique illustrated in Section 5.2 leads to

$$\frac{\partial}{\partial t}\psi(x; t) = \left\{ \frac{k_B T_1}{\gamma}\frac{\partial}{\partial x}\left(1 - \frac{q'(x)^2}{\omega_0^2\Gamma\gamma} \right)\left(\frac{\partial}{\partial x} + \frac{\Phi'_{\text{eff}}}{k_B T} \right) \right\}$$

$$+ \frac{k_B(T_2 - T_1)}{\omega_0^2}\left[\frac{1}{\gamma\Gamma(\gamma + \Gamma)}\frac{\partial}{\partial x}q'(x)\frac{\partial}{\partial x}q'(x) \right.$$

$$\left. \left. + \frac{1}{\gamma^2(\gamma + \Gamma)}\frac{\partial^2}{\partial x^2}q'(x)^2 \right] \right\}\psi(x; t). \qquad (5.5.12)$$

Note that the same energy pumping structure as that of Section 5.4 (see (5.4.9)) is here recovered as generated from the natural transfer of energy from the hot (T_2) into the cold (T_1) 'thermal bath'.

Note that the result of (5.5.12) cannot be recovered by any linear combination of the Itô structure of (5.1.17c) with the Stratonovich structure of (5.1.17e) (the counterpart of the result of (5.4.9)). This shows the complete failure of the naïve method of elimination of fast variables illustrated in Section 5.1. Similar conclusions have been reached by Theiss and Titulaer (1985).

5.6   The inertialess case

In the totally inertialess case the system of (5.1.16b) reads

$$\dot{x} = \varphi(x) + g(x)y(t)$$
$$\dot{y} = -\Gamma y(t) + f_y(t) \tag{5.6.1}$$

(where $\varphi(x) \equiv -(\partial V/\partial x)/\gamma$, $g(x) \equiv -q(x)/\gamma$). For the sake of simplicity we assumed that the additive stochastic force $f(t)$ of (5.1.16a) vanishes. This means that the system of (5.1.16b) can be thought of as being an inertialess anharmonic oscillator at the temperature $T = 0$ excited via a totally incoherent radiation field.

Let us adopt the second-order result of Section 5.2 to determine the equation of motion of $\sigma(x;t) \equiv \int dy \rho(x, y; t)$. We obtain

$$\frac{\partial}{\partial t}\sigma(x;t) = \left\{ -\frac{\partial}{\partial x}\varphi(x) + D\frac{\partial}{\partial x}g(x)\frac{\partial}{\partial x}\Phi(x) \right\}\sigma(x;t), \tag{5.6.2}$$

where $D \equiv \langle y^2 \rangle/\Gamma$, and the function $\Phi(x)$ satisfies the following differential equation:

$$g(x) = \Phi(x) + \tau(\varphi'(x)\Phi(x) - \varphi(x)\Phi'(x)). \tag{5.6.3}$$

This is the result of the straightforward application to the one-dimensional case of the method widely illustrated through the two-dimensional case of Section 5.4 (see (5.4.1a)–(5.4.4)).

Equation (5.6.2), supplemented by (5.6.3), is precisely the best Fokker–Planck approximation derived by Lindenberg and West (1983) and Sancho *et al.* (1982). It has been remarked (Grigolini, 1986; Masoliver, Lindenberg and West, 1987) that both Fox's theory (Fox, 1986) and the mean-field approximation (Hänggi, Mroczowski, Moss and McClintock, 1985) can be thought of as being approximated solutions of (5.6.3). This is, however, somewhat misleading if not supplemented by the following important remark. A careful treatment at the fourth order in $\mathcal{L}_1$ (for the sake of conciseness we omit reporting the tedious calculation behind this result (Faetti *et al.*, 1988, and P. Grigolini *et al.*, manuscript in preparation), but it should be enough to say that the basic rules to apply are precisely those illustrated in Section 5.2) shows that the nonstandard diffusion term

$$\tfrac{3}{2}D^2\tau^2\frac{\partial}{\partial x}g(x)\frac{\partial}{\partial x}g(x)\frac{\partial}{\partial x}g(x)^2\prod{}'(x)$$

$$\prod(x) \equiv (\varphi'(x)g(x) - g'(x)\varphi(x))$$

(5.6.4)

appears (note that the relevance of nonstandard diffusion terms such as that of (5.6.4) has already been pointed out by Hänggi, Marchesoni and Grigolini (1984). When use is made of the renormalization technique of Faetti *et al.* (1988) and Festa *et al.* (1986) this term is replaced by the 'equivalent' standard diffusion term $(3D\tau^2/2)\,(\partial g(x)/\partial x)\varphi(x)\Pi'(x)$, which corrects the predictions of the best Fokker–Planck approximation, (5.6.2) and (5.6.3), at the order τ^2. It is easy to show, however, that in some cases where $\Pi'(x)$ happens to vanish the best Fokker–Planck approximation is proven to lead to the exact equilibrium distribution. The first case is that where $\varphi(x) = -\gamma x$ and $g(x) = 1$ (linear additive case). In this case $\Pi = \gamma$, thereby resulting in $\Pi' = 0$, and the best Fokker–Planck approximation is proven to coincide with the exact steady-state Fokker–Planck equation (Grigolini, 1986). A second very interesting case is that where $g(x) = x^3$, $\varphi(x) = \gamma x - \delta x^3$. In this case $\Pi = -2\gamma$, thereby resulting in $\Pi' = 0$. Via the nonlinear transformation (Suzuki, 1980) $\xi = x^{-2}$, (5.6.1) is transformed into the linear additive case, the steady-state Fokker–Planck solution of which is known to be exact (the preceding case). On coming back to the variable x, we obtain (according to J. M. Sancho, private communication (some caution should be exerted in using the steady-state distribution provided by this equation))

$$\frac{\partial}{\partial t}\sigma(x;t) = \left\{ -\frac{\partial}{\partial x}(\gamma x - \delta x^3) + \frac{2\langle y^2\rangle}{\gamma}\right.$$

$$\left. \times \frac{1}{(1+2\tau\gamma)}\frac{\partial}{\partial x}x^3\frac{\partial}{\partial x}x^3\right\}\sigma(x;t).$$

(5.6.5)

It is straightforward to check that in this case (5.6.3) is given the exact solution $\Phi(x) = x^3/(1+2\tau\gamma)$, which makes the best Fokker–Planck approximation coincident with (5.6.5). A further interesting case is that where $\varphi = 0$ (we have again $\Pi' = 0$). In this case the second-order approximation to (5.2.17) is straightforwardly proven to lead to the same equation of motion for $\langle x^n(t)\rangle$ as the Kubo cumulant theory (Kubo, 1963) which is known to provide the exact solution of this case. Faetti and Grigolini (1988) have very recently shown that this is a general property: whenever $\Pi'(x) = 0$, (5.6.2) turns out to be exact.

References

Ben-Jacob, E., Bergman, D., Carmeli, B. and Nitzan, A. 1981. NBS Special Publication 614. *Proceedings of the Sixth International Conference on Noise in Physical Systems*, pp. 381–3.

Büttiker, M. and Landauer, R. 1982. In *Nonlinear Phenomena at Phase Transitions and Instabilities* (T. Riste, ed.), pp. 111–43. NATO ASI Series. New York: Plenum.

Carmeli, B. and Nitzan, A. 1985. *Phys. Rev. A* **32**, 2439–54.

Dykman, M. I. and Krivoglaz, M. A. 1984. In *Physics Reviews (Soviet Scientific Reviews, Section A)* I. M. Khalatnikov, ed.), p. 265–442. Chur-New York: Harwood Academic.

Faetti, S., Festa, C., Fronzoni, L., Grigolini, P. and Martano, P. 1984. *Phys. Rev. A* **30**, 3252–63.

Faetti, S., Fronzoni, L. and Grigolini, P. 1985. *Phys. Rev. A* **32**, 1150–60.

Faetti, S., Fronzoni, L., Grigolini, P. and Mannella, R. 1988. *J. Stat Phys.* (in press).

Faetti, S. and Grigolini, P. 1987. *Rapid Communication of Phys. Rev. A* **36A**, 441–4.

Faetti, S., Grigolini, P. and Marchesoni, F. 1982. *Z. Phys. B* **47**, 353–63.

Fedchenia, I. I. and Usova, N. A. 1983. *Z. Phys. B* **50**, 263–7.

Ferrario, M., Grigolini, P., Tani, A., Vallauri, R. and Zambon, B. 1985. *Adv. Chem. Phys.* **62**, 225–75.

Festa, C., Fonseca, T., Fronzoni, L., Grigolini, P. and Papini, A. 1986. *Phys. Lett.* **113A**, 57–61.

Fonseca, T. and Grigolini, P. 1986. *Phys. Rev. A* **33**, 1122–33.

Fonseca, T., Fronzoni, L. and Grigolini, P. 1985. *Phys. Lett.* **113A**, 143–6.

Fonseca, T., Grigolini, P. and Pareo, D. 1985. *J. Chem. Phys.* **83**, 1039–48.

Fox, R. F. 1972. *J. Math. Phys.* **13**, 1196–207.

Fox, R. F. 1978. *Phys. Rep.* **48**, 179–283.

Fox, R. F. 1986. *Phys. Rev. A* **33**, 467–76.

Fronzoni, L. and Grigolini, P. 1984. *Phys. Lett.* **106A**, 289–92.

Fronzoni, L., Grigolini, P., Mannella, R. and Zambon, B. 1985. *J. Stat. Phys.* **41**, 553–79.

Fronzoni, L., Grigolini, P., Mannella, R. and Zambon, B. 1986a. *Phys. Rev. A* **34**, 3293–303.

Fronzoni, L., Grigolini, P., Hänggi, P., Moss, F., Mannella, R. and McClintock, P. V. E. 1986b. *Phys. Rev. A* **33**, 3320–7.

Giannozzi, P., Grosso, G., Moroni, S. and Pastori-Parravicini, G. 1988. *J. Appl. Math.* (in press).

Grabert, H. 1982. *Projection Operator Techniques in Nonequilibrium Statistical Mechanics.* Berlin: Springer-Verlag.

Graham, R. and Schenzle, A. 1982. *Phys. Rev. A* **26**, 1676–85.

Grigolini, P. 1974. *J. Chem. Phys.* **61**, 1874–81.

Grigolini, P. 1975. *Molec. Phys.* **30**, 1229–51.

Grigolini, P. 1976. *Molec. Phys.* **31**, 1717–48.

Grigolini, P. 1981. *Il Nuovo Cimento* **63B**, 174–206.

Grigolini, P. 1985. *Adv. Chem. Phys.* **62**, 1–29.

Grigolini, P. 1986. *Phys. Lett.* **119**, 157–62.

Grigolini, P. and Marchesoni, F. 1985. *Adv. Chem. Phys.* **62**, 29–80.

Hänggi, P., Marchesoni, F. and Grigolini, P. 1984. *Z. Phys. B* **56**, 333–9.

Hänggi, P., Mroczowski, T. J., Moss, F. and McClintock, P. V. E. 1985. *Phys. Rev. A* **32**, 695–8.

Horsthemke, W. and Lefever, R. 1984. *Noise Induced Transitions, Theory and*

Applications in Physics, Chemistry and Biology. Berlin: Springer-Verlag.

Janssen, J. A. M. 1986. *Physica* **137A**, 477–501.

Jung, P. and Risken, H. 1985. *Z. Phys. B* **61**, 367–79.

Kramers, H. A. 1940. *Physica* **7**, 284–304.

Kubo, R. 1957. *J. Phys. Soc. Jpn.* **12**, 570–86.

Kubo, R. 1963. *J. Math. Phys.* **4**, 174–83.

Kubo, R. 1966, *Rep. Prog. Phys.* **29**, 255–84.

Landauer, R. 1962. *J. Appl. Phys.* **33**, 2209–16.

Lee, M. H. 1982. *Phys. Rev. B* **26**, 2547–51.

Lindenberg, K. and Seshadri, V. 1981. *Physica* **109A**, 483–99.

Lindenberg, K. and West, B. J. 1983. *Physica* **119A**, 485–503.

Mannella, R., Faetti, S., Grigolini, P., McClintock, P. V. E. and Moss, F. E. 1986. *J. Phys. A* **19**, L699–L704.

Marchesoni, F. and Grigolini, P. 1983. *Physica* **121A**, 269–85.

Marchesoni, F., Sparpaglione, M., Ruffo, S. and Grigolini, P. 1982. *Phys. Lett.* **88A**, 113–16.

McClintock, P. V. E. and Moss, F. 1985. *Phys. Lett.* **107A**, 367–70.

Masoliver, J., Lindenberg, K. and West, B. J. 1987. *Phys. Rev. A* **35**, 3086–94.

Moss, F. and McClintock, P. V. E. 1985. *Z. Phys. B* **61**, 381–6.

Mukamel, S., Oppenheim, I. and Ross, J. 1978. *Phys. Rev. A* **17**, 1988–98.

Muus, L. T. 1972. In *Electron Spin Relaxation in Liquids* (L. T. Muus and P. W. Atkins, eds), pp. 1–24. New York: Plenum Press.

Nordholm, S. and Zwanzig, R. 1975. *J. Stat. Phys.* **13**, 347–71.

Popper, K. R. 1959. *The Logic of Scientific Discovery.* London: Hutchinson.

Ramshaw, J. D. 1985. *J. Stat. Phys.* **38**, 669–80.

Ramshaw, J. D. and Lindenberg, K. 1986. *J. Stat. Phys.* **45**, 295–307.

Sancho, J. M., Saugés, F. and San Miguel, M. 1986. *Phys. Rev. A* **33**, 3399–403.

Sancho, J. M., San Miguel, M., Katz, S. L. and Gunton, J. D. 1982. *Phys. Rev.* **26A**, 1589–609.

San Miguel, M. and Sancho, J. M. 1980. *J. Stat. Phys.* **22**, 605–24.

Schenzle, A. and Brand, H. 1979. *Phys. Rev. A* **20**, 1628–47.

Schenzle, A. and Tél, T. 1985. *Phys. Rev. A* **32**, 596–605.

Smythe, J., Moss, F. and McClintock, P. V. E. 1983. *Phys. Rev. Lett.* **51**, 1062–5.

Smythe, J., Moss, F., McClintock, P. V. E. and Clarkson, D. 1983. *Phys. Lett.* **97A**, 95–8.

Stratonovich, R. L. 1963. *Topics in the Theory of Random Noise.* New York: Gordon and Breach.

Suzuki, M. 1980. *Suppl. Theor. Phys.* **69**, 160–73.

Theiss, W. and Titulaer, U.M. 1985. *Physica* **130A**, 123–42, 143–54.

Uhlenbeck, G. E. and Ornstein, L. S. 1930. *Phys. Rev.* **36**, 823–41.

Van Kampen, N. G. 1971. *Phys. Norv.* **5**, 279–84.

Van Kampen, N. G. 1976. *Phys. Report* **24C**, 171–228.

Van Kampen, N. G. 1981. *J. Stat. Phys.* **24**, 175–87.

Van Kampen, N. G. 1985. *Phys. Report* **124**, 69–160.

Van Kampen, N. G. and Oppenheim, I. 1986. *Physica* **138A**, 231–48.

Wang, M. C. and Uhlenbeck, G. E. 1945. *Rev. Mod. Phys.* **17**, 323–42.

West, B. J., Rovner, G. and Lindenberg, K. 1983. *J. Stat. Phys.* **30**, 632.

Zwanzig, R. 1960. *J. Chem. Phys.* **33**, 1338–41.

Zwanzig, R. 1961. *Phys. Rev.* **124**, 983–92.

6 Methods for solving Fokker–Planck equations with applications to bistable and periodic potentials

H. RISKEN and H. D. VOLLMER

6.1 Introduction

The Fokker–Planck equation (FPE) deals with those fluctuations of continuous variables which stem from many tiny disturbances, each of which changes the variables in an unpredictable but small way. One therefore has to add noise forces to the deterministic differential equations describing the motion of the variables of the system. This can be done by including these noise forces as additive terms on the right hand sides of the equations of motion (additive noise) or by including them as fluctuating prefactors of some terms (multiplicative noise). Thus, instead of deterministic systems, one has to deal with stochastic differential equations. The properties of the noise forces are important for defining the stochastic differential equations. If the noise forces are δ correlated in time one speaks of white noise, otherwise the term colored noise is used. For white noise the stochastic differential equations are called Langevin equations.

To obtain numeric solutions of these equations one may simulate them on a digital computer or build an analog circuitry simulating the equations electronically. Analytic solutions can be given if the Langevin equations are linear and have additive noise terms. Then, using Green's function of the deterministic system, one can write down the solution of the Langevin equation. If the Langevin equation is nonlinear, it is often impossible to solve it analytically. The best way of treating a nonlinear Langevin equation consists in transforming it to an FPE. The FPE is an equation of motion for the probability distribution function of the system variables. For readers not familiar with FPEs, see, for instance, Bharucha-Reid (1960), Chandrasekhar (1943), Gardiner (1983), Goel and Richter-Dyn (1974), Graham (1973), Haken (1975, 1983a, b), Horsthemke and Lefever (1984), Lax (1960, 1966a, b), Risken (1984), Stratonovich (1963, 1967), Uhlenbeck and Ornstein (1930), Van Kampen (1981), Wang and Uhlenbeck (1945), Wax (1954). To derive an FPE one has to assume that one has Gaussian white noise. As will be shown in Section 6.2 for simple Gaussian colored-noise forces one can also derive an FPE if one or more auxiliary variables are introduced. The FPE is a linear

191

partial differential equation for the probability distribution function of the variables. For nonlinear Langevin equations the drift coefficients of the FPE become nonlinear but the differential equation remains linear. In these cases the solution of the FPE is not easily obtained either.

First in this chapter we review methods by which FPEs can be solved. Then we treat in more detail the FPE describing the Brownian motion of a particle in a potential. This equation is also called the Klein–Kramers equation in honor of O. Klein (Klein, 1922) and H. A. Kramers (Kramers, 1940), who did pioneering work in these investigations. We also discuss methods for solving this Klein–Kramers equation. In particular the matrix-continued-fraction (MCF) method is treated in some detail because it seems to be the method that best allows for calculation of the solutions of more general two-dimensional FPEs. We then apply these methods to the Brownian motion problem in a double-well potential and in a tilted periodic potential. The Brownian motion problem in a double-well potential serves as a model for describing fluctuations in such different fields as chemical reaction theory (Blomberg, 1977; Brinkman, 1956a, b; Grote and Hynes, 1980; Hänggi and Mojtabai, 1982; Kramers, 1940; Larson and Kostin, 1978, 1980; Mangel, 1980; Northrup and Hynes, 1978a, b, 1980; Schulten, Schulten and Szabo, 1981; Skinner and Wolynes, 1978, 1980; Visscher, 1976), bistable nonlinear oscillators (Bixon and Zwanzig, 1971; Dykman and Krivoglaz, 1984; Dykman, Soskin and Krivoglaz, 1985; Hänggi, 1986; Landauer, 1962; Risken and Voigtlaender, 1985; Voigtlaender and Risken, 1985), second-order phase transitions (Krumhansl and Schrieffer, 1975). The Brownian motion in a tilted periodic potential is applicable to superionic conductors (Dieterich, Fulde and Peschel, 1980; Fulde, Pietronero, Schneider and Strässler, 1975; Geisel, 1979), Josephson tunneling junctions (Barone and Paterno, 1982; Josephson, 1962; Solymar, 1972), rotation of dipoles (Coffey, 1980; Coffey, Evans and Grigolini, 1984; McConnel, 1980; Praestgard and Van Kampen, 1981), phase-locked loops (Lindsey, 1972; Viterbi, 1966), a simple model of charge-density waves (Toombs, 1978; Wonneberger and Breymayer, 1981).

6.2 Fokker–Planck equation

As mentioned in the introduction the Fokker–Planck equation (FPE) is an equation of motion for a distribution function of a continuous variable under the influence of noise. Without fluctuations the system variables $\{x\} = x_1, \ldots, x_N$ obey the deterministic system of differential equations (t is the time and a dot means differentiation with respect to time)

$$\dot{x}_i = h_i(\{x\}, t). \tag{6.2.1}$$

Including white noise (6.2.1) has to be changed to (*summation convention* is assumed in this section)

$$\dot{x}_i = h_i(\{x\}, t) + g_{ij}(\{x\}, t)\Gamma_j(t). \tag{6.2.2}$$

Here $\Gamma_j(t)$ are δ-correlated Gaussian random forces which are assumed to be of the form

$$\langle \Gamma_i(t)\Gamma_j(t')\rangle = 2\delta_{ij}\delta(t-t'). \tag{6.2.3}$$

The normalization in (6.2.3) can always be accomplished by choosing suitable g_{ij}. If the g_{ij} depend on $\{x\}$ explicitly we speak of multiplicative noise, otherwise the noise is termed additive. The FPE corresponding to (6.2.2) takes the form

$$\partial W/\partial t = \mathbf{L}_{\mathrm{FP}} W \tag{6.2.4}$$

with the Fokker–Planck operator $\mathbf{L}_{\mathrm{FP}}$:

$$\mathbf{L}_{\mathrm{FP}} = -\frac{\partial}{\partial x_i} D_i^{(1)}(\{x\},t) + \frac{\partial^2}{\partial x_i \partial x_j} D_{ij}^{(2)}(\{x\},t). \tag{6.2.5}$$

For the Stratonovich rule (Stratonovich, 1968) the drift and diffusion coefficients are given by

$$D_i^{(1)}(\{x\},t) = h_i(\{x\},t) + g_{kj}(\{x\},t)[(\partial/\partial x_k)g_{ij}(\{x\},t)] \tag{6.2.6}$$

$$D_{ij}^{(2)}(\{x\},t) = g_{ik}(\{x\},t)g_{jk}(\{x\},t). \tag{6.2.7}$$

If the noise is some external physical noise with a finite though very small width the Stratonovich rule should always be used; see, for instance, Risken (1984). For the Itô rule (Itô, 1944) the last term on the right hand side of (6.2.6) has to be omitted. This term, i.e. the difference between the Stratonovich and the Itô rule, is called a noise-induced drift coefficient or spurious drift coefficient. For purely additive noise, where the g_{ij} do not depend on x explicitly, the spurious drift term is absent. Obviously the Fokker–Planck operator is not an Hermitian operator. In some cases (one-variable case in Section 6.3.1, detailed balance in Section 6.3.4) it can be transformed to an Hermitian operator. Then the eigenvalues and eigenfunctions of $\mathbf{L}_{\mathrm{FP}}$ are real. In the general case, however, $\mathbf{L}_{\mathrm{FP}}$ cannot be transformed to an Hermitian operator, and eigenvalues as well as eigenfunctions may become complex.

The FPE, (6.2.4) and (6.2.5), may be written in the form of a continuity equation

$$\partial W/\partial t + \partial S_i/\partial x_i = 0, \tag{6.2.8}$$

where the probability current components S_i are defined by

$$S_i = D_i^{(1)}(\{x\},t)W - (\partial/\partial x_j)D_{ij}^{(2)}(\{x\},t)W. \tag{6.2.9}$$

For further investigations we assume that the drift and diffusion coefficients do not depend explicitly on the time t. Furthermore we assume that the diffusion matrix is positive definite or at least positive semidefinite. In quantum optics there occur FPEs where the diffusion matrix may not be positive (semi)definite (Drummond and Walls, 1980; Gardiner, 1983). Such an FPE cannot be

H. RISKEN and H. D. VOLLMER

interpreted as describing the Brownian motion of a particle obeying the Langevin equation (6.2.2). It was termed quantum FPE. As was shown recently for the model of Drummond and Walls describing dispersive optical bistability such a quantum FPE can nevertheless be solved with the matrix-continued-fraction method discussed in Section 6.4 (Vogel and Risken, 1987).

Stationary solution and transition probability

By solving FPEs one obtains distribution functions from which any averages of variables, or averages of functions of these variables, can be obtained. The stationary distribution function $W_{st}(\{x\})$ (i.e. where $\partial W/\partial t = 0$) is important for stationary one-time expectation values. Stationary two-time expectation values can be derived from the stationary joint-distribution function, which can be expressed by

$$W_2(\{x\}, t; \{x'\}, t') = P(\{x\}, t|\{x'\}, t') W_{st}(\{x'\}). \qquad (6.2.10)$$

Here the transition probability P is the solution of the FPE (6.2.4) with the initial condition

$$P(\{x\}, t'|\{x'\}, t') = \delta(\{x\} - \{x'\}). \qquad (6.2.11)$$

Multi-time correlation functions can be obtained from higher-order joint-distribution functions. In these higher-order joint-distribution functions additional products of transition probabilities have to be placed in front of the right hand side of (6.2.10). For a process described by the Langevin equations (6.2.2) (i.e. a Markov process) all information of the process is contained in the joint-distribution function W_2. Thus the main task is to obtain the stationary solution W_{st} and the transition probability distribution P. The stationary solution is obtained from the transition probability if we take the limit $t \to \infty$ i.e.

$$W_{st}(\{x\}) = \lim_{t-t' \to \infty} P(\{x\}, t|\{x'\}, t'). \qquad (6.2.12)$$

Boundary conditions

For the solution of a partial differential equation like the FPE, initial and boundary conditions must be specified. The initial condition (6.2.11) for the transition probability was already mentioned. If nothing else is said about the boundary conditions, one usually assumes that the variables $\{x\}$ of the process are defined in the whole region ($-\infty < x_i < \infty$), and at infinity the distribution function and its probability current vanish (this is often called a natural boundary condition), so that the function can be normalized to unity

$$\int_{-\infty}^{\infty} W(\{x\}, t) d^N x = 1. \qquad (6.2.13)$$

Because of (6.2.8) the normalization is conserved. There may be, however, other boundary conditions. For the motion in periodic potentials, as

194

presented in Section 6.6, we usually require that the function $W(x, v, t)$ is periodic in x with period 2π, i.e.

$$W(x + 2\pi, v, t) = W(x, v, t) \tag{6.2.14}$$

and that we have natural boundary conditions with respect to v.

6.2.1 *Examples of Fokker–Planck equations for white noise*

Klein–Kramers equation

As mentioned in the introduction the Klein–Kramers equation is a special FPE describing the Brownian motion of a particle in a potential. The Langevin equations are given by

$$\dot{x} = v, \quad \dot{v} = -\gamma v - f'(x) + (\gamma kT)^{1/2}\Gamma(t), \tag{6.2.15}$$

where x is the position and v the velocity of the particle, $f(x)$ is the potential, γ the friction constant, T the temperature and k Boltzmann's constant (the mass is unity by normalization). The Klein–Kramers equation, i.e. the FPE for the distribution function $W(x, v, t)$ in phase space corresponding to (6.2.15), reads

$$\frac{\partial W}{\partial t} = \left[-\frac{\partial}{\partial x} v + \frac{\partial}{\partial v}(\gamma v + f'(x)) + \gamma kT \frac{\partial^2}{\partial v^2} \right] W. \tag{6.2.16}$$

Smoluchowski equation

If the damping constant γ is large we may set $\dot{v} = 0$ in the second equation in (6.2.15). By eliminating v we then obtain the corresponding FPE for the distribution function $W(x, t)$ in position only, i.e. $(\tau = t/\gamma)$

$$\frac{\partial W}{\partial \tau} = \left[\frac{\partial}{\partial x} f'(x) + kT \frac{\partial^2}{\partial x^2} \right] W. \tag{6.2.17}$$

This equation is called the Smoluchowski equation. The derivation of (6.2.17) from the Klein–Kramers equation (6.2.16) is also possible and higher correction terms to it ($1/\gamma$ expansion) can be calculated, see Section 6.5.

6.2.2 *Example of an FPE describing simple colored noise*

Colored noise can also be treated with the help of the FPE. If, for instance, one has an exponentially correlated Gaussian noise force $\varepsilon(t)$

$$\langle \varepsilon(t)\varepsilon(t') \rangle = \gamma D \exp\{-\gamma|t - t'|\}. \tag{6.2.18}$$

We introduce $\varepsilon(t)$ as a new virtual variable driven by the normalized white noise $\Gamma(t)$ according to

$$\dot{\varepsilon} = -\gamma\varepsilon + \gamma D^{1/2}\Gamma(t). \tag{6.2.19}$$

195

In the stationary state this equation guarantees that the correlation function (6.2.18) is found. The Langevin equation (6.2.19) has to be added to the other equation(s). Using (6.2.6) and (6.2.7) it is then an easy matter to write down the corresponding FPE for the distribution function of the variables $\{x\}$ and ε. For the example of the overdamped motion of a particle in the potential $-x^2/2 + x^4/4$ this equation reads (Jung and Risken, 1985)

$$\frac{\partial W}{\partial t} = \left[-\frac{\partial}{\partial x}(x - x^3 + \varepsilon) + \gamma \frac{\partial}{\partial \varepsilon}\varepsilon + \gamma^2 D \frac{\partial^2}{\partial \varepsilon^2} \right] W. \qquad (6.2.20)$$

The distribution for the variables $\{x\}$ is finally obtained by integration of $W(\{x\}, \varepsilon, t)$ over ε. For more complicated correlation functions of the noise force $\varepsilon(t)$ more additional variable have to be introduced; see, for instance, Guardia, Marchesoni and San Miguel (1984), Hänggi (1984) and Marchesoni, Grigolini and Marin (1982), where also more complicated damping terms are treated.

6.3 Some methods for solving Fokker–Planck equations

Mathematically the FPE is a linear partial differential equation, the coefficients of which depend on the variables. In the general case it is very hard to solve the FPE. For some classes of FPEs methods of solving them are known. These classes are distinguished according to such different features as number of variables, stationary solutions, special drift and diffusion coefficients, boundary conditions. These classes may of course overlap. We now give a review of some of the standard methods.

6.3.1 Stationary solutions for one variable

First we look for solutions of those FPEs which have only one system variable. The simplest solution is the stationary-state distribution. Here the probability current S according to (6.2.9) must be constant, i.e.

$$S = D^{(1)}(x) W_{st}(x) - (\partial/\partial x) D^{(2)}(x) W_{st}(x) = \text{const.} \qquad (6.3.1)$$

If we have natural boundary conditions S should be zero at $x \to \pm \infty$. Thus S is zero for every x and we immediately obtain

$$W_{st}(x) = \frac{N}{D^{(2)}(x)} \exp\left\{ \int_{x_0}^{x} \frac{D^{(1)}(x')}{D^{(2)}(x')} dx' \right\} = N \exp\{-\Phi(x)\}. \qquad (6.3.2)$$

The constant N follows from the normalization condition (6.2.13).

If the probability current is not zero, we can still integrate (6.3.1) leading to

$$W_{st}(x) = N \exp\{-\Phi(x)\} - S \exp\{-\Phi(x)\} \int_{x_0}^{x} \frac{\exp\{\Phi(x')\}}{D^{(2)}(x')} dx',$$
$$(6.3.3)$$

where $\Phi(x)$ is defined in (6.3.2). The two integration constants N and S are determined by the normalization condition (6.2.13) and by the boundary condition. If, for instance, $W_{st}(x)$ is periodic with period 2π, we have to use the boundary condition (6.2.14).

6.3.2 Transformation to a Schrödinger equation and eigenfunction expansion

Here we assume that $D^{(2)}(x) = D$ is constant. One may show (Section 5.1 of Risken, 1984) that this can always be achieved by using the transformed variable

$$y = y(x) = \int_{x_0}^{x} [D/D^{(2)}(x')]^{1/2} \, dx' \qquad (6.3.4)$$

as the new independent variable. If the potential

$$f(x) = -\int_{x_0}^{x} D^{(1)}(x') \, dx' \qquad (6.3.5)$$

is introduced, the one-variable FPE takes the form

$$\frac{\partial W}{\partial t} = \left[\frac{\partial}{\partial x} f'(x) + D \frac{\partial^2}{\partial x^2} \right] W = \mathbf{L}_{FP} W, \qquad (6.3.6)$$

which is for $t \to \tau$, $D \to kT$ identical to the Smoluchowski equation (6.2.17).

In the following we assume that we have *natural boundary conditions* and that a stationary solution exists which is then given by

$$W_{st}(x) = N \exp\{-f(x)/D\} \equiv \psi_0^2(x). \qquad (6.3.7)$$

The operator $\mathbf{L}_{FP}$ can be transformed to the Hermitian operator $\mathbf{L}$ of a Schrödinger equation according to

$$\mathbf{L} = D(\partial^2/\partial x^2) - V_S(x) = [\psi_0(x)]^{-1} \mathbf{L}_{FP} \psi_0(x). \qquad (6.3.8)$$

The potential V_S of the Schrödinger equation can be expressed by the potential $f(x)$ of the Smoluchowski equation (6.3.6) according to

$$V_S(x) = [f'(x)]^2/(4D) - f''(x)/2. \qquad (6.3.9)$$

Thus the problem of solving the FPE, (6.2.4) and (6.2.5), for the one-variable case is reduced to the problem of solving a Schrödinger equation. We now assume that we have only discrete eigenvalues λ_n of $\mathbf{L}$ and eigenfunctions $\psi_n(x)$, i.e.

$$\mathbf{L}\psi_n = -\lambda_n \psi_n. \qquad (6.3.10)$$

Using the completeness relation

$$\sum_n \psi_n(x)\psi_n(x') = \delta(x - x') \qquad (6.3.11)$$

197

we can express the transition probability of the FPE in terms of the λ_n and ψ_n. This is seen as follows. We first observe from (6.3.8) and (6.3.10) that $\psi_0\psi_n$ is an eigenfunction of L_{FP}

$$L_{FP}\psi_0\psi_n = -\lambda_n\psi_0\psi_n. \tag{6.3.12}$$

By applying the formal solution of (6.3.6) for the transition probability P we thus obtain (L_{FP} is an operator with respect to x)

$$P(x,t|x',t') = e^{L_{FP}(t-t')}\delta(x-x')$$

$$= e^{L_{FP}(t-t')}\frac{\psi_0(x)}{\psi_0(x')}\delta(x-x')$$

$$= e^{L_{FP}(t-t')}\sum_n\frac{\psi_0(x)}{\psi_0(x')}\psi_n(x)\psi_n(x')$$

$$= \sum_n e^{-\lambda_n(t-t')}\frac{\psi_0(x)}{\psi_0(x')}\psi_n(x)\psi_n(x')$$

$$= \frac{\psi_0(x)}{\psi_0(x')}\sum_n\psi_n(x)\psi_n(x')e^{-\lambda_n(t-t')}. \tag{6.3.13}$$

In going from the third to the fourth line we have employed the fact that a function of an operator applied to an eigenfunction is the function of the eigenvalue times the eigenfunction. If a continuous eigenvalue spectrum occurs the sums have to be replaced by proper integrations.

If the drift coefficient $D^{(1)}(x) = -f'(x)$ in (6.3.6) is linear (Ornstein–Uhlenbeck process) we can solve the FPE analytically; see the next subsection for the N-variable case. The Schrödinger equation then specializes to that one of the harmonic oscillator, see for instance Section 5.5.1 of Risken (1984) for details. For arbitrary potentials one may use the shooting method to obtain eigenvalues and eigenfunctions. Here a numerical integration technique is used starting at large positive and negative x with some trial initial conditions, and a trial eigenvalue and the arbitrary parameters are determined in such a way that both solutions meet smoothly at some intermediate point.

6.3.3   Ornstein–Uhlenbeck processes

For Ornstein–Uhlenbeck processes the drift coefficient is linear and the diffusion coefficient constant, i.e.

$$D_i^{(1)} = -\gamma_{ij}x_j; \tag{6.3.14}$$

where γ_{ij} and $D_{ij}^{(2)} = D_{ji}^{(2)}$ are constant matrix elements.

As seen from (6.2.2) the variables and forces of the corresponding Langevin equations are coupled by linear equations. Therefore the distribution functions of the stochastic variables must be Gaussian distributions, since the

Langevin forces are Gaussian distributed. The solution for the transition probability reads (see, for instance, Section 6.5 of Risken, 1984) using the *summation convention* further on in this section

$$P(\{x\},t|\{x'\},t') = (2\pi)^{-N/2}[\text{Det}\,\sigma(t-t')]^{-1/2}$$

$$\times \exp\{-[\sigma^{-1}(t-t')]_{ij}[x_i - G_{ik}(t-t')x'_k]$$

$$\times [x_j - G_{jl}(t-t')x'_l]/2\}. \tag{6.3.15}$$

Here G_{ij} and σ_{ij} are solutions of the following two systems of differential equations plus initial conditions

$$\dot{G}_{ik} = -\gamma_{ij}G_{jk}, \quad G_{ij}(0) = \delta_{ij}. \tag{6.3.16}$$

$$\dot{\sigma}_{ij} = -\gamma_{il}\sigma_{lj} - \gamma_{ji}\sigma_{li} + 2D_{ij}, \quad \sigma_{ij}(0) = 0. \tag{6.3.17}$$

6.3.4 Potential conditions

If the drift and diffusion coefficients satisfy certain conditions the stationary solution can be found by quadratures. These conditions are termed potential conditions, and it is said that detailed balance would be valid. We first explain the essence of this procedure for the special scalar diffusion coefficient

$$D_{ij}^{(2)} = D\delta_{ij}; \quad D = \text{const}, \tag{6.3.18}$$

and for the case that the reversible part of the drift coefficient is absent. Later on we present the conditions for the general case. For the special diffusion coefficient (6.3.18) the probability current (6.2.9) is given by

$$S_i(\{x\}) = D_i^{(1)}(\{x\})W - D\,\partial W(\{x\})/\partial x_i$$

$$= W(\{x\})[D_i^{(1)}(\{x\}) - D\,\partial \ln W(\{x\})/\partial x_i]. \tag{6.3.19}$$

In the one-variable case the probability current was always constant, and was zero for natural boundary conditions. This is no longer true for the N-variable case. Thus, for natural boundary conditions, we do not always have $S_i(\{x\}) = 0$ for FPEs with $N \geqslant 2$. We may, however, first look for the special solutions with $S_i(\{x\}) = 0$. Because $W(\{x\}) \neq 0$, $S_i(\{x\})$ can only vanish everywhere if $D_i^{(1)}(\{x\})$ is the gradient of a potential Φ, i.e.

$$D_i^{(1)}(\{x\}) = -D\,\partial\Phi(\{x\})/\partial x_i. \tag{6.3.20}$$

Necessary and sufficient conditions for the existence of $\Phi(\{x\})$ are the *potential conditions*

$$\partial D_i^{(1)}(\{x\})/\partial x_j = \partial D_j^{(1)}(\{x\})/\partial x_i. \tag{6.3.21}$$

If (6.3.21) is fulfilled the stationary solution of (6.2.4) and (6.2.5) reads

$$W_{\text{st}}(\{x\}) = N\exp\{-\Phi(\{x\})\}, \tag{6.3.22}$$

where $\Phi(\{x\})$ is given by the line integral

$$\Phi(\{x\}) = -D^{-1} \int_{\{x_0\}}^{\{x\}} D_i^{(1)}(\{x'\})\,dx_i'. \tag{6.3.23}$$

If (6.3.21) is not fulfilled $S_i(\{x\})$ cannot be zero everywhere.

General case

In the general case we may split up the drift coefficient $D_i(\{x\})$ into two parts:

$$D_i^{(1)}(\{x\}) = D_i^{(s)}(\{x\}) + D_i^{(a)}(\{x\}), \tag{6.3.24}$$

where the first and the second part should obey the relations

$$D_i^{(s)}(\{x\}) = \partial D_{ij}^{(2)}(\{x\})/\partial x_j - D_{ij}^{(2)}(\{x\})\,\partial\Phi(\{x\})/\partial x_j \tag{6.3.25}$$

$$(\partial/\partial x_i)[D_i^{(a)}(\{x\})\exp\{-\Phi(\{x\})\}] = 0. \tag{6.3.26}$$

Here $\Phi(\{x\})$ is the negative logarithm of the stationary solution, see (6.3.22). If the inverse of $D_{ij}^{(2)}(\{x\})$ exists we may write

$$\partial\Phi(\{x\})/\partial x_i \equiv A_i(\{x\})$$

$$= [(D^{(2)}(\{x\}))^{-1}]_{ij}[\partial D_{jk}^{(2)}(\{x\})/\partial x_k - D_j^{(s)}(\{x\})]. \tag{6.3.27}$$

From (6.3.27) we thus have the generalized potential conditions

$$\partial A_i(\{x\})/\partial x_j = \partial A_j(\{x\})/\partial x_i. \tag{6.3.28}$$

If the FPE has a stationary solution the decomposition (6.3.24) with (6.3.25) and (6.3.26) always exists. Without knowing the stationary distribution, however, the decomposition (6.3.24) is generally hard to find. In some special cases one may find it by some guesswork. The first try would be $D_i^{(a)}(\{x\}) = 0$. If the potential condition (6.3.28) is fulfilled we obtain $\Phi(\{x\})$ and therefore also the stationary solution by the line integral

$$\Phi(\{x\}) = \int_{\{x_0\}}^{\{x\}} A_i(\{x'\})\,dx_i'. \tag{6.3.29}$$

If the detailed-balance condition holds we can always find the decomposition (6.3.24). The drift coefficient $D_i^{(s)}(\{x\})$ has to be replaced by the irreversible part of the drift coefficient and $D_i^{(a)}(\{x\})$ has to be replaced by the reversible part; see, for instance, Risken (1984, Section 6.4) for a definition of reversible and irreversible parts and for a discussion of the detailed-balance condition.

6.3.5 Other methods

The MCF method will be discussed in the next section. There exist a variety of other methods. Some of these are listed below:

(i) analytical solutions for model potentials;
(ii) variational methods;

(iii) WKB approximations (limit solutions for vanishing diffusion);
(iv) adiabatic elimination (see Section 6.4 for application);
(v) transformation to integral equations;
(vi) first-passage-time methods;
(vii) path-integral methods;
(viii) digital and analog simulation methods.

6.4   Matrix-continued-fraction method

The method will be explained for an FPE with two variables x and y. Though the method may in principle be applied to an FPE with an arbitrary number of variables, it becomes less practical for more than two variables. The method also works very well for the one-variable case (Risken, 1984, 1987) but this will not be discussed here any further. Applications in laser theory are reviewed in Risken and Vollmer (1985).

6.4.1   *Expansion into complete sets*

We first assume that we have two complete orthonormalized sets $\phi^q(x)$ and $\psi_n(y)$ which satisfy the boundary conditions in x and y, respectively. If x and y extend from minus infinity to plus infinity (natural boundary conditions) we may use, for instance, Hermite functions $\phi^q(x) = N^q H_q(\alpha x) \exp(-\alpha^2 x^2/2)$ and $\psi_n(y) = N_n H_n(\beta y) \exp(-\beta^2 y^2/2)$ with suitable scaling factors α and β. Complete sets which are more adapted to the considered problem may also be used. We may, for instance, construct a system of polynomials orthogonal to a certain weight function (Blackmore and Shizgal, 1985; Shizgal, 1981). The weight function should be connected to the investigated problem. For the Klein–Kramers equation in Section 6.5, for instance, we may apply as weight function the stationary solution of the Smoluchowski equation.

Because the sets are complete the probability density can be expanded according to

$$W(x, y, t) = F(x, y) \sum_{q,n} c_n^q(t) \, \phi^q(x) \, \psi_n(y). \qquad (6.4.1)$$

This procedure is sometimes called the Bubnow–Galerkin method (Michlin and Smolizki, 1969). The choice of the function $F(x, y)$ will be discussed later on. If we insert (6.4.1) into the FPE, we obtain an infinite system of coupled differential equations for the expansion coefficients c_n^q. One choice for this system is given by (assuming $F(x, y) \neq 0$)

$$\dot{c}_n^q = \sum_{p,m} [F^{-1} \mathbf{L}_{\mathrm{FP}} F]_{nm}^{qp} c_m^p. \qquad (6.4.2)$$

Here we have denoted the matrix element of an operator $\mathbf{A}$ by

$$A_{nm}^{qp} = \iint [\phi^q(x)\psi_n(y)]^* \, \mathbf{A} \, \phi^p(x)\psi_m(y) \, \mathrm{d}x \, \mathrm{d}y. \qquad (6.4.3)$$

If we do not divide by $F(x, y)$ we obtain the more complicated system

$$\sum_{p,m} F_{nm}^{qp} \dot{c}_m^p = \sum_{p,m} [\mathbf{L}_{\mathrm{FP}} F]_{nm}^{qp} c_m^p, \tag{6.4.4}$$

which may, however, allow for generalizations; see below.

If the infinite system is truncated at $q = Q$ and $n = N$ we may solve it numerically, leading to approximate solutions of the FPE. Because the number of equations in the truncated system is of the order $Q \times N$, we can use only low Q and N values.

6.4.2 Tridiagonal vector-recurrence relations

Sometimes the structure of the system (6.4.2) is such that only nearest neighbors with respect to the lower index are coupled. By properly choosing $F(x, y)$ and the sets many equations may be brought to that form. If such a coupling occurs for the upper index we may, of course, change notation. If we introduce the column vector

$$\mathbf{c}_n = (c_n^1, \ldots, c_n^Q) \tag{6.4.5}$$

the system (6.4.2) has the form of the tridiagonal vector-recurrence relation

$$\dot{\mathbf{c}}_n = \mathbf{Q}_n^- \mathbf{c}_{n-1} + \mathbf{Q}_n \mathbf{c}_n + \mathbf{Q}_n^+ \mathbf{c}_{n+1} \tag{6.4.6}$$

with matrices $\mathbf{Q}_n^+, \mathbf{Q}_n^-, \mathbf{Q}_n$ built from the matrix elements (6.4.3) with $\mathbf{A} = F^{-1} \mathbf{L}_{\mathrm{FP}} F$. If two nearest neighbors with respect to the lower index are coupled, we have to use the $2 \times Q$-dimensional column vector

$$\mathbf{c}_n = (c_{2n}^1, \ldots, c_{2n}^Q, c_{2n+1}^1, \ldots, c_{2n+1}^Q), \tag{6.4.7}$$

and we can again cast (6.4.2) into the form (6.4.6); see, for instance, Risken (1984) for the M-nearest-neighbor case.

We now explain the main idea for solving (6.4.6) for the stationary case, i.e. for

$$\mathbf{Q}_n^- \mathbf{c}_{n-1} + \mathbf{Q}_n \mathbf{c}_n + \mathbf{Q}_n^+ \mathbf{c}_{n+1} = 0. \tag{6.4.8}$$

The time-dependent problem as well as the eigenvalue problem can be solved similarly. We first introduce the matrices $\mathbf{S}_n$ which connect $\mathbf{c}_{n+1}$ and $\mathbf{c}_n$:

$$\mathbf{c}_{n+1} = \mathbf{S}_n \mathbf{c}_n. \tag{6.4.9}$$

Insertion into (6.4.8) leads to

$$\mathbf{Q}_n^- \mathbf{c}_{n-1} + (\mathbf{Q}_n + \mathbf{Q}_n^+ \mathbf{S}_n) \mathbf{c}_n = 0,$$

i.e. to

$$\mathbf{c}_n = -(\mathbf{Q}_n + \mathbf{Q}_n^+ \mathbf{S}_n)^{-1} \mathbf{Q}_n^- \mathbf{c}_{n-1} = \mathbf{S}_{n-1} \mathbf{c}_{n-1}.$$

By comparison with (6.4.9) we conclude (changing n to $n+1$)

$$\mathbf{S}_n = [-\mathbf{Q}_{n+1} - \mathbf{Q}_{n+1}^+ \mathbf{S}_{n+1}]^{-1} \mathbf{Q}_{n+1}^- \tag{6.4.10}$$

or by iteration of (6.4.10) we obtain the infinite matrix continued fraction

$$\mathbf{S}_n = [-\mathbf{Q}_{n+1} - \mathbf{Q}_{n+1}^+ [-\mathbf{Q}_{n+2} - \mathbf{Q}_{n+2}^+ [-\mathbf{Q}_{n+3} - \cdots]^{-1}$$
$$\times \mathbf{Q}_{n+3}^-]^{-1} \mathbf{Q}_{n+2}^-]^{-1} \mathbf{Q}_{n+1}^-. \tag{6.4.11}$$

If we insert $\mathbf{c}_1 = \mathbf{S}_0 \mathbf{c}_0$ into the first equation ($n = 0$) of (6.4.8) we thus have

$$\mathbf{R}_0 \mathbf{c}_0 = 0 \quad \text{or} \quad \sum_{p=0}^{Q} R_0^{qp} c_0^p = 0, \tag{6.4.12}$$

with

$$\mathbf{R}_0 = -\mathbf{Q}_0 - \mathbf{Q}_0^+ \mathbf{S}_0. \tag{6.4.13}$$

Insertion of $\mathbf{S}_0$ from (6.4.11) leads to

$$\mathbf{R}_0 = -\mathbf{Q}_0 - \mathbf{Q}_0^+ [-\mathbf{Q}_1 - \mathbf{Q}_1^+ [-\mathbf{Q}_2 - \cdots]^{-1} \mathbf{Q}_2^-]^{-1} \mathbf{Q}_1^-. \tag{6.4.14a}$$

If the matrix inversions are written by fraction lines $\mathbf{R}$ takes the form

$$\mathbf{R}_0 = -\mathbf{Q}_0 - \mathbf{Q}_0^+ \cfrac{\mathbf{I}}{-\mathbf{Q}_1 - \mathbf{Q}_1^+ \cfrac{\mathbf{I}}{-\mathbf{Q}_2 - \cdots} \mathbf{Q}_2^-} \mathbf{Q}_1^-. \tag{6.4.14b}$$

For a discussion of the convergence of MCFs, see Denk and Riederle (1982). It follows from (6.4.6) and $c_0^0 = 1$, $\dot{c}_0^0 = 0$ that R^{0p} is also zero. Therefore we may write the second equation of (6.4.12) in the form

$$\sum_{p=1}^{Q} R_0^{qp} c_0^p = -R_0^{q0} c_0^0 = -R^{q0}, \quad q = 1, \ldots, Q. \tag{6.4.15}$$

From this equation we obtain c_0^p, i.e. the vector $\mathbf{c}_0$. By iteration according to (6.4.9) (the $\mathbf{S}_n$ need not be calculated separately, but they appear in intermediate steps in the calculation of $\mathbf{R}_0$) we thus obtain all $\mathbf{c}_n$:

$$\mathbf{c}_n = \mathbf{S}_{n-1} \mathbf{S}_{n-2} \cdots \mathbf{S}_0 \mathbf{c}_0. \tag{6.4.16}$$

If $\mathbf{c}_1$ is obtained from (6.4.16) the other $\mathbf{c}_n$ may be obtained by upiteration of (6.4.8). In contrast to (6.4.16), however, this upiteration is numerically unstable. In the actual calculations one also has to truncate the expansion in ψ_n at $n = N$. This means that we only take into account N iterations.

The MCF, (6.4.14a) and (6.4.14b), can be easily translated into an algorithm for a digital computer. In order to do this we use the recursion

$$\mathbf{R}_n = -\mathbf{Q}_n - \mathbf{Q}_n^+ [\mathbf{R}_{n+1}]^{-1} \mathbf{Q}_{n+1}^- \tag{6.4.14c}$$

starting with $\mathbf{R}_N = -\mathbf{Q}_N$ at $n = N$ and iterating (6.4.14c) to lower n till we arrive at $\mathbf{R}_0$. This downward or 'tail to head' iteration of (6.4.14a) and (6.4.14b) seems to be easier to use than a possible upward or 'head to tail' iteration; see section 9.5.2 of Risken (1984) for a discussion of the advantages and disadvantages of both procedures.

The truncation indices N and Q must be determined in such a way that a further increase of N and Q does not change the final result beyond a given accuracy. Thus the accuracy of the final result can easily be controlled. For the limit $Q \to \infty$ and $N \to \infty$ the results become exact. If the parameters are not too extreme the method usually works very well, see Sections 6.6 and 6.7 for examples. For very small diffusion coefficients (or for some other extreme parameters), however, the distribution function gets more peaked and therefore more coefficients in the expansion have to be taken into account. Thus the dimension of the matrices to be inverted also increases. For numerical calculations the method is then no longer tractable.

At a first glance the M-nearest-neighbor coupling leading to the tridiagonal recurrence relation (6.4.6), which in turn leads to the MCF solution described above, seems to occur only for very special FPEs. An M-nearest-neighbor coupling, however, is valid for a large class of Fokker–Planck operators. If, for instance, we have natural boundary conditions, if we use Hermite functions, and if $D_i^{(1)}$ and $D_{ij}^{(2)}$ are given by polynomials in y of finite order, we obtain a finite number of nearest-neighbor couplings in the lower index for $F = 1$. (The coefficients of the polynomial may still depend on x.) By using a suitable notation the equation of motion can then be cast into the tridiagonal vector-recurrence relation (6.4.2). If the drift and diffusion coefficients are rational functions in y (i.e. quotient of two polynomials in y) and if we choose F to be the common denominator, we can cast the equation for the coefficients again into a tridiagonal form. Generally this form is more complicated than (6.4.2) because the time derivatives on the left hand side also have a tridiagonal coupling. For the eigenvalue problem as well as for the stationary solution, the above method can still be applied.

It should be stressed that the MCF method does not require that some potential conditions are fulfilled or that the Fokker–Planck operator can be brought to an Hermitian form.

6.5   Methods for solving the Klein–Kramers equation

In this section we discuss methods for solving the Klein–Kramers equation (6.2.16). The Klein–Kramers equation is a special FPE describing the Brownian motion in a potential. The corresponding Langevin equations are given in (6.2.15). For intermediate friction constants γ the full two-variable FPE (6.2.16) has to be solved. For high and low damping constants γ, only one variable is relevant, and special methods can be applied. In the following subsections we first discuss these two limit-solution methods and then the continued-fraction (CF) method for the general case.

6.5.1   High-friction case

If we eliminate the velocity v in (6.2.15) we have

$$\ddot{x} + \gamma \dot{x} + f'(x) = (\gamma kT)^{1/2} \Gamma(t). \qquad (6.5.1)$$

For high friction constants we may neglect the second-derivative term in (6.5.1), thus obtaining the first-order Langevin equation

$$\dot{x} + (1/\gamma)f'(x) = (kT/\gamma)^{1/2}\Gamma(t). \tag{6.5.2}$$

The corresponding FPE is the Smoluchowski equation (6.2.17). As discussed in Section 6.3.2 it can be transformed to a Schrödinger equation, and thus all the well-known methods for solving Schrödinger equations may be applied to the Smoluchowski equation (6.2.17).

The reduction of the system of Langevin equations (6.2.15) to (6.5.2) is done by adiabatic elimination. This means that the variable v, which is moving much faster than the variable x, follows the variable x adiabatically and is finally eliminated in the equation for x. Haken terms this principle the slaving principle (Haken, 1975, 1983a). The fast (or irrelevant) variable v is slaved by the slow (or relevant) variable x. The derivation of the Smoluchowski equation (6.2.17) directly from the FPE is also possible. Generally one obtains corrections to the Smoluchowski equation, the $1/\gamma$ expansion; see, for instance, Titulaer (1978) and Wilemski (1976). By expanding the MCFs of Section 6.5.3 one may also obtain this $1/\gamma$ expansion (Risken, Vollmer and Mörsch, 1981).

6.5.2   Low-friction case

Without any friction (i.e. $\gamma = 0$) the damping and fluctuating terms in (6.5.1) are absent. The energy

$$E = \dot{x}^2/2 + f(x) \tag{6.5.3}$$

is then a constant of motion. For small friction constants the energy E will become the slow or relevant variable whereas both x and v are fast or irrelevant variables. Therefore we should use the energy or a function of the energy like the action integral as the relevant variable. In order to obtain this one-variable equation one has to transform first the x and v variables to x and E variables (Stratonovich, 1963, 1967). To retain the full information of the distribution function $W(x,v,t)$ we have to introduce the two distribution functions $\tilde{W}_+(x,E,t)$ and $\tilde{W}_-(x,E,t)$ for the positive and the negative sign of the velocity $v = \pm\{2[E - f(x)]\}^{1/2}$, respectively. In this way one gets two coupled equations for $\tilde{W}_+$ and $\tilde{W}_-$. Next one has to perform an average over the fast variable x. Here we have to distinguish whether we are near to or far away from a critical trajectory. The critical trajectories are those solutions of the deterministic equations (6.2.1) (Langevin equations without noise and damping) where the form of the solution changes. For the double-well potential in Section 6.7 with $d_2 = -1$ and $d_4 = 1$, for instance, the critical trajectory is the '8'-shaped curve $E = v^2/2 - x^2/2 + x^4/4$ in the phase-space. Near a critical trajectory the x-dependence must be taken into account in a small region around it. The width of the region is proportional to $\gamma^{1/2}$. Outside this boundary layer only the E dependence needs to be taken into account and the differences between $\tilde{W}_+$ and $\tilde{W}_-$ vanish. By the boundary layer conditions the functions $\tilde{W}(E,t)$ and their derivatives are coupled. Thus solving the one-variable FPEs for $\tilde{W}(E,t)$ outside the critical regions separately, and connecting them by the boundary layer theory, leads to the solution in the low-friction case. The lowest nonzero eigenvalue for the double-well potential, for instance, has the following dependence on γ:

$$\lambda = a\gamma + b\gamma^{3/2} + \cdots = \gamma(a + b\gamma^{1/2} + \cdots). \tag{6.5.4}$$

The first term proportional to γ stems from the equation inside the well with the boundary condition that the distribution function vanishes at the critical trajectory, whereas the last term stems from the boundary-layer theory (Büttiker and Landauer, 1984; Risken, Vogel and Vollmer, 1988; Risken and Voigtlaender, 1985; Voigtlaender and Risken, 1984. See also Mel'nikov, 1984, and Mel'nikov and Meshkov, 1986). For the tilted periodic-potential problem treated in Section 6.6 the critical trajectory separates solutions describing the steady motion in one direction, and the locked solutions describing the particles oscillating within the wells (for details see Section 6.6). A detailed calculation shows that the drift velocity has a γ dependence of the form (Section 11.4 of Risken, 1984; Risken and Vollmer, 1979b)

$$\langle v \rangle = (1/\gamma)(\tilde{a} + \tilde{b}\gamma^{1/2} + \cdots). \tag{6.5.5}$$

The last term ($\sim \gamma^{-1/2}$) in (6.5.5) stems from the boundary-layer theory. For a further discussion of the boundary-layer theory, see Risken and Voigtlaender (1985).

6.5.3 Solutions in terms of matrix continued fractions

Brinkman's hierarchy

We first expand the distribution function according to

$$W(x,v,t) = \psi_0(v) \sum_{n=0}^{\infty} c_n(x,t)\psi_n(v). \tag{6.5.6}$$

Here $\psi_0(v)\psi_n(v)$ are eigenfunctions of the operator appearing in the Klein–Kramers equation (6.2.16)

$$L_v = \gamma[(\partial/\partial v)v + kT\partial^2/\partial v^2], \tag{6.5.7a}$$

i.e. of the equation

$$L_v\psi_0\psi_n = -n\gamma\,\psi_0\psi_n. \tag{6.5.7b}$$

The functions $\psi_n(v)$ are the Hermite functions

$$\psi_n(v) = h_n(v/(2kT)^{1/2})(kT)^{-1/4} \tag{6.5.8}$$

$$h_n(\xi) = (n!2^n)^{-1/2}(2\pi)^{-1/4}H_n(\xi)\exp(-\xi^2/2) \tag{6.5.9}$$

known also from the oscillator eigenfunctions of quantum mechanics. The eigenfunctions (6.5.8) obey the orthonormality relation

$$\int_{-\infty}^{\infty} \psi_n(v)\,\psi_m(v)\,dv = \delta_{nm}. \tag{6.5.10}$$

Inserting (6.5.6) into the Klein–Kramers equation (6.2.16), multiplying it by $\psi_m(v)/\psi_0(v)$ and using (6.5.8) and (6.5.10) and the recurrence relations for H_n we obtain the following infinite system of coupled differential equations $(n \geqslant 0, c_{-1} = 0)$

$$\partial c_n/\partial t = -n^{1/2}\hat{\mathbf{D}}c_{n-1} - n\gamma c_n - (n+1)^{1/2}\mathbf{D}c_{n+1}, \tag{6.5.11}$$

where the operators $\mathbf{D}$ and $\hat{\mathbf{D}}$ are defined by

$$\mathbf{D} = (kT)^{1/2}\partial/\partial x; \quad \hat{\mathbf{D}} = (kT)^{1/2}\partial/\partial x + f'(x)/(kT)^{1/2}. \tag{6.5.12}$$

The system (6.5.11) is sometimes called Brinkman's hierarchy (Brinkman, 1956a, b).

Derivation of the Smoluchowski equation

If we truncate the expansion (6.5.6) after the first two terms $(c_2 = c_3 = \ldots = 0)$ Brinkman's hierarchy (6.5.11) reduces to

$$\begin{aligned}\partial c_0/\partial t &= -\mathbf{D}c_1\\ \partial c_1/\partial t &= -\gamma c_1 - \hat{\mathbf{D}}c_0.\end{aligned} \tag{6.5.13}$$

For large γ we may further neglect the time derivative in the last equation of (6.5.13). Elimination of c_1 then leads to the Smoluchowski equation (6.2.17), which for large γ is the correct equation for the distribution in x only, i.e. for

$$c_0(x,t) = \int_{-\infty}^{\infty} W(x,v,t)\,dv. \tag{6.5.14}$$

For decreasing friction constants increasingly more terms of the expansion (6.5.6) have to be taken into account. Thus one has to solve the system (6.5.11) for a large number of coefficients.

Tridiagonal vector-recurrence relations

Because of the tridiagonal structure of Brinkman's hierarchy the solution of (6.5.11) can be obtained in terms of matrix continued fractions. Expanding the coefficients $c_n(x,t)$ into a complete set with respect to x, i.e.

$$c_n(x,t) = \sum_q c_n^q \phi^q(x), \tag{6.5.15}$$

inserting (6.5.15) into (6.5.11) and using the orthonormality condition

$$\int [\phi^p(x)]^* \phi^q(x)\,dx = \delta_{pq}, \tag{6.5.16}$$

207

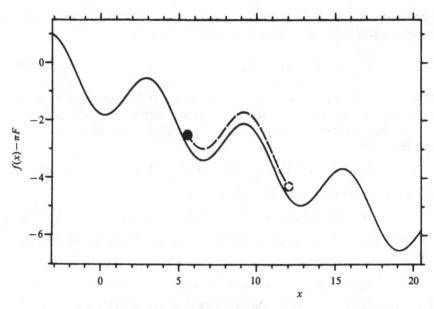

Figure 6.1. The potential (6.6.1) as a function of x for $d = 1$, $F = 0.25$. (Taken from Risken, 1984.)

we arrive at the tridiagonal recurrence relation (6.4.6). The infinite expansion (6.5.15) has to be truncated at an upper limit Q for performing the calculations. The column vectors $\mathbf{c}_n$ then have the components c_n^q, i.e.

$$\mathbf{c}_n = (c_n^0, c_n^1, \ldots, c_n^Q) \quad \text{or} \quad \mathbf{c}_n = (c_n^{-Q}, \ldots, c_n^0, \ldots, c_n^Q). \qquad (6.5.17)$$

The first form has to be used if the index q runs from 0 to Q, the second one if q runs from $-Q$ to $+Q$. The matrices $\mathbf{Q}_n^{\pm}$, $\mathbf{Q}_n$ are given by

$$\mathbf{Q}_n^+ = -(n+1)^{1/2}\mathbf{D}, \quad \mathbf{Q}_n = -n\gamma\mathbf{I}, \quad \mathbf{Q}_n^- = -n^{1/2}\hat{\mathbf{D}} \qquad (6.5.18)$$

where $\mathbf{I}$ is the unit matrix. The matrices $\mathbf{D}$ and $\hat{\mathbf{D}}$ are the matrices corresponding to the operators $\mathbf{D}$ and $\hat{\mathbf{D}}$. The matrix elements of $\mathbf{D}$, for example, are the expressions

$$D^{pq} = \int [\phi^p(x)]^* \mathbf{D}\, \phi^q(x)\, dx. \qquad (6.5.19)$$

We therefore can solve the matrix form of Brinkman's hierarchy by the MCF method as described in Section 6.4.

6.6 Brownian motion in a tilted periodic potential

In this section we apply the methods described in the previous section to the Brownian-motion problem in the tilted cosine potential (Figure 6.1)

$$f(x) = -d\cos x - Fx. \qquad (6.6.1)$$

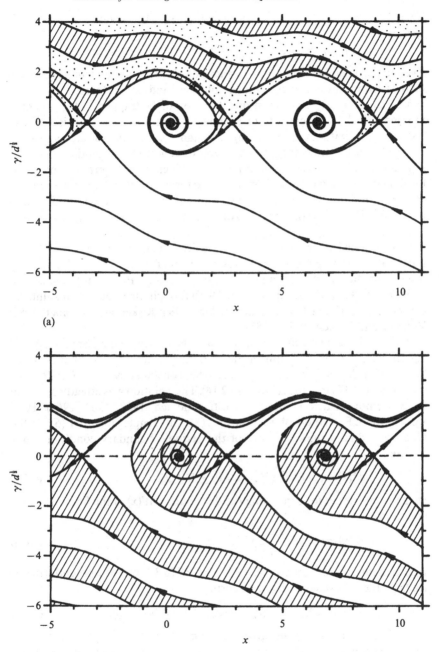

Figure 6.2. The trajectories of the equation (6.2.15) going through the saddle points for the potential (6.6.1) and without noise term, for $\gamma/d^{1/2} = 0.25$ and (a) $F/d = 0.25$; (b) $F/d = 0.5$. (Taken from Risken, 1984.)

209

Therefore the force $-f'(x)$ appearing in the Langevin equation (6.5.1) specializes to

$$-f'(x) = -d \sin x + F. \qquad (6.6.2)$$

Even the problem without noise cannot be solved in closed form. Depending on the friction constant γ, potential amplitude d and external force F, solutions can be continuously moving in one direction or be trapped in the potential minima (if $F/d < 1$). Figures 6.2(a, b) show two examples for the trajectories of the equation without noise going through one of the saddle points. In Figure 6.2(b) the trajectory going in the direction to the saddle point is a separatrix which separates the trajectories which finally spiral in one of the focal points (shaded regions) from the trajectories which end up at the running solution (unshaded regions). In Figure 6.2(a) all the trajectories spiral in one of the focal points. Areas that spiral to different points are alternately shaded and unshaded.

As mentioned in the introduction the Brownian-motion problem in the potential (6.6.1) is applicable in a large number of fields. In this section we only discuss the stationary solution of the corresponding FPE (6.2.16) (Risken and Vollmer, 1979a; Vollmer and Risken, 1979) though time-dependent solutions can be treated similarly (Jung and Risken, 1984; Risken and Vollmer, 1982; Vollmer and Risken, 1982, 1983).

In the *high-friction limit* one has to solve the corresponding Smoluchowski equation (6.2.17). Its solution in the stationary state is given by (6.3.2) and (6.3.3) with $D^{(1)} = -f'(x)$, $D^{(2)} = kT$. The constants N and S follow by insertion of (6.3.3) into (6.2.13) and (6.2.14). This solution was already obtained by Stratonovich a long time ago (Stratonovich, 1958, 1963, 1967). Here we use the CF method of Section 6.4. For the simple potential (6.6.1) the CFs are ordinary ones. Because of the periodic boundary conditions in x we use

$$\phi^n(x) = (2\pi)^{-1/2} e^{inx}, \quad n = 0, \pm 1, \pm 2, \dots \qquad (6.6.3)$$

as the complete set for expanding the distribution $W(x)$ of the Smoluchowski equation

$$W(x) = (2\pi)^{-1/2} \sum_n c_n e^{inx}. \qquad (6.6.4)$$

Insertion into the integrated one-variable FPE (6.3.1) with the potential (6.6.1) leads to the two-sided tridiagonal form

$$(F - inkT)c_n + \tfrac{1}{2}id(c_{n-1} - c_{n+1}) = \gamma S(2\pi)^{1/2} \delta_{n0}. \qquad (6.6.5)$$

For $n \geqslant 1$ these equations give an infinite CF for $c_1/c_0 = c^*_{-1}/c_0$. The normalization requires $c_0 = (2\pi)^{-1/2}$. Thus the average of $\sin x$ is easily evaluated by calculating the CF

$$\langle \sin x \rangle = (2\pi)^{1/2}(c_1^* - c_1)/2i$$

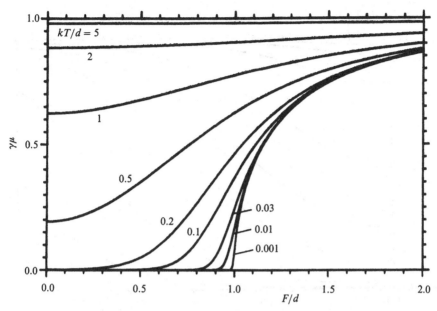

Figure 6.3. The mobility μ multiplied by the damping constant γ as a function of F/d for various temperatures kT/d for the cosine potential (6.6.1). The zero-temperature-limit curve agrees with the curve for $kT/d = 0.001$ within the linewidth of the drawing. (Taken from Risken, 1984.)

$$(2\pi)^{1/2} c_1 = \langle e^{-ix} \rangle = 2 \cfrac{0.25}{1(kT/d) + i(F/d) + \cfrac{0.25}{2(kT/d) + i(F/d) + \cdots}}.$$
(6.6.6)

The average $\langle \sin x \rangle$ is needed to obtain the mean drift velocity $\langle v \rangle$ and the mobility μ

$$\langle v \rangle = \mu F = \langle \dot{x} \rangle = \gamma^{-1}(F - d \langle \sin x \rangle).$$
(6.6.7)

In Figure 6.3 we show the mobility μ times the damping constant γ as a function of the external field F/d. It is seen that for larger F/d values and also for high temperatures the mobility approaches the free particle mobility $\gamma\mu = 1$. For low temperatures we approximately obtain the mobility of the equation without noise, i.e. it is zero for $F/d < 1$ and it is $\gamma\mu = [1 - (d/F)^2]^{1/2}$ for $F/d > 1$.

In the *low-friction limit* we have to use the energy as the relevant variable, see Section 6.5.2. In the limit $\gamma \to 0$ a stationary solution can only exist if the external force F is proportional to the friction constant γ, i.e. if F/γ is finite. Thus for defining the energy variable we omit the potential energy term due to the external force $-Fx$ in the asymptotic limit $\gamma \to 0$. In the stationary state we can solve the FPE analytically. The final expression is quite complicated. In it

211

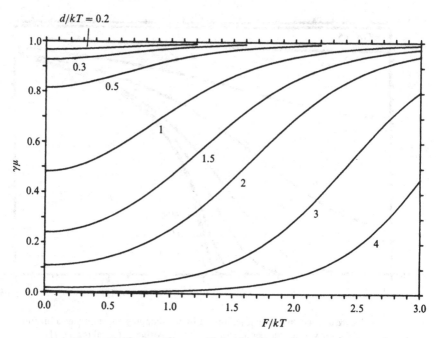

Figure 6.4. The mobility μ multiplied by the damping constant γ as a function of F/kT for various ratios d/kT for the cosine potential (6.6.1) for $\gamma/(kT)^{1/2} = 1$. (Taken from Risken and Vollmer, 1979a.)

occur integrals over the complete elliptic functions. By taking into account the boundary layer we get the square-root dependence of the mobility according to (6.5.5), see also Figure 6.5.

For *intermediate friction constants* γ we have to solve Brinkman's hierarchy (6.5.11) or its matrix representation (6.4.8) with (6.5.18). As expansion set in x we use (6.6.3). As explained in Section 6.5.3 we only need the matrix elements of the operators $\mathbf{D}$ and $\hat{\mathbf{D}}$ defined in (6.5.12). They are easily evaluated and read

$$D^{pq} = \frac{(kT)^{1/2}}{2\pi} \int_0^{2\pi} e^{-ipx} \frac{\partial}{\partial x} e^{iqx} \, dx = i(kT)^{1/2} q \delta_{pq} \qquad (6.6.8)$$

$$\hat{D}^{pq} = \frac{(kT)^{1/2}}{2\pi} \int_0^{2\pi} e^{-ipx} \left(\frac{\partial}{\partial x} + \frac{d \sin x - F}{kT} \right) e^{iqx} \, dx$$

$$= (kT)^{1/2} \{ [iq - F/(kT)] \delta_{pq} - i(\delta_{p,q+1} - \delta_{p,q-1}) d/(2kT) \}. \qquad (6.6.9)$$

The normalization requires $c_0^0 = (2\pi)^{-1/2}$. Thus the coefficients c_n^p can be obtained by using (6.4.14a)–(6.4.16). For the drift velocity we obtain

$$\langle v \rangle = \int_0^{2\pi} \int_{-\infty}^{\infty} v W(x, v) \, dx \, dv = (2\pi kT)^{1/2} c_1^0. \qquad (6.6.10)$$

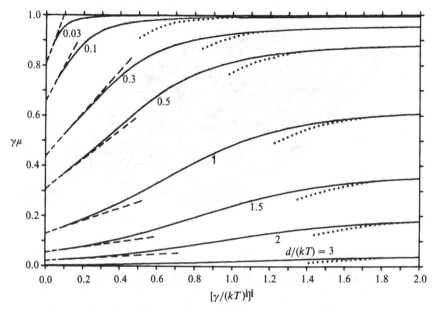

Figure 6.5. The mobility μ multiplied by the damping constant γ in linear response as a function of $[\gamma/(kT)^{1/2}]^{1/2}$ for various ratios d/kT for the cosine potential (6.6.1). The full curves are calculated by the MCF method, the broken lines follow from the low-friction theory, and the dotted lines are results of the inverse-friction expansion, valid for large γ. (Taken from Risken and Vollmer, 1979b.)

Thus the main task in determining the drift velocity is calculating the MCF (6.4.14a) where the matrices $\mathbf{Q}$ are expressed by (6.5.18) which in turn are given by (6.6.8) and (6.6.9). To evaluate this infinite CF it is approximated by its Nth approximant, i.e. by putting $\mathbf{S}_N = \mathbf{0}$. The number N and the truncation number Q of the Fourier series have been determined such that a further increase of N and Q does not alter the result beyond a given accuracy. It turned out that the number N for giving results accurate to three decimal places was of the order $N = 20kT^{1/2}/\gamma$, and $Q = 12$ was sufficient for d values up to $d/(kT) = 4$. In Figure 6.4 the mobility μ times the damping constant γ is shown as a function of $F/(kT)$ for various amplitudes $d/(kT)$ of the cosine potential. For large forces the effect of the periodic potential becomes negligible and therefore one obtains the result $\gamma\mu = 1$ which is also valid for vanishing force $f' = 0$. (This result is immediately obtained by averaging (6.5.1) in the stationary state, i.e. $\langle \ddot{x} \rangle = 0, \langle \dot{x} \rangle = \langle v \rangle = F/\gamma$.) The result for a small external force F (linear response) is shown in Figure 6.5 as a function of $[\gamma/(kT)]^{1/2}$. (To evaluate the expression for $F = 0$ a small F must be used, because otherwise the expression will become infinite.) It is seen that for small damping constants the plots fit the low-friction approximation very accurately, and for large damping constants they fit the high-friction approximation.

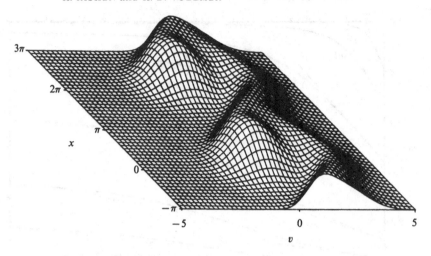

Figure 6.6. Perspective plots of the stationary probability distribution of the Klein–Kramers equation (6.2.16) for the cosine potential (6.6.1) for $\gamma/(kT)^{1/2} = 0.25$, $d/kT = 1$, $F/(kT) = 0.5$. (Taken from Vollmer and Risken, 1979.)

By calculating all S_n we can find all the other coefficients c_n^q by upiteration according to (6.4.16) and thus obtain the stationary distribution itself. Figures 6.6 and 6.7 show some typical distribution functions for the Brownian motion problem in an inclined cosine potential obtained by Vollmer and Risken (1979) using the MCF method.

6.7   Brownian motion in a double-well potential

In this section we discuss the Brownian motion in the double-well potential (Landau potential)

$$f(x) = d_2 x^2/2 + d_4 x^4/4, \quad d_2 < 0 \quad d_4 > 0. \tag{6.7.1}$$

As mentioned in the introduction the Brownian motion in the bistable potential (6.7.1) serves as a model for a transition from one state to another in quite a number of fields. The equations of motion without the noise term are best discussed by looking at those trajectories in phase space which go through the saddle point $x = v = 0$, see Figure 6.8. The other singular points are $x = x_\pm$, $v = 0$. Notice that trajectories not going through the saddle point cannot cross each other. Trajectories starting at the shaded region go to the left well, the others to the right well. Adding noise has the effect that the particles do not always stay in the same region of Figure 6.8, but they will mix up and may also cross a critical trajectory, so that particles leaving from the left well have a chance to reach the right well and vice versa.

For a detailed investigation of the problem we have to solve the corresponding FPE. Because the detailed-balance condition is valid here, the *stationary*

214

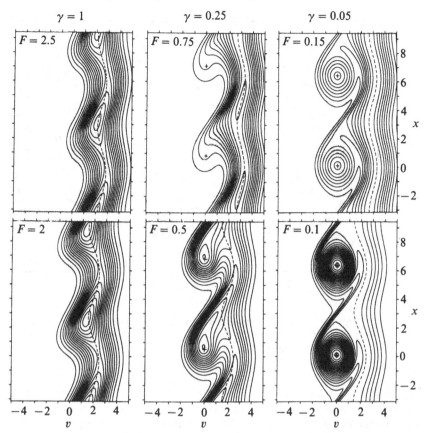

Figure 6.7. Contour lines of the stationary probability distribution of the Klein–Kramers equation (6.2.16) with (6.6.1) for the indicated F and γ values for $kT = 1$, $d = 1$. (Taken from Vollmer and Risken, 1979.)

solution of the corresponding FPE, (6.2.16), can be expressed by an integral. It is the well known Boltzmann distribution

$$W_{st}(x, v) = N \exp \left\{ - E/(kT) \right\}, \tag{6.7.2}$$

where E is the energy of the particle

$$E = v^2/2 + f(x). \tag{6.7.3}$$

It is much more difficult to obtain the *time-dependent solutions*. These solutions may be expanded into the eigenfunctions of the FPE. The eigenvalues determine the decay rates due to the approach to the stationary state. If the potential difference between the maximum of the potential (6.7.1) at $x = 0$ and its minimum at $x = x_{\pm}$ is much larger than kT, i.e.

$$\Delta E = d_2^2/(4d_4) \gg kT, \tag{6.7.4}$$

215

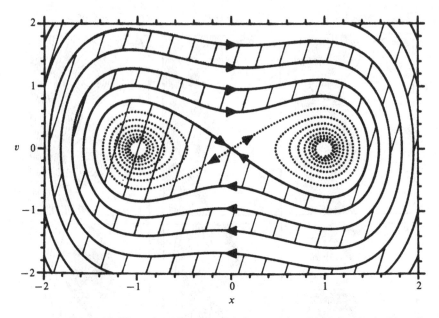

Figure 6.8. The trajectories of the noiseless equation (6.2.15) for the double-well potential (6.7.1), going through the saddle point for $d_4 = 1$, $d_2 = -1$, $\gamma = 0.1$. (Taken from Voigtlaender and Risken, 1985.)

the lowest nonzero eigenvalue becomes very small and is well separated from the higher ones. This lowest nonzero eigenvalue thus determines the decay of the eigenfunction to the stationary distribution (6.7.2) for large times. It is equal to the sum of the transition rates from the left to the right well and vice versa (for high and intermediate friction).

To apply the MCF method we follow the procedure in Section 6.5.3. Here we have to expand the coefficients $c_n(x, t)$ into a complete set with respect to x for which we choose the Hermite functions (6.5.9)

$$\phi^p(x) = h_n(x/(2^{1/2}x_s))/x_s^{1/2}. \tag{6.7.5}$$

Here x_s is a scaling length, by which the convergence of the expansion can be improved. For the expansion of the coefficients we use

$$c_n(x, t) = \exp\left\{-f(x)/(2kT)\right\} \sum_{p=0}^{\infty} c_n^p(t)\phi^p(x). \tag{6.7.6}$$

To get a symmetric form of the matrices $\mathbf{D}$ and $\hat{\mathbf{D}}$ the exponential factor was added in front of (6.7.6). Inserting (6.7.6) in Brinkman's hierarchy (6.5.11), multiplying the equation by $\exp\left\{f(x)/(2kT)\right\}\phi^q(x)$ and integrating the expression over x, we obtain the tridiagonal recurrence relation (6.4.6) with

(6.5.18), where the matrices $\mathbf{D}$ and $\hat{\mathbf{D}}$ are now given by

$$D^{pq} = \frac{(kT)^{1/2}}{2x_s} \left\{ \frac{d_4 x_s^4}{kT} [q(q-1)(q-2)]^{1/2} \delta^{p,q-3} \right.$$

$$+ \left[1 - \frac{3qd_4 x_s^4}{kT} - \frac{d_2 x_s^2}{kT} \right] q^{1/2} \delta^{p,q-1}$$

$$- \left[1 + \frac{3pd_4 x_s^4}{kT} + \frac{d_2 x_s^2}{kT} \right] p^{1/2} \delta^{p,q+1}$$

$$\left. - \frac{d_4 x_s^4}{kT} [p(p-1)(p-2)]^{1/2} \delta^{p,q+3} \right\} \tag{6.7.7a}$$

$$\hat{D}^{pq} = \frac{(kT)^{1/2}}{2x_s} \left\{ \frac{d_4 x_s^4}{kT} [q(q-1)(q-2)]^{1/2} \delta^{p,q-3} \right.$$

$$+ \left[1 + \frac{3qd_4 x_s^4}{kT} + \frac{d_2 x_s^2}{kT} \right] q^{1/2} \delta^{p,q-1}$$

$$- \left[1 - \frac{3pd_4 x_s^4}{kT} - \frac{d_2 x_s^2}{kT} \right] p^{1/2} \delta^{p,q+1}$$

$$\left. + \frac{d_4 x_s^4}{kT} [p(p-1)(p-2)]^{1/2} \delta^{p,q+3} \right\}. \tag{6.7.7b}$$

The *eigenvalues* of the Klein–Kramers equation are obtained as follows. Inserting the ansatz

$$\mathbf{c}_n(t) = \hat{\mathbf{c}}_n e^{-\lambda t} \tag{6.7.8}$$

into the tridiagonal recurrence relation (6.4.6) we get (6.4.8), where $\mathbf{Q}_n$ has to be replaced by

$$\mathbf{Q}_n \to \hat{\mathbf{Q}}_n(\lambda) = \mathbf{Q}_n + \lambda \mathbf{I}. \tag{6.7.9}$$

The same elimination procedure for the $\mathbf{c}_n$ with $n \geqslant 1$ as in Section 6.4 can now be performed leading to

$$\hat{\mathbf{R}}(\lambda)\hat{\mathbf{c}}_0 = 0 \tag{6.7.10}$$

where $\hat{\mathbf{R}}(\lambda)$ is given by (6.4.14) with $\mathbf{Q}_n$ replaced by $\hat{\mathbf{Q}}_n$. Because we are looking for nontrivial solutions of (6.7.10) the eigenvalues are the roots of

$$\mathrm{Det}\{\hat{\mathbf{R}}(\lambda)\} = 0. \tag{6.7.11}$$

In Figures 6.9(a, b) the eigenvalues are shown as a function of the damping constant. Notice that some of the eigenvalues are complex. In Figure 6.10 the lowest nonzero eigenvalue multiplied by the Arrhenius factor $\exp[\Delta E/(kT)]$ is shown as a function of the damping constant for various $\Delta E/(kT)$ ratios. This eigenvalue is real. Furthermore Kramers' escape-rate result as well as the

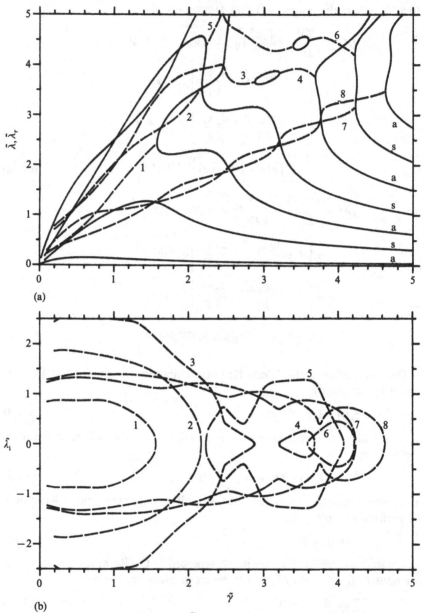

(a)

(b)

Figure 6.9. The eigenvalues $\tilde{\lambda} = \lambda/(|d_2|)^{1/2}$ of the Klein–Kramers equation (6.2.16) for the double-well potential (6.7.1) as a function of the damping constant $\tilde{\gamma} = \gamma/(|d_2|)^{1/2}$ for the barrier height $\Delta E/kT = d_2^2/4d_4 kT = 1$. The real parts $\tilde{\lambda}_r$ are given in (a), the imaginary ones $\tilde{\lambda}_i$ in (b). In (a) the real eigenvalues are shown by solid lines, the real parts of complex ones as broken lines. The numbers indicate corresponding real and imaginary parts. (Taken from Voigtlaender and Risken, 1985.)

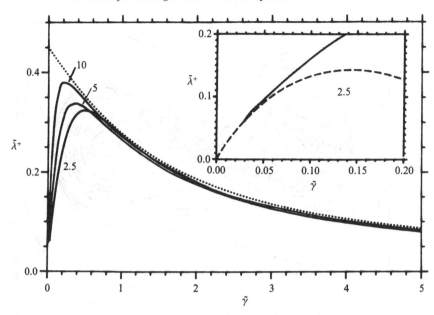

Figure 6.10. The lowest nonzero eigenvalues $\tilde{\lambda}_1 = \lambda_1/(|d_2|)^{1/2}$ multiplied by the Arrhenius factor, i.e. $\tilde{\lambda}^+ = \tilde{\lambda}_1 \exp(\Delta E/kT)$ of the Klein–Kramers equation (6.2.16) for the double-well potential (6.7.1) as a function of the damping constant $\tilde{\gamma} = \gamma/(|d_2|)^{1/2}$ for the barrier heights $\Delta E/kT = d_2^2/4d_4kT = 2.5$, 5 and 10. The high-friction asymptotic expression for high barriers is shown by a dotted line, the zero-friction-limit result by a broken line. (Taken from Voigtlaender and Risken, 1985.)

low-friction results are shown in this figure. Figure 6.11 shows the altitude chart of the eigenfunction belonging to the lowest nonzero eigenvalue. (For a possible normalization of this eigenfunction see appendix C of Risken and Voigtlaender (1985).) This solution can be interpreted as follows. Suppose that we finally have a small deviation of the distribution from its stationary form. It is then described by the stationary distribution plus a small correction of the form of this eigenfunction. This resulting distribution shows an increase in probability at one well and a decrease at the other, with corrections to the shape of the stationary distribution where the transitions between the wells take place.

Fourier transforms of the correlation functions can also be obtained by the MCF method. As explained in Voigtlaender and Risken (1985) one has to evaluate an MCF of the form (6.4.14a) with $\mathbf{Q}_n$ replaced by $\tilde{\mathbf{Q}}_n$ according to

$$\tilde{\mathbf{Q}}_n = \mathbf{Q}_n - i\omega \mathbf{I}. \tag{6.7.12}$$

In Figures 6.12(a, b) the real part $\tilde{K}'_{xx}$ of the one-sided Fourier transform of

219

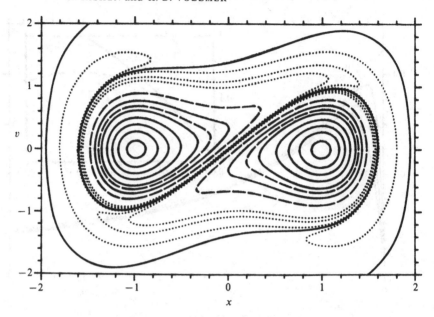

Figure 6.11. Altitude chart of the (antisymmetric) eigenfunction ψ_1 belonging to the lowest nonzero eigenvalue λ_1 of the Klein–Kramers equation (6.2.16) for the bistable potential (6.7.1) in suitable normalization for $\gamma/(|d_2|)^{1/2} = 0.25$, $\Delta E/kT = 1$. The contour lines are shown for heights $c = 0$, ± 0.1, ± 0.2,..., ± 0.6 (———), $c = \pm 0.05$, ± 0.15 (-----), and $c = \pm 0.001$, ± 0.005, ± 0.01 (·······). (Taken from Voigtlaender and Risken, 1985.)

the correlation function

$$\tilde{K}_{xx} = \tilde{K}'_{xx} - i\tilde{K}''_{xx} = \int_0^\infty \langle x(t)x(0)\rangle e^{-i\omega t}\, dt \tag{6.7.13}$$

is shown. It is quite remarkable that $\tilde{K}_{xx}$ can be obtained by MCFs for such low damping constants so that even the transition to the low-friction limit can be seen. The low-friction limit was investigated by Dykman, Soskin and Krivoglaz (1985), Onodera (1970), and the appendix of Voigtlaender and Risken (1985).

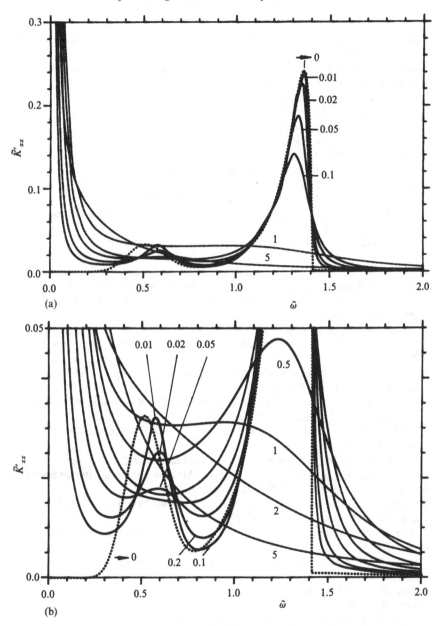

Figure 6.12. The real part of the half-sided Fourier transform of the correlation function of the coordinate, for the Brownian motion in the normalized ($d_2 = -1$, $d_4 = 1$) double-well potential (6.7.1) with $\Delta E/kT = 1/4kT = 5$, for the indicated friction constants (——) and for the zero-friction limit (······) as a function of $\tilde{\omega} = \omega/(|d_2|)^{1/2}$. The plot (b) is an enlarged version of (a). (Taken from Voigtlaender and Risken, 1985.)

References

Barone, A. and Paterno, G. 1982. *Physics and Applications of the Josephson Effect.* New York: Wiley.

Bharucha-Reid, A. T. 1960. *Elements of the Theory of Markov Processes and Their Applications.* New York: McGraw-Hill.

Bixon, M. and Zwanzig, R. 1971. *J. Stat. Phys.* 3, 245.

Blackmore, R. and Shizgal, B. 1985. *Phys. Rev. A* 31, 1855.

Blomberg, C. 1977. *Physica* 86A, 49.

Brinkman, H. C. 1956a. *Physica* 22, 29.

Brinkman, H. C. 1956b. *Physica* 22, 149.

Büttiker, M. and Landauer, R. 1984. *Phys. Rev. Lett.* 52, 1250.

Chandrasekhar, S. 1943. *Rev. Mod. Phys.* 15, 1.

Coffey, W. T. 1980. *Adv. Molec. Relax. Int. Proc.* 17, 169.

Coffey, W. T., Evans, M. W. and Grigolini, P. 1984. *Molecular Diffusion and Spectra.* New York: Wiley.

Denk, H. and Riederle, M. 1982. *J. Approx. Theory* 35, 355.

Dieterich, W., Fulde, P. and Peschel, I. 1980. *Adv. Phys.* 29, 572.

Drummond, P. D. and Walls, D. F. 1980. *J. Phys. A* 13, 725.

Dykman, M. I. and Krivoglaz, M. A. 1984. In *Physics Reviews (Soviet Scientific Reviews, Section A)* (I. M. Khalatnikov, ed.), p. 265–442. Chur-New York: Harwood Academic.

Dykman, M. I., Soskin, S. M. and Krivoglaz, M. A. 1985. *Physica* 133A, 53.

Fulde, P., Pietronero, L., Schneider, W. R. and Strässler, S. 1975. *Phys. Rev. Lett.* 35, 1776.

Gardiner, C. W. 1983. *Handbook of Stochastic Methods.* Springer Series in Synergetics 13. Berlin: Springer.

Geisel, T. 1979. In *Physics of Superionic Conductors* (M. B. Salamon, ed.), p. 201. Topics in Current Physics 15. Berlin: Springer.

Goel, N. S. and Richter-Dyn, N. 1974. *Stochastic Models in Biology.* New York: Academic Press.

Graham, R. 1973. In *Quantum Statistics in Optics and Solid-State Physics* (G. Höhler, ed.), pp. 1–97. Springer Tracts in Modern Physics 66. Berlin: Springer.

Grote, R. F. and Hynes, J. T. 1980. *J. Chem. Phys.* 73, 2715.

Guardia, E., Marchesoni, F. and San Miguel, M. 1984. *Phys. Lett.* 100A, 15.

Haken, H. 1975. *Rev. Mod. Phys.* 47, 67.

Haken, H. 1983a. *Synergetics.* Springer Series in Synergetics 1, 3rd edn. Berlin: Springer.

Haken, H. 1983b. *Advanced Synergetics.* Springer Series in Synergetics 20. Berlin: Springer.

Hänggi, P. 1984. *Phys. Lett.* 103A, 97.

Hänggi, P. 1986. *J. Stat. Phys.* 42, 105.

Hänggi, P. and Mojtabai, F. 1982. *Phys. Rev. A* 26, 1168.

Horsthemke, W. and Lefever, R. 1984. *Noise-Induced Transitions.* Springer Series in Synergetics 15. Berlin: Springer.

Itô, K. 1944. *Proc. Imp. Acad.* 20, 519.

Josephson, B. D. 1962. *Phys. Lett.* 1, 251.

Jung, P. and Risken, H. 1984. *Z. Phys. B* 54, 357.

Methods for solving Fokker–Planck equations

Jung, P. and Risken, H. 1985. *Z. Phys. B* **61**, 367.
Klein, O. 1922. *Arkiv for Mathematik, Astronomi och Fysik* **16**, (5), 1.
Kramers, H. A. 1940. *Physica* **7**, 284.
Krumhansl, J. A. and Schrieffer, J. R. 1975. *Phys. Rev. B* **11**, 3535.
Landauer, R. 1962. *J. Appl. Phys.* **33**, 2209.
Larson, R. S. and Kostin, M. D. 1978. *J. Chem. Phys.* **69**, 4821.
Larson, R. S. and Kostin, M. D. 1980. *J. Chem. Phys.* **72**, 1392.
Lax, M. 1960. *Rev. Mod. Phys.* **32**, 25.
Lax, M. 1966a. *Rev. Mod. Phys.* **38**, 359.
Lax, M. 1966b. *Rev. Mod. Phys.* **38**, 541.
Lindsey, W. C. 1972. *Synchronization Systems in Communication and Control.* Englewood Cliffs, NJ: Prentice Hall.
McConnel, J. 1980. *Rotational Brownian Motion and Dielectric Theory.* London: Academic.
Mangel, M. 1980. *J. Chem. Phys.* **72**, 6602.
Marchesoni, F., Grigolini, P. and Marin, P. 1982. *Chem. Phys. Lett.* **87**, 451.
Mel'nikov, V. I. 1984. *Sov. Phys. JETP* **60**, 380.
Mel'nikov, V. I. and Meshkov, S. V. 1986. *J. Chem. Phys.* **85**, 1018.
Michlin, S. G. and Smolizki, C. L. 1969. *Näherungsmethoden zur Lösung von Differential- und Integralgleichungen.* Leipzig: Teubner.
Northrup, S. C. and Hynes, J. T. 1978a. *J. Chem. Phys.* **69**, 5246.
Northrup, S. C. and Hynes, J. T. 1978b. *J. Chem. Phys.* **69**, 5261.
Northrup, S. C. and Hynes, J. T. 1980. *J. Chem. Phys.* **73**, 2700.
Onodera, Y. 1970. *Prog. Theor. Phys.* **44**, 1477.
Praestgard, E. and Van Kampen, N. G. 1981. *Molec. Phys.* **43**, 33.
Risken, H. 1984. *The Fokker–Planck Equation.* Springer Series in Synergetics 18. Berlin: Springer.
Risken, H. 1987. Solutions of Fokker–Planck equations in terms of matrix continued fractions. In *Lasers and Synergetics* (R. Graham and A. Wunderlin, eds.), p. 148. Springer Proceedings in Physics 19. Berlin: Springer.
Risken, H., Vogel, K. and Vollmer, H. D. 1988. *IBM J. Res. Develop.* **32**, 112.
Risken, H. and Voigtlaender, K. 1985. *J. Stat. Phys.* **41**, 825.
Risken, H. and Vollmer, H. D. 1979a. *Z. Phys. B* **33**, 297.
Risken, H. and Vollmer, H. D. 1979b. *Z. Phys. B* **35**, 177.
Risken, H. and Vollmer, H. D. 1982. *Mol. Phys.* **46**, 555.
Risken, H. and Vollmer, H. D. 1985. *Acta Physica Austr.* **57**, 75.
Risken, H., Vollmer, H. D. and Mörsch, M. 1981. *Z. Physik B* **40**, 343.
Schulten, K., Schulten, Z. and Szabo, A. 1981. *J. Chem. Phys.* **74**, 4426.
Shizgal, B. 1981. *J. Comp. Phys.* **41**, 309.
Skinner, J. L. and Wolynes, P. G. 1978. *J. Chem. Phys.* **69**, 2143.
Skinner, J. L. and Wolynes, P. G. 1980. *J. Chem. Phys.* **72**, 4913.
Solymar, L. 1972. *Superconductive Tunneling and Applications.* London: Chapman and Hall.
Stratonovich, R. L. 1958. *Radiotekhnica; elektronika* **3**, 497.
Stratonovich, R. L. 1963. *Topics in the Theory of Random Noise*, vol. I. New York: Gordon and Breach.
Stratonovich, R. L. 1967. *Topics in the Theory of Random Noise*, vol. II. New York: Gordon and Breach.

H. RISKEN and H. D. VOLLMER

Stratonovich, R. L. 1968. *Conditional Markov Processes and Their Application to the Theory of Optimal Control*, New York: Elsevier.

Titulaer, U. M. 1978. *Physica* **91A**, 321.

Toombs, G. A. 1978. *Phys. Rev.* **40C**, 181.

Uhlenbeck, G. E. and Ornstein, L. S. 1930. *Phys. Rev.* **36**, 823.

Van Kampen, N. G. 1981. *Stochastic Processes in Physics and Chemistry*. Amsterdam: North-Holland.

Visscher, P. B. 1976. *Phys. Rev. B* **14**, 347.

Viterbi, A. J. 1966. *Principles of Coherent Communication*. New York: McGraw-Hill.

Vogel, K. and Risken, H. 1987. *Optics Comm.* **62**, 45.

Voigtlaender, K. and Risken, H. 1984. *Chem. Phys. Lett.* **105**, 506.

Voigtlaender, K. and Risken, H. 1985. *J. Stat. Phys.* **40**, 397.

Vollmer, H. D. and Risken, H. 1979. *Z. Phys. B* **34**, 313.

Vollmer, H. D. and Risken, H. 1982. *Physica* **110A**, 106.

Vollmer, H. D. and Risken, H. 1983. *Z. Phys. B* **52**, 259.

Wang, M. C. and Uhlenbeck, G. E. 1945. *Rev. Mod. Phys.* **17**, 323.

Wax, N. (ed.) 1954. *Selected Papers on Noise and Stochastic Processes*. New York: Dover Publications.

Wilemski, G. 1976. *J. Stat. Phys.* **14**, 153.

Wonneberger, W. and Breymayer, H. J. 1981. *Z. Phys. B* **43**, 329.

7 Macroscopic potentials, bifurcations and noise in dissipative systems

ROBERT GRAHAM

7.1 Introduction

The two most prominent branches of macroscopic physics, reversible classical mechanics and reversible equilibrium thermodynamics, are both characterized by extremum principles – the principle of stationary action for a classical trajectory, and the principle of maximum entropy for thermodynamic equilibrium in a closed system. It is well known how both of these extremum principles arise by taking the macroscopic limit of more fundamental underlying microscopic theories, in which these extremum principles do not hold: quantum mechanics and statistical mechanics, respectively. In these microscopic theories the extremum principles are violated by the occurrence of fluctuations; quantum fluctuations by which finite probability amplitudes are assigned to nonclassical trajectories, and classical fluctuations which assign nonzero probabilities to states with less than the maximum entropy. Hence, there is a deep connection in physics between fluctuation phenomena and extremum principles, the former providing a mechanism allowing the system to explore a neighbourhood of the extremizing state and thereby to identify the extremum.

The last decade has seen a considerable increase of interest in non-equilibrium phenomena in macroscopic systems, as it was realized that simple and rather general mechanisms of self-organization exist in such systems (cf., e.g. Haken, 1977, 1983; Nicolis and Prigogine, 1977), leading to the selection, formation and competition of patterns in space and (or) time. The evolution equations governing these nonequilibrium phenomena neither belong to the realm of equilibrium thermodynamics, therefore thermodynamic extremum principles are not applicable, nor to the realm of reversible classical mechanics covered by the principle of least action. As a consequence an extremum principle allowing us to characterize the time-dependent or time-independent solutions of such evolution equations is not readily available.

Classical fluctuations, of course, also exist in nonequilibrium systems, and are either of thermal origin or due to other stochastic perturbations of a less universal nature, connected with the particular system under consideration. Clearly, therefore, just as in a state of thermodynamic equilibrium, not only the deterministic trajectory or state is explored in such nonequilibrium systems, but also a small neighborhood of it, and extremum principles analogous to the ones introduced before must also hold in this case. In cases where one is able to identify the minimized potential function, one may characterize stable steady states (attractors) of macroscopic evolution equations by an extremum condition, which is what one desires, e.g. in order to understand pattern selection in self-organizing systems.

Fluctuations play a role not only in identifying local extrema but also in choosing the absolute extremum among the local ones. The fluctuations must lift the system over barriers of the minimized potential in order to go from one local minimum to another; therefore, the quantitative question of the size of the fluctuations, the height of the potential barriers, and the relevant time scales is very important. In macroscopic nonequilibrium systems the different states among which the system may choose are macroscopically distinguishable, e.g. by different macroscopic patterns, presumably separated by macroscopic potential barriers. Thermal fluctuations are therefore generally much too small to allow a unique selection in a physically relevant time scale, except for systems which are very close to bifurcation points. Therefore, unless there are stronger stochastic fluctuations present one can only expect to find the steady state in a local minimum of some potential, and not necessarily in a global one. The problem of pattern selection can in this case not be solved completely by a minimum principle alone, because which local minimum is selected depends on the system's prehistory, just as for metastable states (e.g. diamond) in thermodynamic equilibrium.

After emphasizing similarities let us now also note the most important difference between macroscopic nonequilibrium systems and reversible classical mechanics or equilibrium thermodynamics, namely the lack of reversibility in time in the former. In a stochastic description this lack of reversibility manifests itself by the absence of the detailed balance property with respect to the microscopically defined transformation of time reversal. The possibility to distinguish reversible from irreversible processes is, of course, at the heart of equilibrium thermodynamics, and is already essential for the thermodynamic definition of entropy. It enables us to obtain entropy and all other thermodynamic potentials from the equations of state by quadratures, without ever having to refer to the statistical meaning of entropy. (Indeed, that statistical meaning was discovered by Boltzmann only long after equilibrium thermodynamics had been established.)

In the absence of detailed balance in nonequilibrium systems we cannot hope that a similar short cut to the potential function should exist there as well. Therefore, in order to determine the potential function to be minimized, one is forced to consider its statistical meaning, i.e. to solve a problem of nonequilibrium statistical physics, at least approximately. This is often difficult and often requires a considerable knowledge of the behavior of the nonequilibrium system. In a given application this may, of course, be a serious disadvantage. However, a similar situation is not at all unusual in the statistical calculation of equilibrium entropies. (For example, the physical nature of the superconducting state had to be understood completely before its thermodynamic potential could be calculated.)

In equilibrium statistical mechanics detailed balance enforces a compatibility condition between the dynamics of a system and its thermodynamic potential (potential conditions). The absence of this property has important consequences for the analytical properties of potential functions in nonequilibrium systems. It turns out that the potential functions are continuous and differentiable near attractors, but may be only piecewise differentiable at a finite distance from attractors with surfaces of non-differentiability piling up around saddles.

The purpose of the present chapter is to review the properties and to present methods of construction of macroscopic potentials for nonequilibrium systems (nonequilibrium potential Φ). The presentation closely follows and extends two recent articles by the author (Graham, 1987a, b). The detailed plan of the article is as follows. In Section 7.2 some basic properties of stochastic equilibrium thermodynamics are recalled, which are needed as a background. In Section 7.3 nonequilibrium potentials are defined generally and their governing equations are obtained. Sections 7.4–7.6 give some elementary examples, where Φ can be obtained exactly, without difficulty. In Section 7.7 the influence of the noise bandwidth on Φ is determined asymptotically for small bandwidth. This section also gives a first example of how Φ can be obtained in a nontrivial case by a systematic power series expansion in a small parameter. Further examples of this kind are treated in Sections 7.8 and 7.9, where nonequilibrium potentials are explicitly constructed for the Van der Pol oscillator and a case of codimension-two bifurcation. In Section 7.10 it is shown for the example of codimension-two bifurcation how nonequilibrium potentials can be used to calculate mean first passage times out of the domain of an attractor due to weak noise. In Section 7.11 we begin the discussion of the peculiar analytical properties which nonequilibrium potentials may have due to their lack of simple time-reversal symmetries; Section 7.12 gives a simple example. In Section 7.13 we briefly review available methods for identifying special cases with smooth nonequilibrium potentials. The general case is taken up again in Section 7.14, where simple asymptotic properties of functional integrals are used to formulate general integral expressions for Φ which show that Φ, in general, is not everywhere continuously differentiable. In Sections 7.15 and 7.16 further applications are given, in Section 7.15 to bifurcations in current driven resistively shunted Josephson junctions and, finally, in Section 7.16 to the strange attractor in the Lorenz model.

7.2   Stochastic equilibrium thermodynamics

Later in this chapter, the basic structure of equilibrium thermodynamics will be generalized to more general nonequilibrium systems. Therefore we start by recalling equilibrium thermodynamics (De Groot and Mazur, 1962; Landau and Lifshitz, 1958). We shall adopt a mezoscopic point of view, in between the microscopic viewpoint of statistical mechanics and the macroscopic viewpoint of thermodynamics proper (Green, 1952). On this level a system is described by macroscopic variables q^v, $v = 1, \ldots, n$, fluctuating around their mean values due to thermodynamic fluctuations. (For simplicity we shall always consider discrete sets of variables.) The time-dependence of these fluctuations follows first order stochastic differential equations (Gardiner, 1983; Haken, 1977, 1983; Hänggi and Thomas, 1982; Van Kampen, 1981; Risken, 1983; Schuss, 1980; Stratonovich, 1963, 1967) of the form

$$\dot{q}^v = K^v(q) + g_i^v(q)\xi^i(t) \tag{7.2.1}$$

stated in the sense of Itô, where the $\xi^i(t)(i = 1, \ldots, m \leqslant n)$ are mutually statistically independent sources of Gaussian white noise

$$\langle \xi^i(t) \rangle = 0, \langle \xi^i(t)\xi^j(t') \rangle = \eta \delta^{ij} \delta(t - t'). \tag{7.2.2}$$

Repeated indices are summed over. The parameter η is used as a formal device to discuss the weak noise limit $\eta \to 0$. In the present case it may be identified with Boltzmann's constant k_B. There exists a stochastically equivalent Fokker–Planck equation corresponding to (7.2.1) of the form

$$\frac{\partial P}{\partial t} = -\frac{\partial}{\partial q^v} K^v(q)P + \tfrac{1}{2}\eta \frac{\partial^2}{\partial q^v \partial q^\mu} Q^{v\mu}(q)P. \tag{7.2.3}$$

$P(q, t)$ is the probability density to observe q at time t and

$$Q^{v\mu}(q) = g_i^v(q)g_i^\mu(q) \tag{7.2.4}$$

is the non-negative, symmetric matrix of transport coefficients of the system. The stochastic process (7.2.1) is assumed to be ergodic, i.e. for $t \to \infty$ the solutions of (7.2.3) approach the equilibrium distribution

$$P_\infty(q)d^n q = N \exp\left(-\frac{\Phi(q)}{\eta} \right)dV_q, \tag{7.2.5}$$

where N is a normalization constant. $\Phi(q)$ is the negative of the entropy of the total system formally closed by including all reservoirs, and constrained in such a way that the variables q^v are fixed at arbitrary sharp values, and

$$dV_q = \frac{d^n q}{(g(q))^{1/2}} \tag{7.2.6}$$

is the invariant volume element in q-space (which reduces to $d^n q$ in extensive variables (Grabert, Graham and Green, 1980)). The function $g(q)$ is the determinant of the contravariant metric tensor describing the adopted metric in the space of the macroscopic variables. In the Euclidean metric $g = 1$.

Equation (7.2.5) may be considered as the statistical definition of entropy $-\Phi(q)$ in the present mezoscopic formulation. Due to the necessity to artificially fix the variables q^{ν} when defining $\Phi(q)$, this quantity is often called a coarse-grained thermodynamic potential. It does not share the usual convexity properties of a thermodynamic potential defined microscopically as the logarithm of a partition function in which all fluctuations (including those of the q^{ν}) are summed over.

Independent from (7.2.5), there is a second way to define the negative entropy $\Phi(q)$ in equilibrium thermodynamics, which is based on reversibility properties. It proceeds by splitting the drift $K^{\nu}(q)$ into its reversible part $r^{\nu}(q)$ and irreversible part $d^{\nu}(q)$ under the (microscopically defined) transformation of time reversal

$$K^{\nu}(q) = d^{\nu}(q) + r^{\nu}(q). \tag{7.2.7}$$

The reversible part induces a volume preserving flow in q-space and is therefore called conservative

$$(g(q))^{1/2} \frac{\partial}{\partial q^{\nu}} \frac{r^{\nu}(q)}{(g(q))^{1/2}} = 0. \tag{7.2.8}$$

The irreversible part $d^{\nu}(q)$ and the transport matrix $Q^{\nu\mu}(q)$ are sufficient to define $\phi(q)$ by

$$d^{\nu}(q) - \frac{\eta}{2} \frac{\partial Q^{\nu\mu}(q)}{\partial q^{\mu}} = -\tfrac{1}{2} Q^{\nu\mu}(q) \frac{\partial \Phi(q)}{\partial q^{\mu}}. \tag{7.2.9}$$

If the inverse $Q_{\nu\mu}$ of $Q^{\nu\mu}$ exists, we may express $\phi(q)$ by a quadrature

$$\Phi(q) - \Phi(q_0) = -2 \int_{q_0}^{q} dq^{\nu} Q_{\nu\mu}(q) \left[d^{\mu}(q) - \frac{\eta}{2} \frac{\partial Q^{\mu\lambda}(q)}{\partial q^{\lambda}} \right]. \tag{7.2.10}$$

The reversible part $r^{\nu}(q)$ leaves $\Phi(q)$ invariant

$$r^{\nu}(q) \frac{\partial \Phi(q)}{\partial q^{\nu}} = 0, \tag{7.2.11}$$

i.e. $r^{\nu}(q)$ has the form

$$r^{\nu}(q) = -\tfrac{1}{2} A^{\nu\mu}(q) \frac{\partial \Phi(q)}{\partial q^{\mu}} \tag{7.2.12}$$

with some antisymmetric matrix

$$A^{\nu\mu}(q) = - A^{\mu\nu}(q). \tag{7.2.13}$$

For $\eta \to 0$ (7.2.1) reduces to $\dot{q}^{\nu} = K^{\nu}(q)$, and the second law

$$\frac{d\Phi(q)}{dt} = K^{\nu}(q) \frac{\partial \Phi(q)}{\partial q^{\nu}} = -\tfrac{1}{2} Q^{\nu\mu}(q) \frac{\partial \Phi}{\partial q^{\nu}} \frac{\partial \Phi}{\partial q^{\mu}} \leqslant 0 \tag{7.2.14}$$

follows from (7.2.7), (7.2.9) and (7.2.11).

ROBERT GRAHAM

Parenthetically we note that for $\eta > 0$ the second law is not obeyed with respect to the coarse-grained negative entropy $\Phi(q)$, since fluctuations may violate it in the mezoscopic description. The second law for $\eta > 0$ only holds for the fully averaged convex Gibbsian entropy defined by

$$S(t) = -\eta \int d^n q \, P(q,t) \ln \frac{P(q,t)}{P_\infty(q)} + \text{const.} \qquad (7.2.15)$$

For a proof of the H-theorem for the entropy (7.2.15) see Lebowitz and Bergmann (1957).

The definition of $\Phi(q)$ provided by (7.2.9) or (7.2.10) gives back the $\Phi(q)$ as defined by (7.2.5). This is easily seen by inserting (7.2.7) with (7.2.9) in (7.2.3) and solving for P_∞ by using (7.2.8) and (7.2.11). But why does (7.2.9) hold for any choice of $\Phi(q)$? The reason is the principle of detailed balance of thermodynamic equilibrium, described by (7.2.5), with respect to the same time-reversal transformation as used in (7.2.7) to define $d^\nu(q)$. From this principle the relation (7.2.9) between $d^\nu(q)$ and $\Phi(q)$ follows as a compatibility condition (cf., e.g., Graham, 1973a).

The central problem now to be discussed in the following is the generalization of at least a part of this structure to evolution equations of nonequilibrium systems for which restrictions due to detailed balance with respect to microscopical time reversal do not apply.

7.3   Nonequilibrium potentials

Nonequilibrium potentials have their origin in independent developments in physics and mathematics, at least dating back to the 1960s. Haken (1964, 1970) in his theory of lasers made use of free-energy-like potentials, and the physical use of such potentials was later generalized, e.g. in Graham (1973a, b), Graham and Haken (1971a, b), Kitahara (1975), Kubo, Matsuo and Kitahara (1973), and Tomita and Tomita (1974). Independent mathematical work using such potentials was carried out in Cohen and Lewis (1967), Freidlin and Ventzel (1984), Jona-Lasinio (1985), Ludwig (1974, 1975), and Ventzel and Freidlin (1970).

Let us consider a general set of evolution equations, which we again write in the form (7.2.1)

$$\dot{q}^\nu = K^\nu(q) + g_i^\nu(q)\xi^i(t) \qquad (7.3.1)$$

with the further definitions (7.2.2)–(7.2.4) implied. We assume the class of evolution equations to be restricted in such a way that a unique steady state distribution $P_\infty(q)$ is approached for $t \to \infty$ from a given initial probability density $P(q,0)$. Then a statistical definition of a nonequilibrium potential function is made by

$$\Phi(q) = -\lim_{\eta \to 0} \eta \ln P_\infty(q). \qquad (7.3.2)$$

230

In other words, a steady state distribution $P_\infty(q, \eta)$ for small η of the form

$$P_\infty(q, \eta) = Z(q) \exp\left[-\frac{\Phi(q)}{\eta} + O(\eta) \right] \tag{7.3.3}$$

is anticipated, and the weak noise limit $\eta \to 0$ is used to extract the form of $\Phi(q)$. The prefactor $Z(q)$ is then defined as the limit

$$\ln Z(q) = \lim_{\eta \to 0}\left[\ln P_\infty(q, \eta) + \frac{1}{\eta}\Phi(q) \right]. \tag{7.3.4}$$

Equations governing $\Phi(q)$ and $Z(q)$ are easily obtained from the Fokker–Planck equation (7.2.3). The latter may be written in the form

$$\frac{\partial g^\nu(q, \eta)}{\partial q^\nu} = 0, \tag{7.3.5}$$

with the probability current density

$$g^\nu(q, \eta) = -K^\nu(q) P_\infty(q, \eta) + \frac{\eta}{2}\frac{\partial}{\partial q^\mu} Q^{\nu\mu}(q) P_\infty(q, \eta). \tag{7.3.6}$$

$\Phi(q)$ satisfies a nonlinear first order partial differential equation independent of $Z(q)$

$$K^\nu(q)\frac{\partial\Phi}{\partial q^\nu} + \tfrac{1}{2}Q^{\nu\mu}(q)\frac{\partial\Phi}{\partial q^\nu}\frac{\partial\Phi}{\partial q^\mu} = 0, \tag{7.3.7}$$

while $Z(q)$ satisfies a linear first order partial differential equation depending on $\Phi(q)$

$$\left(K^\nu + Q^{\nu\mu}\frac{\partial\Phi}{\partial q^\mu} \right)\frac{\partial Z}{\partial q^\nu}$$

$$+ \left(\frac{\partial K^\nu}{\partial q^\nu} + \frac{\partial Q^{\nu\mu}}{\partial q^\mu}\frac{\partial\Phi}{\partial q^\mu} + \tfrac{1}{2}Q^{\nu\mu}\frac{\partial^2\Phi}{\partial q^\nu\partial q^\mu} \right)Z = 0. \tag{7.3.8}$$

We now define the streaming velocity $r^\nu(q, \eta)$ of the probability flow in the steady state by

$$r^\nu(q, \eta) = \frac{g^\nu(q, \eta)}{P_\infty(q, \eta)}. \tag{7.3.9}$$

Taking the limit $\eta \to 0$, and using (7.3.2) and (7.3.6), we find

$$r^\nu(q) = \lim_{\eta \to 0} r^\nu(q, \eta) = K^\nu(q) + \tfrac{1}{2}Q^{\nu\mu}(q)\frac{\partial\Phi}{\partial q^\mu}, \tag{7.3.10}$$

which we adopt as a new definition of $r^\nu(q)$ which is independent of time reversal and therefore more general than the definition given by (7.2.7). The comparison of (7.3.10) and (7.3.7) yields

$$r^\nu(q)\frac{\partial\Phi(q)}{\partial q^\nu} = 0. \tag{7.3.11}$$

ROBERT GRAHAM

Hence, the statistical definition of $\Phi(q)$ by (7.3.2) allows one to split the drift $K^v(q)$ appearing in (7.3.1) into two parts like in (7.2.7)

$$K^v(q) = d^v(q) + r^v(q),\qquad(7.3.12)$$

where the part r^v conserves the potential $\Phi(q)$ just as in (7.2.10), while the other part d^v, according to (7.3.10), may be written as

$$d^v(q) = -\tfrac{1}{2}Q^{v\mu}(q)\frac{\partial\Phi}{\partial q^\mu}\qquad(7.3.13)$$

just as in (7.2.9) for $\eta \to 0$.

It follows from (7.3.11)–(7.3.13) that (7.2.14) is still satisfied. Equation (7.3.2) and the normalizability condition $P_\infty(q,\eta)$ ensure that $\Phi(q)$ is bounded from below. Therefore the monotonic decrease of Φ under the deterministic dynamics $\dot{q}^v = K^v(q)$ implies that Φ is a Lyapunov function of this dynamics which takes a minimum on attractors. Moreover, Φ must be constant on all extended attractors like limit cycles, higher dimensional tori or even strange attractors, since these are ω-limit sets describing dynamical steady states of $\dot{q}^v = K^v(q)$ for $t \to \infty$ in which $d\Phi(q)/dt = 0$.

In summary, an important part of the structure of equilibrium thermodynamics may be transferred to general evolution equations under the assumptions made. This is the part based on the statistical definition of Φ. However, r^v and d^v can, in general, no longer be identified with the reversible and irreversible components of K^v, i.e. this part of equilibrium thermodynamics cannot be generalized. In general, Φ and r^v can only be obtained via their statistical definitions, i.e. by solving (7.3.7), a problem considered later in this chapter. However, there are simple but important special cases where either the reversibility properties of equilibrium thermodynamics are preserved in the nonequilibrium case, or where it is possible to specify a new macroscopically defined transformation of time reversal with respect to which r^v and d^v are reversible and irreversible, respectively (Graham, 1980). In these cases, the analogy of the nonequilibrium dynamics to equilibrium thermodynamics is complete, r^v and d^v are easily identified by their time-reversal signatures, and Φ is directly obtained from (7.3.13). Examples will be discussed next.

7.4 Neighborhood of monostable steady states

As the simplest possible first example for the construction of Φ in a nonequilibrum system (Graham, 1973b, 1981), assume that $q^v = 0$, $v = 1,\dots,n$ is a stable fixed point of the drift K^v, expand around this point:

$$K^v(q) = B^v_\mu q^\mu + O((q)^2)$$
$$Q^{v\mu}(q) = Q^{v\mu} + O(q)\qquad(7.4.1)$$
$$g(q) = 1 + O(q),$$

232

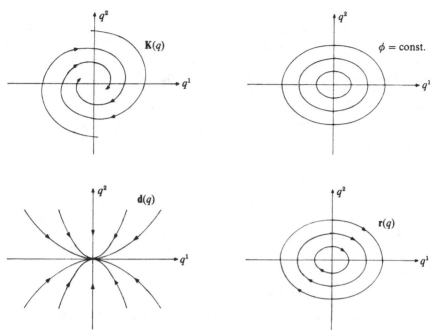

Figure 7.1. The draft **K** and its two components **r** and **d** and the equipotential surfaces of ϕ near a stable fixed point. (Schematic plot, after Graham, 1981.)

and enter (7.3.7) with the ansatz

$$\Phi(q) = \tfrac{1}{2}(\sigma^{-1})_{\nu\mu} q^{\nu} q^{\mu} + \mathrm{O}((q)^3). \tag{7.4.2}$$

One finds that $\sigma^{\nu\mu}$ must satisfy

$$B^{\alpha}_{\mu} \sigma^{\mu\beta} + B^{\beta}_{\mu} \sigma^{\alpha\mu} = -Q^{\alpha\beta}, \tag{7.4.3}$$

which has a solution. The two parts d^{ν} and r^{ν} of K^{ν} are now easily found. One also easily checks that in this simple case r^{ν} even satisfies (7.2.8) neglecting the corrections in (7.4.1) and (7.4.2).

In Figure 7.1 an example is shown with $n = 2$. Thus, for linear evolution equations the potential Φ is a quadratic form and can be constructed without difficulty. This shows that Φ exists as a smooth function at least in a small neighborhood of the attractors. Equation (7.3.8) is also easily solved in the present case by $Z = \text{const}$. The same result is, of course, obtained by solving the Fokker–Planck equation of the linearized Gaussian process near $q = 0$ and using (7.2.3).

7.5 Hopf and pitchfork bifurcations

As another simple example we consider a system in the neighborhood of a supercritical Hopf bifurcation. The single-mode laser near threshold is an important case (DeGiorgio and Scully, 1970; Graham, 1973a, b; Graham and Haken, 1970; Haken, 1964, 1970, 1977, 1983; Risken, 1965). Two results of center manifold theory (Guckenheimer and Holmes, 1983) tell us that (i) a reduced description of the system in a two-dimensional configuration space (center manifold) is possible on a sufficiently long time scale $\Delta t = O(1/\varepsilon)$, where $\varepsilon = |\mu - \mu_c|/\mu_c \ll 1$ and μ, μ_c are, respectively, the control parameter and its critical value where the Hopf bifurcation occurs, and (ii) the deterministic dynamics on the two-dimensional manifold is governed by an amplitude equation, which for sufficiently small amplitudes takes the normal form

$$\dot{\beta} = [(\mu - \mu_c) - i\omega]\beta - (b - i\Omega)|\beta|^2\beta + O(\beta^5). \tag{7.5.1}$$

Here β is the complex amplitude of the oscillation appearing at the bifurcation point, and ω, b, and Ω are real parameters. If the set of stochastic differential equations describing the system is known, then, for $\varepsilon \ll 1$, (7.5.1) can be derived from them together with an additive complex noise term $(2\eta Q)^{1/2}\xi(t)$, with the only nonvanishing cumulant

$$\langle \xi(t)\xi^*(t') \rangle = \delta(t - t'). \tag{7.5.2}$$

In addition, the parameters ω, b and Ω are related to the parameters of the model in such a derivation. Like the deterministic part of the amplitude equation, the noise part also is only a lowest order contribution, which would receive corrections if the expansion is carried to higher orders in ε. Super-critical Hopf bifurcations require that $b > 0$, which we shall assume from now on. (A subcritical Hopf bifurcation occurs if $b < 0$ and would require that we carry the expansion at least to fifth order terms in β in order to achieve overall stability.) A potential satisfying (7.3.7) can immediately be written down for (7.5.1)

$$\Phi = \frac{1}{Q}\left[-(\mu - \mu_c)|\beta|^2 + \frac{b}{2}|\beta|^4 \right]. \tag{7.5.3}$$

The prefactor Z determined from (7.2.8) is simply $Z = \text{const.}$ From

$$K_\beta = d_\beta + r_\beta, \tag{7.5.4}$$

with K_β defined by the right hand side of (7.5.1), we find

$$d_\beta = -Q\frac{\partial\Phi}{\partial\beta^*}, \quad r_\beta = -i\omega\beta + i\Omega|\beta|^2\beta. \tag{7.5.5}$$

A reasonable choice of the metric on the center manifold near $\beta = 0$ is the Euclidean metric, where $g(\beta, \beta^*) = 1$. We note that r_β as given by (7.5.5) even satisfies (7.2.8) in addition to (7.2.11), i.e. the field lines of r_β, r_{β^*} are closed (in fact circles). There exists a transformation

$$t \to -t, \quad \beta \to \beta^*, \quad i \to -i, \tag{7.5.6}$$

which we call time reversal, under which d_β transforms as an irreversible drift, i.e. with opposite sign compared to $\dot\beta$, and r_β transforms as a reversible drift, i.e. in the same way as $\dot\beta$. In the example of the laser the transformation (7.5.6) coincides with the microscopically defined transformation of time reversal, but in general this need not be the case. According to (7.5.5) potential conditions hold for the part of the drift K_β that is irreversible under the transformation (7.5.6), and consequently there is detailed balance in the steady state with respect to this time reversal transformation. The analogy of (7.5.3) to the Landau theory of phase transitions is apparent (DeGiorgio and Scully, 1970; Graham and Haken, 1970).

A very similar analysis applies to pitchfork bifurcations. An example is the Bénard instability of thermal convection (Cross, 1982; Graham, 1973c, 1974, 1975; Haken, 1973, 1975; Newel and Whitehead, 1969; Schlüter, Lortz and Busse, 1965; Segel, 1969; Swift and Hohenberg, 1977). In spatially strongly confined systems (small aspect ratio) the center manifold is one-dimensional, i.e. β is real and $\omega = \Omega = 0$ in (7.5.1) and (7.5.3). In very large aspect ratio systems β is again complex and must be treated as spatially varying in the horizontal (x, y)-plane (Newel and Whitehead, 1969; Segel, 1969). From the point of view of nonequilibrium potentials this case has been considered in Cross (1982), Graham (1973c, 1974, 1975), Haken (1973, 1975) and Swift and Hohenberg (1977). Nonequilibrium potentials for Hopf and pitchfork bifurcations in chemical systems including space dependence have been derived in Fraikin and Lemarchand (1985) and in Lemarchand (1984) and Lemarchand and Nicolis (1984), respectively.

7.6 Optical bistability

Optical bistability is another example where, at least in a special case, it is easy to write down a nonequilibrium potential (Graham and Schenzle, 1983). Here, two attractors compete, and a nonequilibrium potential permits a quantification and comparison of the stability of the two attractors according to the relative depth of their potential wells.

The evolution equations, in this case, are given by (Graham and Schenzle, 1983)

$$\dot\beta = K(\beta, \beta^*) + (2\eta Q)^{1/2}\xi(t)$$

$$K(\beta, \beta^*) = -(1 + i\delta)\beta - \Gamma^2\frac{1 + i\Delta}{1 + |\beta|^2}\beta + E_0 \tag{7.6.1}$$

where β is a complex field amplitude, $\xi(t)$ is the same as in (7.5.2), Γ, δ, Δ and Q are real parameters, and E_0 is a real externally applied field amplitude which serves as a control parameter. Equation (7.3.7) takes the form

$$K\frac{\partial\Phi}{\partial\beta} + K^*\frac{\partial\Phi}{\partial\beta^*} + 2Q\frac{\partial\Phi}{\partial\beta}\frac{\partial\Phi}{\partial\beta^*} = 0. \tag{7.6.2}$$

235

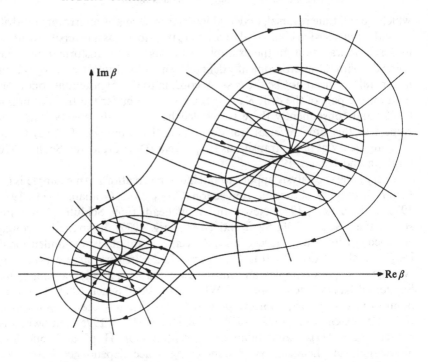

Figure 7.2. Equipotential surfaces $\phi = \text{const.}$ of (7.6.4) and field lines of $\mathbf{r}$, (7.6.5), and field lines of $\mathbf{d}$, (7.6.5), in the complex β-plane. Regions where ϕ is below its saddle point value are shaded.

For $E_0 = 0$ the solution of (7.6.2) is

$$\Phi_0 = \frac{1}{Q}[|\beta|^2 + \Gamma^2 \ln(1 + |\beta|^2)]. \tag{7.6.3}$$

The remarks concerning time reversal made after (7.5.5) also apply here. For $E_0 \neq 0$ and $\delta = \Delta$ an exact solution of (7.6.2) was found by Graham and Schenzle (1983), namely

$$\Phi = \Phi_0 - \frac{E_0}{Q(1 + \delta^2)}[(1 + i\delta)\beta + (1 - i\delta)\beta^*] \tag{7.6.4}$$

with

$$d_\beta = -Q\frac{\partial\Phi}{\partial\beta^*}$$

$$r_\beta = -i\delta\left(1 + \frac{\Gamma^2}{1 + |\beta|^2}\right)\beta + i\delta\frac{1 - i\delta}{1 + \delta^2}E_0. \tag{7.6.5}$$

Choosing a Euclidean metric in the complex β-plane we have $g = 1$ in (7.2.6) and (7.2.8), and (7.2.8) is satisfied in addition to (7.2.11), i.e. the field lines of r_β, r_{β^*} are closed curves. The prefactor Z obtained from (7.2.8) is simply $Z = \text{const.}$ In Figure 7.2 the equipotential surfaces of and the field lines of (7.6.4)

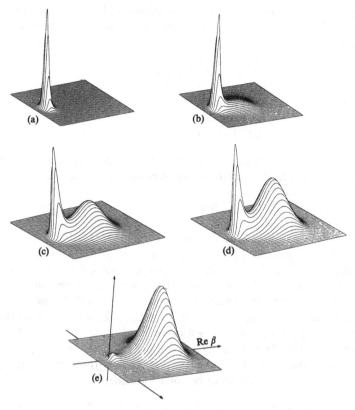

Figure 7.3. The probability density (7.2.5) for the potential (7.6.4) for $\delta = 0$, $\eta Q = 1$, $\Gamma = 5$, $E_0 = 6$ (*a*), 9.3 (*b*), 10.6 (*c*), 10.8 (*d*), 11.5 (*e*). (From Schenzle, 1981.)

and (7.6.5) are sketched. The relative depths of the potential wells can now serve as a measure for relative stability of their minima. The special case $\delta = \Delta = 0$ of (7.6.4) was first obtained by Schenzle and Brand (1978). In Figures 7.3(a–e) the probability density (7.2.5) is plotted for this case for increasing values of E_0 (Schenzle, 1981), displaying the regime of bistability. If the depth of the potential wells is large compared to η both local minima are stable on physically acceptable time scales. The knowledge of Φ, (7.6.4), allows a calculation of the mean first passage time from one well to the other (cf. below). The explicit calculation for the present example can be found in Talkner and Hänggi (1984).

7.7   Influence of colored noise

Let us assume that a stochastic model of the form (7.3.1), subject to Gaussian white noise, has the nonequilibrium potential Φ and the prefactor Z satisfying (7.3.7) and (7.3.8), respectively. What is the effect on the shape of the potential if the white noise is changed into colored noise? For the sake of concreteness we shall assume that the colored noise is of Ornstein–Uhlenbeck type with bandwidth ε^{-2}

$$\langle y^\nu(t) y^\mu(t') \rangle = \frac{\eta}{2} Q^{\nu\mu} \exp\left(-\frac{|t-t'|}{\varepsilon^2} \right) \tag{7.7.1}$$

satisfying the equation of motion

$$\dot{y}^\nu = -\frac{1}{\varepsilon^2} y^\nu + \frac{\eta^{1/2}}{\varepsilon} g_i^\nu \xi^i(t). \tag{7.7.2}$$

For simplicity we shall here assume g_i^ν to be independent of q. The colored noise y is assumed coupled to the original system according to

$$\dot{q}^\nu = K^\nu(q) + \frac{1}{\varepsilon} y^\nu(t). \tag{7.7.3}$$

The normalizations are chosen in such a manner that the original white noise process is recovered for $\varepsilon \to 0$. The question raised above was answered by Schenzle and Tél (1985) who, for small ε, derived a Fokker–Planck equation for the reduced single-time probability density of q alone and solved for the time-independent probability density. Here we shall follow a different and somewhat more direct approach, bypassing the derivation of the reduced process and the nontrivial question to what extent it is non-Markovian, by directly solving (7.3.7) and (7.3.8) for the enlarged process (q, y). From the nonequilibrium potential $\Phi(q, y)$ the potential $\bar{\Phi}(q)$ for q alone is then extracted at the end. The equations to be solved are then

$$\frac{1}{2} Q^{\nu\mu} \frac{\partial\Phi}{\partial y^\nu} \frac{\partial\Phi}{\partial y^\mu} - y^\nu \frac{\partial\Phi}{\partial y^\nu} + \varepsilon y^\nu \frac{\partial\Phi}{\partial q^\nu} + \varepsilon^2 K^\nu \frac{\partial\Phi}{\partial q^\nu} = 0 \tag{7.7.4}$$

and

$$\left(-y^\nu + Q^{\nu\mu} \frac{\partial\Phi}{\partial y^\mu} \right) \frac{\partial \ln Z}{\partial y^\nu} + (\varepsilon y^\nu + \varepsilon^2 K^\nu) \frac{\partial \ln Z}{\partial q^\nu}$$
$$+ \frac{1}{2} Q^{\nu\mu} \frac{\partial^2 \Phi}{\partial y^\nu \partial y^\mu} - n + \varepsilon^2 \frac{\partial K^\nu}{\partial q^\nu} = 0. \tag{7.7.5}$$

In (7.7.5) n denotes the dimension of the q-space. Since we are interested in modifications of the white noise case, ε will be considered as a small parameter and the solutions of (7.7.4) and (7.7.5) can be constructed as the power series

$$\Phi = \Phi_0 + \varepsilon \Phi_1 + \varepsilon^2 \Phi_2 + \cdots$$
$$\ln Z = \psi_0 + \varepsilon \psi_1 + \varepsilon^2 \psi_2 + \cdots. \tag{7.7.6}$$

We briefly indicate how this construction can be carried out step by step and present Φ in order ε^2 and $\ln Z$ in order ε. The equation for Φ_0 in order ε^0 has the general solution

$$\Phi_0 = Q_{\nu\mu} y^\nu y^\mu + \tilde{\Phi}_0(q), \tag{7.7.7}$$

where $Q_{\nu\mu}$ is the inverse of $Q^{\nu\mu}$ whose existence we must assume here, and $\tilde{\Phi}_0(q)$ is still an arbitrary function of q. Such arbitrary functions appear in any order ε^l and are determined by solubility conditions for the equations appearing in order ε^{l+2}. The equation for Φ_1 in order ε^1 has the general solution

$$\Phi_1 = - y^\nu \frac{\partial \tilde{\Phi}_0(q)}{\partial q^\nu} + \tilde{\Phi}_1(q), \tag{7.7.8}$$

where $\tilde{\Phi}_1(q)$ is again not yet fixed. The equation for Φ_2 in order ε^2 has the general solution

$$\Phi_2 = - y^\nu \frac{\partial \tilde{\Phi}_1(q)}{\partial q^\nu} + \tfrac{1}{2} y^\nu y^\mu \frac{\partial^2 \tilde{\Phi}_0(q)}{\partial q^\nu \partial q^\mu} + \tilde{\Phi}_2(q)$$

$$+ \left[\tfrac{1}{2} Q^{\nu\mu} \frac{\partial \tilde{\Phi}_0}{\partial q^\nu} \frac{\partial \tilde{\Phi}_0}{\partial q^\mu} + K^\nu(q) \frac{\partial \tilde{\Phi}_0}{\partial q^\nu} \right] \ln |y^\kappa|. \tag{7.7.9}$$

Now the first solubility condition appears from the following requirements. In order to avoid the unphysical logarithmic singularity of Φ_2 at $y = 0$ the bracket $[\cdots]$ in (7.7.9) has to vanish, and we obtain a solubility condition which fixes $\tilde{\Phi}_0(q)$

$$\tfrac{1}{2} Q^{\nu\mu} \frac{\partial \tilde{\Phi}_0}{\partial q^\nu} \frac{\partial \tilde{\Phi}_0}{\partial q^\mu} + K^\nu(q) \frac{\partial \tilde{\Phi}_0}{\partial q^\nu} = 0. \tag{7.7.10}$$

We see from (7.7.10) that $\tilde{\Phi}_0(q)$ must be just the original white noise potential of the system, which had to be expected for the solution in order ε^0. Similar solubility conditions now appear by the avoidance of logarithmic singularities in all higher orders. In orders ε^3 and ε^4 they read, respectively,

$$\left(Q^{\nu\mu} \frac{\partial \tilde{\Phi}_0}{\partial q^\nu} + K^\nu(q) \right) \frac{\partial \tilde{\Phi}_1}{\partial q^\nu} = 0 \tag{7.7.11}$$

$$\left(Q^{\nu\mu} \frac{\partial \tilde{\Phi}_0}{\partial q^\mu} + K^\nu(q) \right) \frac{\partial}{\partial q^\nu} \left(\tilde{\Phi}_2 - \tfrac{1}{2} Q^{\kappa\lambda} \frac{\partial \tilde{\Phi}_0}{\partial q^\kappa} \frac{\partial \tilde{\Phi}_0}{\partial q^\lambda} \right) = 0, \tag{7.7.12}$$

and they are solved by $\tilde{\Phi}_1(q) = 0$ and $\tag{7.7.13}$

$$\tilde{\Phi}_2(q) = \tfrac{1}{2} Q^{\kappa\lambda} \frac{\partial \tilde{\Phi}_0}{\partial q^\kappa} \frac{\partial \tilde{\Phi}_0}{\partial q^\lambda}.$$

As a result, to order ε^2 we find

$$\Phi(q, y) = \tilde{\Phi}_0(q) + \frac{\varepsilon^2}{2} Q^{\nu\mu} \frac{\partial \tilde{\Phi}_0}{\partial q^\nu} \frac{\partial \tilde{\Phi}_0}{\partial q^\mu} - \varepsilon y^\nu \frac{\partial \tilde{\Phi}_0}{\partial q^\nu}$$

$$+ y^\nu y^\mu \left(Q_{\nu\mu} + \frac{\varepsilon^2}{2} \frac{\partial^2 \tilde{\Phi}_0}{\partial q^\nu \partial q^\mu} \right). \tag{7.7.14}$$

The prefactor Z can be determined by the same method, but now the result for Φ has to be used as an input in the equations determining $\ln Z$ in each order.

239

We content ourselves with the result to first order in ε, which reads

$$\ln Z = \tilde{\Psi}_0(q) - \varepsilon y^\nu \frac{\partial \tilde{\Psi}_0(q)}{\partial q^\nu}, \tag{7.7.15}$$

where $\tilde{\Psi}_0(q)$ follows from the solubility condition in order ε^2, which reads

$$\left(Q^{\nu\mu} \frac{\partial \tilde{\Phi}_0}{\partial q^\mu} + K^\nu \right) \frac{\partial \tilde{\Psi}_0}{\partial q^\nu} + \frac{\partial K^\nu}{\partial q^\nu} + \tfrac{1}{2} Q^{\nu\mu} \frac{\partial^2 \tilde{\Phi}_0}{\partial q^\nu \partial q^\mu} = 0. \tag{7.7.16}$$

Thus, by comparison with (7.3.8), $\tilde{\Psi}_0$ is just the logarithm of the prefactor Z in the white noise case, again as expected. We remark that no first order correction $\tilde{\Psi}_1(q)$ modifying $\tilde{\Psi}_0(q)$ appears in (7.7.15) which follows from the solubility condition in order ε^3 with an identically vanishing solution. The joint probability density of q and y of the form (7.3.3) is now easily integrated over y, and we obtain for the reduced potential $\bar{\Phi}(q)$ and prefactor $\ln \bar{Z}(q)$, respectively,

$$\bar{\Phi}(q) = \tilde{\Phi}_0(q) + \frac{\varepsilon^2}{4} Q^{\nu\mu} \frac{\partial \tilde{\Phi}_0}{\partial q^\nu} \frac{\partial \tilde{\Phi}_0}{\partial q^\mu} + \mathrm{O}(\varepsilon^4) \tag{7.7.17}$$

$$\ln \bar{Z}(q) = \tilde{\Psi}_0(q) + \mathrm{O}(\varepsilon^2).$$

To the order considered this result agrees with that derived by Schenzle and Tél (1985). As discussed by these authors, the main physical effect of the correction in (7.7.17) is to sharpen the minima of the white noise potential $\bar{\Phi}_0$ in such a way that this sharpening is the strongest for those minima which are sharpest in the white noise case. On the other hand, maxima of $\bar{\Phi}_0$ are broadened by the correction in (7.7.17). Again the sharpest maxima are affected the most strongly.

7.8   Van der Pol oscillator

The Van der Pol oscillator is a simple but very important example of a system undergoing a Hopf bifurcation from a fixed point to a limit cycle as a parameter is varied. It is therefore also an important example which one would like to study from the point of view of nonequilibrium potentials. The equations of the model are given by

$$\left. \begin{array}{l} \dot{x} = v \\ \dot{v} = -x + v(\alpha - \gamma x^2 - \delta v^2) + \xi(t) \end{array} \right\}, \tag{7.8.1}$$

with the Gaussian white noise source $\xi(t)$

$$\langle \xi(t)\xi(t') \rangle = \eta \delta(t - t'). \tag{7.8.2}$$

The Hopf bifurcation in the absence of noise occurs at $\alpha = 0$. Nonequilibrium potentials for the model described by (7.8.1) and (7.8.2) have recently been derived by a number of authors. Ebeling and Schimansky-Geier (1985) gave separate treatments of the special case $\gamma = 0, \delta = 1$ in the two limits $\alpha^2 \gg \eta$ and $\alpha^2 \ll \eta$. For $\alpha^2 \ll \eta$ this calculation was later extended to arbitrary δ and γ by Ebeling, Herzel, Richert and Schimansky-Geier (1986). Subsequently it was shown by T. Tél (1988) that after the rescaling

$$x \rightarrow \alpha^{1/2} x, \quad v \rightarrow \alpha^{1/2} v \qquad (7.8.3)$$

a nonequilibrium potential and the prefactor can be constructed by an expansion in α (assuming $\eta/\alpha \ll 1$). This expansion uses the same technique employed in the following section to study codimension-two bifurcations (Graham and Tél, 1987). As the expansion in α is made after the expansion in η has been performed, the assumption is made that $\eta \ll \alpha \ll 1$. However, Tél's results not only reproduced (and slightly extended) the earlier results by Ebeling *et al.* for $\eta \ll \alpha^2$, but within the same calculation also reproduced their result for $\alpha^2 \ll \eta$, if his calculation was formally extended to that case. This strongly indicates that the limits $\alpha \rightarrow 0$ and $\eta \rightarrow 0$ commute. Here, we shall confirm this conjecture by avoiding the expansion in α altogether, leaving α arbitrary, but instead expanding in the difference $\gamma - \delta$. For this purpose it is useful to rescale the variables by

$$x \rightarrow \left(\frac{2}{\gamma + \delta}\right)^{1/2} x, \quad v \rightarrow \left(\frac{2}{\gamma + \delta}\right)^{1/2} v, \quad \eta \rightarrow \frac{2}{\gamma + \delta} \eta, \qquad (7.8.4)$$

and to introduce the parameter

$$\varepsilon = \frac{\gamma - \delta}{\gamma + \delta}. \qquad (7.8.5)$$

The rescaled equations of motion then read

$$\left. \begin{array}{l} \dot{x} = v \\ \dot{v} = -x + v(\alpha - x^2 - v^2 - \varepsilon(x^2 - v^2)) + \xi(t). \end{array} \right\} \qquad (7.8.6)$$

The equations satisfied by Φ and $\ln Z$ then take the form

$$\frac{1}{2}\left(\frac{\partial \Phi}{\partial v}\right)^2 + (-x + v(\alpha - x^2 - v^2 - \varepsilon(x^2 - v^2)))\frac{\partial \Phi}{\partial v}$$

$$+ v\frac{\partial \Phi}{\partial x} = 0 \qquad (7.8.7)$$

and

$$v\frac{\partial \ln Z}{\partial x} + \left(-x + v(\alpha - x^2 - v^2 - \varepsilon(x^2 - v^2)) + \frac{\partial \Phi}{\partial v}\right)\frac{\partial \ln Z}{\partial v}$$

$$= -\alpha + x^2 + 3v^2 + \varepsilon(x^2 - 3v^2) - \frac{1}{2}\frac{\partial^2 \Phi}{\partial v^2}. \qquad (7.8.8)$$

The solutions are constructed by an expansion in ε:

$$\Phi = \Phi_0 + \varepsilon \Phi_1 + \cdots$$

$$\ln Z = \ln Z_0 + \varepsilon \ln Z_1 + \cdots. \qquad (7.8.9)$$

In fact, all the earlier results mentioned can already be recovered in first order in ε, and we shall therefore content ourselves with this accuracy. Equations (7.8.7) and (7.8.8) to zero order in ε are solved trivially by

$$\left.\begin{array}{l} \Phi_0 = -\alpha(x^2 + v^2) + \tfrac{1}{2}(x^2 + v^2)^2 \\ \ln Z_0 = \text{const.} \end{array}\right\} \tag{7.8.10}$$

The equations appearing in first order are

$$(-x - \alpha v + v(x^2 + v^2))\frac{\partial \Phi_1}{\partial v} + v\frac{\partial \Phi_1}{\partial x}$$

$$= -2\alpha v^2(x^2 - v^2) + 2v^2(x^4 - v^4) \tag{7.8.11}$$

and

$$(-x - \alpha v + v(x^2 + v^2))\frac{\partial \ln Z_1}{\partial v} + v\frac{\partial \ln Z_1}{\partial x}$$

$$= x^2 - 3v^2 - \frac{1}{2}\frac{\partial^2 \Phi_1}{\partial v^2}. \tag{7.8.12}$$

In view of the form of the inhomogeneity of these equations Φ_1 and $\ln Z_1$ will have the form of power series in x and v

$$\Phi_1 = \sum_{n,m} C_{n,m} v^n x^m \tag{7.8.13}$$

and

$$\ln Z_1 = \sum_{n,m} D_{n,m} v^n x^m \tag{7.8.14}$$

with $C_{nm} = D_{nm} = 0$ if $n < 0$ or $m < 0$.

From (7.8.11) and (7.8.12) we obtain the following recursion relations:

$$-\alpha k C_{k,l} - (k+1)C_{k+1,l-1} + (l+1)C_{k-1,l+1}$$

$$+ kC_{k,l-2} + (k-2)C_{k-2,l}$$

$$= 2\alpha(\delta_{k,l}\delta_{l0} - \delta_{k2}\delta_{l2}) + 2(\delta_{k2}\delta_{l4} - \delta_{k6}\delta_{l0}) \tag{7.8.15}$$

and

$$-\alpha k D_{k,l} - (k+1)D_{k+1,l-1} + (l+1)D_{k-1,l+1}$$

$$+ kD_{k,l-2} + (k-2)D_{k-2,l}$$

$$= -3\delta_{k2}\delta_{l0} + \delta_{k0}\delta_{l2} - \tfrac{1}{2}(k+1)(k+2)C_{k+2,l}, \tag{7.8.16}$$

which can easily be solved recursively. We merely give the results for Φ_1 (and $\ln Z_1$) up to powers of x and v of fourth (and second) order, respectively,

$$\Phi_1 = \frac{1}{4 + 3\alpha^2}\left[\left(-1 + \frac{3\alpha^2}{2}\right)(x^2 + v^2)^2\right.$$

$$\left. + 4\alpha x v^3 - 3\alpha^2 v^4 + O(6)\right] \tag{7.8.17}$$

and

$$\ln Z_1 = -\frac{3}{4+3\alpha^2}(2xv + 3\alpha x^2) + O(4), \tag{7.8.18}$$

where we indicated the order of the omitted terms. For $\alpha^2 \ll 1$ our results agree with those of Tél (1988). It is now straightforward to see that no singularities arise in Φ or $\ln Z$ if we take $\alpha = 0$. This confirms Tél's conjecture. In order to obtain the nonequilibrium potential to sufficient accuracy to characterize the limit cycle as the minimum of Φ, it is necessary to treat α as of the order of x^2, v^2 and xv, i.e. together with the term αxv^3 we have also to include powers of x and v of sixth order. Here we only state the result for Φ and $\ln Z$ obtained for small α (Tél, 1988)

$$\left.\begin{aligned}
\Phi_1 &= -\tfrac{1}{4}(x^2+v^2)^2 + (\alpha - x^2 - v^2)xv^3 + O(8)\\
\ln Z_1 &= -\tfrac{3}{2}xv - \tfrac{9}{4}\alpha x^2 + 4x^4\\
&\quad + \tfrac{23}{4}x^2v^2 - \tfrac{23}{32}(x^2+v^2)^2 + O(6).
\end{aligned}\right\} \tag{7.8.19}$$

It can be now easily be checked that, $\alpha > 0$, Φ is minimal on the limit cycle

$$x^2 + v^2 - \varepsilon xv^3 = \alpha\left(1 + \frac{\varepsilon}{2}\right) + O(\varepsilon^2, \alpha^2). \tag{7.8.20}$$

While Φ is degenerate on the limit cycle, $\ln Z$ is of course not, i.e. the probability density (7.2.3) varies along the limit cycle and, in general, will have local maxima at isolated points. It would be wrong, however, to infer from this fact, as is sometimes done in the literature, that the noise has qualitatively changed the bifurcation behavior of the model from Hopf bifurcation to some different bifurcation involving only 'fixed points'. Physically, maxima of the probability density with velocity $v \neq 0$ do not correspond to fixed points, since $\dot{x} = v \neq 0$. Rather, the change of P_∞ along the limit cycle is merely due to a nonconstant but finite velocity along the limit cycle. For $\eta \to 0$ the resulting localized maxima of P_∞ do not become infinitely sharp. In fact, already without noise the probability density along the limit cycle is proportional to the inverse of the absolute value of $(\dot{x}^2 + \dot{v}^2)^{1/2}$ and therefore not constant, leading to isolated extrema of the probability density not associated with fixed points. The preceding discussion underlines the fact that in order to obtain a potential which is minimal on the deterministic attractors (and serves as a Lyapunov function for the deterministic dynamics) the weak noise limit $\eta \to 0$ in (7.3.2) is absolutely essential. As a potential $\Phi(q)$ has useful properties which are not shared by $\ln P_\infty(q, \eta)$.

7.9 Codimension-two bifurcations and noise

Let us now consider bifurcations of a more complicated type than those of Section 7.5. Generally speaking, bifurcations of the limit sets (i.e. attractors, repellors and saddles) of equations of the form $\dot{q}^v = K^v(q)$, $v = 1,\ldots,n$ only occur on special surfaces in parameter space, characterized by their codimension, i.e. by the number of independent relations between the parameters they imply. Codimension-one bifurcations are those occurring on codimension-one surfaces in parameter space. They are classified as saddle-node, transcritical, pitchfork, or Hopf bifurcations (Guckenheimer and Holmes, 1983). Two such bifurcations coincide at the intersection of two codimension-one surfaces and then constitute a bifurcation of codimension two. Codimension-two bifurcations of a fixed point, chosen as the origin, can be classified by the behavior of the linearized dynamics (Guckenheimer and Holmes, 1983). After diagonalizing the linearized dynamics one may have either (i) two vanishing eigenvalues, (ii) a purely imaginary pair of eigenvalues and a vanishing eigenvalue, or (iii) two purely imaginary pairs of eigenvalues. Nonequilibrium potentials have been determined for all three types of codimension-two bifurcations by Graham and Tél (1987); for a special case see also Sulpice, Lemarchand and Lemarchand (1987). Here we shall restrict our attention to type (i) for simplicity, which has, e.g., relevance for recent experiments on thermal convection of binary mixtures in a porous medium heated from below (Rehberg and Ahlers, 1985).

The local normal form (near $x = v = 0$) of the equations of motion on the two-dimensional center manifold (Guckenheimer and Holmes, 1983) to lowest nontrivial order takes the form

$$\left.\begin{array}{l} \dot{x} = v \\ \dot{v} = \mu_1 x + \mu_2 v - ax^3 + bx^2 v + \cdots . \end{array}\right\} \qquad (7.9.1)$$

Here μ_1, μ_2, a and b are parameters, and $\mu_1 = \mu_2 = 0$ defines the bifurcation point. For convection the normal form including fluctuation terms can be derived from a basic set of hydrodynamic equations using a multiple time scale technique or by other equivalent methods of adiabatic elimination of fast variables in the presence of noise. The derivation of normal forms for codimension-two bifurcations in the presence of noise was discussed by Knobloch and Wiesenfeld (1983) and Wunderlin and Haken (1981). The normal form for the case of binary convection in a porous medium (without noise) was derived in Brand and Steinberg (1983). As was shown there, $a < 0$, $b < 0$ is the case of physical interest, in this case. Equation (7.9.1) can be reduced to a conservative system, weakly perturbed by dissipative terms (Guckenheimer and Holmes, 1983) by introducing the scaling, for $\varepsilon \to 0$

$$\left.\begin{array}{ll} \mu_1 = \varepsilon^2 v_1, & \mu_2 = \varepsilon^2 v_2 \\ x = \varepsilon \bar{x} & v = \varepsilon^2 \bar{v} \\ t = \bar{t}/\varepsilon. \end{array}\right\} \qquad (7.9.2)$$

Henceforth, the bars from the scaled variables will be omitted, for simplicity.

244

Macroscopic potentials, bifurcations and noise

Equation (7.9.1) then takes the form

$$\dot{x} = \frac{\partial H}{\partial v}, \quad \dot{v} = -\frac{\partial H}{\partial x} + \varepsilon v g(x) \tag{7.9.3}$$

with

$$H = \frac{v^2}{2} + V(x)$$
$$V(x) = -\frac{v_1}{2}x^2 + \frac{a}{4}x^4 + \cdots \tag{7.9.4}$$

and

$$g(x) = v_2 + bx^2 + \cdots. \tag{7.9.5}$$

We now consider the case where a noise term

$$(\eta Q)^{1/2}\xi(t); \quad \langle \xi \rangle = 0; \quad \langle \xi(t)\xi(0) \rangle = \delta(t) \tag{7.9.6}$$

is added to the equation for $\dot{v}$ in (7.9.3) (Knobloch and Wiesenfeld, 1983). The macroscopic potential Φ has to be determined from (7.3.7), which, in the present case, reads

$$\frac{Q}{2}\left(\frac{\partial \Phi}{\partial v}\right)^2 + \frac{\partial H}{\partial v}\frac{\partial \Phi}{\partial x} - \frac{\partial H}{\partial x}\frac{\partial \Phi}{\partial v} + \varepsilon v g(x)\frac{\partial \Phi}{\partial v} = 0. \tag{7.9.7}$$

Its solution can be constructed as a power series in ε (Graham and Tél, 1987)

$$\Phi = \sum_{n=1}^{\infty} \varepsilon^n \Phi_n. \tag{7.9.8}$$

We shall restrict our attention to the lowest nontrivial order, where one finds

$$\Phi_1(x,v) = -\frac{2}{Q}\int_{E_0}^{H(x,v)} \frac{v_g(E)}{\bar{v}(E)}\,dE + \text{const}, \tag{7.9.9}$$

with the abbreviations

$$\bar{v}(E) = \int_{x_1(E)}^{x_2(E)} [2(E - V(x))]^{1/2}\,dx \tag{7.9.10}$$

$$v_g(E) = \int_{x_1(E)}^{x_2(E)} g(x)[2(E - V(x))]^{1/2}\,dx. \tag{7.9.11}$$

In (7.9.10) and (7.9.11) $x_1(E) \leqslant x_2(E)$ are roots of

$$V(x) = E \tag{7.9.12}$$

such that

$$E \geqslant V(x) \ (x_1(E) \leqslant x \leqslant x_2(E)). \tag{7.9.13}$$

The potential $V(x)$ and the equipotential contours $\Phi_1(x,v) = \text{const.}$ are sketched in Figure 7.4 for the case $a < 0$, $v_1 < 0$. In this case the local normal

245

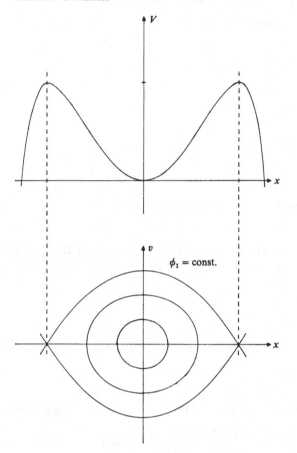

Figure 7.4. The potential $V(x)$ of (7.9.4) and equipotential curves of the nonequilibrium potential ϕ_1 of (7.9.9) for $a < 0$, $v_1 < 0$.

form determines $\Phi \simeq \varepsilon\Phi_1$ only for $0 = E_0 \leqslant E \leqslant V_0$, i.e. only in a neighborhood of the fixed point $x = v = 0$. For $v_1 > 0$ there is no attractor of the normal form near the origin, and a macroscopic potential cannot be locally defined. (For the general connection between the existence of attractors and the existence of Φ, see Section 7.14.) A detailed discussion of the extrema of the nonequilibrium potential can now be performed, whose results are summarized in Figure 7.5. There $\Phi \simeq \varepsilon\Phi_1(H)$ is presented as a function of H in the different regimes of the (v_1, v_2)-plane which have to be distinguished (Graham and Tél, 1987); see also Guckenheimer and Holmes (1983) for a discussion of the bifurcation diagram without the use of a potential.

Having determined Φ to the desired order one may now go on to solve (7.3.8) for the prefactor Z, again expanding in ε. To lowest nontrivial order Z is found

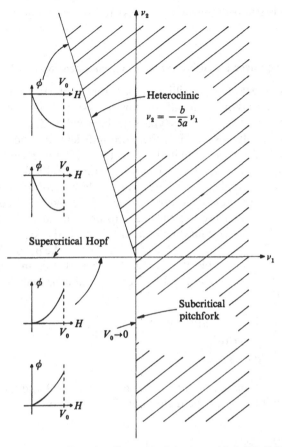

Figure 7.5. Bifurcation diagram and the potential $\phi_1(H)$, (7.9.9) for $a < 0$, $b < 0$.

to be constant. Thus, for the probability density we obtain

$$P_\infty(x, v, \eta) = \text{const} \cdot \exp\left[\frac{\varepsilon}{\eta}\Phi_1(H(x, v)) + O\left(\frac{\varepsilon^2}{\eta}, \varepsilon, \eta\right)\right]. \qquad (7.9.14)$$

So far we have neglected a difficulty apparent in (7.9.14) from the fact that this probability density is not normalizable on the (x, v)-plane. Physically, the reason is that fluctuations make the system (7.9.1) unstable, even in the case where the deterministic system has a stable attractor at or near $x = v = 0$. A steady state distribution $P_\infty(x, v, \eta)$ therefore cannot exist, after all, and the definition (7.9.2) is, strictly speaking, inapplicable. A simple way out of this difficulty is to add stabilizing higher order terms to the normal form (7.9.1). For example, a term $- 6cx^5$, with $c > 0$ on the right hand side of the equation for $\dot{v}$ serves this purpose. Then the potential $V(x)$ in (7.9.4) receives a stabilizing term

247

cx^6; $P_\infty(x, v, \eta)$ is still being given by (7.9.14) and becomes normalizable, and $\Phi(x, v)$ now exists.

7.10   Escape over potential barriers in a codimension-two bifurcation

One important aspect of nonequilibrium potentials is the fact that they build up barriers which may make it difficult for the system to escape from some regions of configuration space. This type of problem will be dealt with at length elsewhere in these volumes, and it is therefore sufficient to indicate here how the results obtained for the potential Φ and the prefactor Z can be used to calculate average mean first passage times asymptotically for weak noise. Then we shall give an application to the results of the preceding section. Let an attractor of the system be a fixed point at the origin and let us consider the average time $\langle \tau \rangle$ which the system needs to cross a closed boundary $\partial\Omega$ surrounding a domain Ω containing the attractor if the system is started somewhere in Ω. This time averaged over Ω depends on the properties of Φ on the boundary $\partial\Omega$. For example, if Φ has local degenerate minima on $\partial\Omega$ in the points q_k, then for $\eta \to 0$ $\langle \tau \rangle$ is given by (see, e.g., Hänggi, 1986, for a review; and Schuss, 1980)

$$\langle \tau \rangle = \frac{(2\pi\eta)^{1/2} Z(0)}{\sum_k \left(\dfrac{H(0)}{H_k} \right)^{1/2} Z(q_k) |K^0(q_k)|} \exp\left(\frac{\Delta\Phi_{min}}{\eta} \right). \qquad (7.10.1)$$

Here $\Delta\Phi_{min}$ is the minimal potential difference between $\partial\Omega$ and the attractor. Furthermore we introduced the definitions (Schuss, 1980)

$$\left. \begin{aligned} H(0) &= \det\left(\frac{\partial^2\Phi}{\partial q^\nu \partial q^\mu} \right)_{q=0} \\ H_k &= \det\left(\frac{\partial^2\Phi}{\partial q^\nu \partial q^\mu} \right)_{q=q_k}^{\partial\Omega} \\ K^0(q) &= n_\nu(q) K^\nu(q). \end{aligned} \right\} \qquad (7.10.2)$$

The notation $(\ )^{\partial\Omega}$ indicates that derivatives are taken only parallel to the surface $\partial\Omega$ and $n_\nu(q)$ is the unit normal vector on $\partial\Omega$ at q. The form of (7.10.1) is modified in rather obvious ways if some of the underlying assumptions have to be modified (Schuss, 1980). For example, if Φ is minimal on $\partial\Omega$, not in isolated points q_k but rather on a whole arc $U \in \partial\Omega$, one has to modify (7.10.1) by the substitution

$$\sum_k \frac{F(q_k)}{(2\pi\eta H_k)^{1/2}} \to \int_{U \in \partial\Omega} \frac{dS(q)}{[(2\pi\eta)^n]^{1/2}} F(q). \qquad (7.10.3)$$

Other, but similarly obvious, modifications are necessary, e.g. if the Hessian

$H(0)$ at the attractor vanishes because one of the eigenvalues of the matrix vanishes.

We now apply these formulas to the example of a codimension-two bifurcation treated in the preceding section (Graham, 1986) and calculate the mean first passage times out of the domain of the fixed point attractor at the origin which exists for $v_1 < 0$, $v_2 < 0$, and out of the domain of the limit cycle

$$2H(x,v) \equiv v^2 + |v_1|x^2 - \tfrac{1}{2}x^4 = 2E_c \tag{7.10.4}$$

which exists for $v_1 < 0, 0 < v_2 < |v_1|/5$. The constant E_c in (7.10.4) is defined as the solution of the equation

$$v_2 \int_0^{x(E_c)} dx (2E_c + v_1 x^2 + \tfrac{1}{2}x^4)^{1/2}$$

$$= \int_0^{x(E_c)} dx\, x^2 (2E_c + v_1 x^2 + \tfrac{1}{2}x^4)^{1/2}. \tag{7.10.5}$$

Let us first consider the exit from the domain of the fixed point attractor. The boundary $\partial\Omega$ of the exit is the curve (the outer equipotential curve in Figure 7.4)

$$H(x,v) = \frac{v_1^2}{4}$$

on which $\Phi(x,v)$ is constant. Thus, the modification (7.10.3) must be used. It is now easy to evaluate all quantities in (7.10.1)–(7.10.3), and one obtains

$$\langle \tau \rangle = \frac{15\pi}{4(2)^{1/2}} \frac{\eta Q}{\varepsilon^2} \frac{\exp\left(\dfrac{\varepsilon}{\eta}\Phi_1\left(\dfrac{v_1^2}{4}\right)\right)}{v_1^2 |v_2|(5|v_2| + |v_1|)}. \tag{7.10.6}$$

Equation (7.10.6) is valid for

$$\frac{\varepsilon}{\eta}\Phi_1\left(\frac{v_1^2}{4}\right) \gg 1 \tag{7.10.7}$$

and sufficiently far from the bifurcation lines at $v_1 = 0$ and $v_2 = 0$. Usually, $\langle \tau \rangle$ will be very long and difficult to observe. However, moving towards the bifurcation line $v_1 \to -0$ (but still observing the condition (7.10.7)) the time $\langle \tau \rangle$ becomes shorter and may be observable. In that limit (7.10.6) can be simplified further and reads

$$\langle \tau \rangle = \frac{3\pi}{8(2^{1/2})} \frac{1}{y} \exp y \tag{7.10.8}$$

with the scaling variable

$$y = \frac{|\mu_2|\mu_1^2}{2\eta Q}. \tag{7.10.9}$$

In (7.10.8) and (7.10.9) we transformed back to unscaled parameters and times. We note again that the validity of (7.10.8) is restricted to $y \gg 1$.

Let us now turn to the domain $v_2 > 0$ and consider the exit from the domain of the limit cycle (7.10.4). In the variables x, H the limit cycle corresponds to $H = E_c$. Formula (7.10.1) is most easily evaluated in these coordinates with the result (Graham, 1986)

$$\langle \tau \rangle = \frac{15}{4} \left(\frac{\pi \eta}{\varepsilon^3} \right)^{1/2} \frac{T(E_c)}{(\Phi_1''(E_c))^{1/2} |v_1|^{3/2} (|v_1| - 5v_2)} \frac{\exp\left[\frac{\varepsilon}{\eta} \Delta \Phi_1 \right]}{}, \tag{7.10.10}$$

where $T(E_c)$ is the round trip time on the limit cycle (in scaled units)

$$T(E_c) = 4 \int_0^{x(E_c)} \frac{dx}{(2E_c + v_1 x^2 + \frac{1}{2} x^4)^{1/2}}, \tag{7.10.11}$$

and

$$\Phi_1''(E_c) = -\frac{2}{Q} \frac{\int_0^{E_c} dx \left[(v_2 - x^2)/(2E_c + v_1 x^2 + \frac{1}{2} x^4)^{1/2} \right]}{\int_0^{E_c} dx (2E_c + v_1 x^2 + \frac{1}{2} x^4)^{1/2}}. \tag{7.10.12}$$

It is interesting to see how $\langle \tau \rangle$ behaves if the homoclinic bifurcation (Guckenheimer and Holmes, 1983) at $v_2 \to \frac{1}{5} |v_1| - 0$ is approached where the limit cycle loses its stability. In this limit (7.10.10) may be considerably simplified by evaluating the quadratures in (7.10.11) and (7.10.12), and we find

$$\langle \tau \rangle \sim \frac{1}{(|\mu_1|y)^{1/2}} \exp y, \tag{7.10.13}$$

with the scaling variable

$$y = \frac{|\mu_1|(\mu_2 - \frac{1}{5}|\mu_1|)^2}{\eta Q \left| \ln \frac{\mu_2 - \frac{1}{5}|\mu_1|}{|\mu_1|} \right|}. \tag{7.10.14}$$

Again, the validity of (7.10.13) is restricted to $y \gg 1$.

Formulas (7.10.6) and (7.10.10) apply to the two different regimes $v_2 < 0$ and $v_2 > 0$ with $v_2^2 > \eta Q/\varepsilon |v_1|$. A simple interpolation can be given (Graham, 1986) which smoothly connects these two results through the Hopf bifurcation line at $v_2 = 0$ where nothing drastic happens to the mean first passage time $\langle \tau \rangle$, contrary to the bifurcation lines at $v_1 = 0$, $v_2 < 0$ and $v_2 < \frac{1}{5}|v_1|$, $v_1 < 0$ discussed above.

7.11   Analytical properties of the weak noise limit in nonequilibrium systems

We now investigate some general properties of the solutions of (7.3.7) (Graham

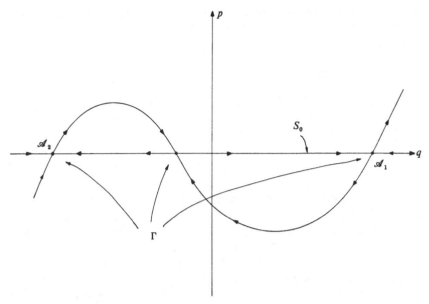

Figure 7.6. Schematic plot of Poincaré cross-section of the Hamiltonian system (7.11.5) and (7.11.6) for $H = 0$ in the case where a smooth separatrix exists.

and Tél, 1984a, b, 1986). For this purpose it is very useful to introduce a formal analogy to mechanics by interpreting (7.3.7) as a time-independent Hamilton Jacobi equation for an action $\Phi(q)$ at energy $H = 0$. The q^ν now play the role of generalized coordinates whose canonically conjugate momenta are defined in terms of the action Φ as

$$p_\nu = \frac{\partial \Phi(q)}{\partial q^\nu}.$$ (7.11.1)

The Hamiltonian H may be read off from (7.3.7) written in the form

$$H\left(q, \frac{\partial \Phi}{\partial q}\right) = 0$$ (7.11.2)

and takes the form

$$H(q, p) = \tfrac{1}{2} Q^{\nu\mu}(q) p_\nu p_\mu + K^\nu p_\nu.$$ (7.11.3)

Continuing with the mechanical analogy we now consider the $(2n - 2)$-dimensional Poincaré cross-section of the formal Hamiltonian system by taking $H = 0$ and a fixed value of one coordinate, say q^n. A schematic example is shown in Figure 7.6. Let us consider a case where several attractors $\mathscr{A}_i$ of the deterministic system

$$\dot{q}^\nu = K^\nu(q)$$ (7.11.4)

251

ROBERT GRAHAM

coexist. The deterministic dynamics (7.11.4) is embedded in the larger Hamiltonian dynamics generated by the Hamiltonian (7.11.3)

$$\dot{q}^\nu = \frac{\partial H}{\partial p_\nu} = Q^{\nu\mu}(q)p_\mu + K^\nu(q) \tag{7.11.5}$$

$$\dot{p}_\nu = -\frac{\partial H}{\partial q^\nu} = -\frac{1}{2}\frac{\partial Q^{\mu\lambda}(q)}{\partial q^\nu}p_\mu p_\lambda + \frac{\partial K^\mu(q)}{\partial q^\nu}p_\mu \tag{7.11.6}$$

as the special case

$$p_\nu \equiv 0 (\nu = 1,\ldots,n). \tag{7.11.7}$$

The conditions (7.11.7) define an n-dimensional surface S_0 in the (q,p) phase space. All the limit sets Γ of (7.11.4) (i.e. its attractors, repellors and saddles) must therefore be located on S_0. At this point we recall that the non-equilibrium potential Φ we are looking for must have vanishing first derivatives in Γ, because Φ is minimal in attractors (see the remark after (7.3.13)), maximal in repellors (which is shown by reversing time in (7.11.4), which turns repellors into attractors, and applying the same arguments as before) and stationary in saddles (which is shown by applying the same arguments as before, but separately along the stable and unstable manifolds of the saddle). The limit sets Γ are singular sets of the deterministic system (7.11.4) and therefore have stable and (or) unstable manifolds on S_0. In the $2n$-dimensional phase space of the Hamiltonian system (7.11.5), (7.11.6) the surface S_0 therefore forms an n-dimensional invariant manifold connecting these limit sets. However, the Γ may also be considered as singular sets of (7.11.5) and (7.11.6) and therefore must have, in addition, invariant manifolds of codimension n (i.e again of dimension n) transverse to S_0 forming separatrices

$$p_\nu = p_\nu^{(0)}(q)(\nu = 1,\ldots,n). \tag{7.11.8}$$

By construction, these separatrices pass through the limit sets Γ, and therefore satisfy

$$p_\nu^{(0)}(\Gamma) = 0. \tag{7.11.9}$$

Being invariant manifolds of the Hamiltonian system (7.11.5), (7.11.6) these separatrices must have the Lagrangian property

$$\frac{\partial p_\nu^{(0)}(q)}{\partial q^\mu} = \frac{\partial p_\mu^{(0)}(q)}{\partial q^\nu}. \tag{7.11.10}$$

Equation (7.11.10) guarantees that one may associate an action Φ with (7.11.8) via

$$\Phi(q) = \int_{q_0}^{q} p_\nu^{(0)}(q)\,dq^\nu + \Phi(q_0), \tag{7.11.11}$$

252

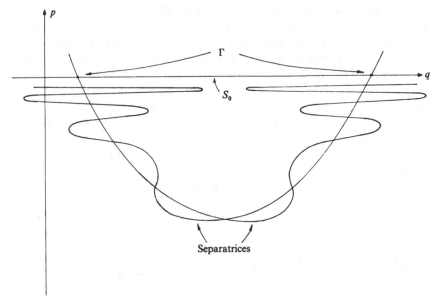

Figure 7.7. Schematic plot of Poincaré cross-section of the Hamiltonian system
(7.11.5), (7.11.6) for $H = 0$ in the generic case.

where q_0 is an arbitrary initial point, which by construction satisfies the
Hamilton Jacobi equation (7.3.7), and whose first derivatives vanish in Γ. It
therefore satisfies necessary requirements of the solution of (7.3.7) we are
looking for. We are now in a position to gain some insight into the analytical
properties of this solution.

First we consider the case where the $p_v^{(0)}(q)$ are everywhere single-valued
functions of q, connecting all the limit sets Γ. This case is schematically drawn
in Figure 7.6. In this case a smooth, continuous, everywhere differentiable
potential $\Phi(q)$ exists and is defined by (7.11.11). In the examples considered in
Sections 7.4–7.10 this was always the case. However, from the present more
general point of view we may now ask how typical this situation is. Drawing on
well known general results of the theory of Hamiltonian systems we have to
observe that smooth separatrices are nongeneric features of such systems.
However, smooth separatrices *are* generic features in the special class of
integrable Hamiltonian systems. Thus, smooth nonequilibrium potentials
exist if the Hamiltonian (7.11.3) is integrable at $H = 0$. However, neither
integrability nor smoothness of separatrices is a structurally stable property,
i.e. smooth separatrices occurring in special cases, e.g. for a special choice of
model parameters, are destroyed by typical small perturbations $K^v \to K^v +
\delta K^v$, $Q^{v\mu} \to Q^{v\mu} + \delta Q^{v\mu}$. Therefore, the typical phase space structure of a
Hamiltonian of the form (7.11.3) looks like Figure 7.7. The separatrices
emanating from the different limit sets Γ transverse to S_0 split up and form
infinitely many heteroclinic intersections forcing the separatrices to oscillate

253

wildly around each other and leading to infinitely many local branches of the separatrix $p_v = p_v^{(0)}(q)$. Why does this behavior not occur in equilibrium thermodynamics? The general reason is that the Hamiltonian systems (7.11.5), (7.11.6) corresponding to equilibrium thermodynamics are nongeneric due to their special time-reversal symmetries. More specifically, as was discussed in Section 7.2 $r^v(q)$ and $d^v(q)$ in equilibrium thermodynamics must satisfy opposite time-reversal properties. Hence $r^v(q)$ and $d^v(q)$ have to be single-valued functions of q if $K^v(q)$ is, because any wild oscillations occurring in r^v and d^v would have to cancel in their sum K^v, which is impossible for two components with opposite time-reversal symmetry. However, if we have to conclude from this that d^v is single valued, and if $Q^{v\mu}(q)$ is given as single valued, then $\partial\Phi/\partial q^v$ must also be single valued, since

$$d^v = -\tfrac{1}{2}Q^{v\mu}(q)\frac{\partial\Phi}{\partial q^\mu}. \tag{7.11.12}$$

The discussion of the generic case of wildly oscillating separatrices will be continued below. However, first we give a simple example.

7.12 Periodically driven overdamped nonlinear oscillator

For the general analytical properties discussed in the preceding section we consider the following example (Graham and Tél, 1984a, b; further examples for phase locking of two oscillations are given in Graham and Tél, 1985a):

$$\frac{1}{\varepsilon}\dot{x} = (x - x^3)(1 + a\cos\omega t) + \eta^{1/2}\xi(t). \tag{7.12.1}$$

with real Gaussian white noise

$$\langle \xi(t)\xi(0)\rangle = \delta(t). \tag{7.12.2}$$

It is convenient to eliminate the explicit time dependence from (7.12.1) by introducing a second variable y via

$$y = \omega t, \quad \dot{y} = \omega. \tag{7.12.3}$$

Rescaling time to put $\omega = 1$, the Hamiltonian (7.11.3) for the present case reads

$$H = \tfrac{1}{2}p_x^2 + \varepsilon p_x(x - x^3)(1 + a\cos y) + p_y. \tag{7.12.4}$$

If $a = 0$, then y is a cyclic variable, i.e. p_y is conserved and therefore the Hamiltonian is integrable. The Poincaré cross-section $H = 0$, $y = 0$ is shown in Figure 7.8. The separatrix (7.11.8) is then given by

$$\left.\begin{array}{l} p_x^{(0)} = -2\varepsilon(x - x^3) \\ p_y^{(0)} = 0 \end{array}\right\} \tag{7.12.5}$$

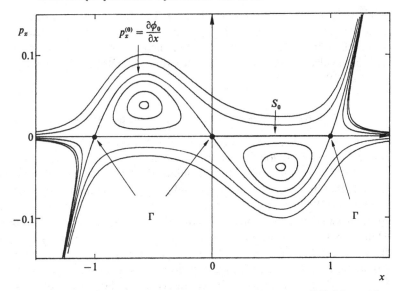

Figure 7.8. Poincaré cross-section $H = 0$, $y = 0$ of the system (7.12.4) for $a = 0$. (From Graham and Tél, 1984b.)

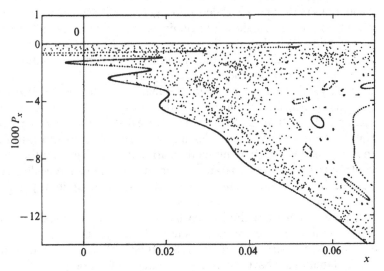

Figure 7.9. Poincaré cross-section $H = 0$, $y = 0$ of the system (7.12.4) for $a = 0.1$. (From Graham and Tél, 1984b.)

and the potential obtained from (7.11.11) is

$$\Phi = -\varepsilon(x^2 - \tfrac{1}{2}x^4).$$ (7.12.6)

For $a \neq 0$ the Hamiltonian (7.12.4) is no longer integrable. This is evident from the Poincaré cross-section of Figure 7.9 for $a = 0.1$, which clearly shows the

ROBERT GRAHAM

wild oscillations close to the saddle at $x = 0$ of the separatrix which emanates from the attractor at $x = 1$. The appearance of these wild oscillations due to heteroclinic intersections can also be studied analytically (Graham and Tél, 1985a; see also Jauslin, 1986, where Melnikov's method (Guckenheimer and Holmes, 1983) is used).

7.13 Search for special cases with smooth potentials

Smooth potentials are exceptional, as was discussed in Section 7.11. It is of interest, therefore, to look for methods which allow one to identify such special cases within a given class of Hamiltonians (7.11.3). Such methods have been investigated by Graham, Roekaerts and Tél, 1985, and are further applied by Hietarinta (1985) and Roekaerts and Schwartz (1987). A simple but in practice quite successful method consists in constructing a perturbative solution of the Hamilton Jacobi equation (7.3.7) (cf., e.g., Sections 7.7–7.9) in the neighborhood of a nontypical case where a potential can be written down immediately. Oscillations in the perturbative solution for Φ then serve as an empirical indicator of oscillations of the separatrix, even though quantitatively the perturbative solution may not be trusted in those regions where oscillations do occur. If a special choice of parameters can be found which eliminates these oscillations in the first few orders of the perturbative solution, in practice this often turns out to be sufficient to eliminate the oscillations to all orders in the perturbation. Whether this is, indeed, the case, must then, of course, be checked by substitution of the smooth solution in the Hamilton Jacobi equation. An example where this strategy is successful is the case of optical bistability studied in Section 7.6 (see Graham and Schenzle, 1983). An obvious potential (7.6.3) for the case $\delta \neq \Delta$ then exists for $E_0 = 0$. A perturbative solution of (7.6.2) around $E_0 = 0$ then reveals that for $\delta = \Delta$ a particularly simple solution exists in first order in E_0 which turns out to be true to all orders in E_0. A possible pitfall of this method arises from the fact that the oscillatory contributions to the separatrix often depend nonanalytically on some parameters of the model and therefore do not show up in any finite order, when expanded in those parameters.

A more systematic method somewhat less based on trial and error makes use of the Painlevé conjecture. According to this conjecture systems which have the Painlevé property are integrable. Solutions of the canonical equations for a given Hamiltonian have the Painlevé property if their only movable singularities in the complex time-plane are poles. Singularities are movable if their location in the complex plane depends on the initial conditions. In practice one checks for the Painlevé property by solving the canonical equations by a Laurent–Taylor series in the neighborhood of a movable singularity, trying to adjust the parameters of the system in such a way that the series contains sufficiently many undetermined coefficients (including the position of the singularity) to have a general solution. If the latter condition can be satisfied

256

for some choice of parameters then the system is conjectured to be integrable for this special case, but this conjecture must still be checked, e.g. by explicit construction of conserved phase space functions. For the application of this method to Fokker–Planck Hamiltonians of the form (7.11.3) we refer to the literature (Graham, Roekaerts and Tél, 1985; Hietarinta, 1985; Roekaerts and Schwartz, 1987).

7.14 Functional integral representation and extremum principle for the potential

We now return to the problem of oscillating separatrices on the weak noise limit described in Section 7.11, and address the question how the non-equilibrium potential Φ can be constructed in this case and what its properties are (Graham and Tél, 1985a, b, 1986). Clearly, the Hamilton Jacobi equation (7.3.7) is not sufficient to answer this question, as it contains only local information about Φ which is no help in deciding which of the infinitely many branches of Φ should be chosen in the case of nonintegrability. At this point we recall that the statistical definition of Φ provided by (7.3.2) contains more than this local information. In order to capture the global information contained in (7.3.2) we make use of the functional integral representation of $P_\infty(q, \eta)$ of the right hand side of (7.3.2) (Freidlin and Ventzel, 1984; Graham, 1973a, 1977; Kubo, Matsuo and Kitahara, 1973; Langouche, Roekaerts and Tirapegui, 1982). The conditional probability density $P(q/q_0, t)$, i.e. the solution of (7.2.3) with initial condition $P(q|q_0, 0) = \delta^{(n)}(q - q_0)$, can be written as the functional integral

$$P(q|q_0, t) = \int D\mu \exp\left[- \int_{q(-t)=q_0}^{q(0)=q} L(q, \dot{q}, \eta) \, d\tau \right], \qquad (7.14.1)$$

where $D\mu$ is a formal functional measure of integration and $L(q, \dot{q}, \eta)$ is a certain 'Lagrangian', depending on η (and the choice of the functional measure $D\mu$). In the weak noise limit (7.14.1) can be evaluated in saddle point approximation. In the limit in (7.3.2) it is sufficient to evaluate the result to leading order in η

$$P(q|q_0, t) \sim \exp\left[- \frac{1}{\eta} \min \int_{q(-t)=q_0}^{q(0)=q} L_0(q, \dot{q}) \, d\tau \right]. \qquad (7.14.2)$$

Here 'min' indicates that the absolute minimum is taken over all paths starting in q_0 at time $(-t)$ and ending in q at time 0, and L_0 is the Lagrangian derived from the Hamiltonian (7.11.3) by the usual Legendre transformation

$$L_0 = \tfrac{1}{2} Q_{\nu\mu}(\dot{q}^\nu - K^\nu(q))(\dot{q}^\nu - K^\mu(q)). \qquad (7.14.13)$$

In (7.11.3) we have to assume that $Q_{\nu\mu}$ exists. For the generalizations necessary if this assumption is not satisfied see Graham and Tél (1985b) and also the example given in Section 7.15. The Lagrangian L_0 has the important property that it vanishes if and only if $\dot{q}^{v} = K^{v}$ holds, i.e. along all deterministic trajectories. For all other trajectories L_0 is positive. A functional integral representation of $P_{\infty}(q)$ can now be obtained from (7.14.2) by taking $t \to \infty$ and noting that $P(q|q_0, \infty) = P_{\infty}(q)$. Using (7.3.2) we then find

$$\Phi(q) = \min \int_{q(-\infty)=q_0}^{q(0)=q} L_0(q, \dot{q}) \, d\tau + \text{const.} \tag{7.14.4}$$

Equation (7.14.4) repeats the statement that $\Phi(q)$ plays the role of a minimal action in our analogy with mechanics. However, it also raises the question of how the dependence on q_0 drops out on the right hand side of (7.14.4), as the left hand side is clearly independent of q_0. The answer is as follows. A minimizing path from q_0 to q within an infinite time interval necessarily proceeds first along a deterministic trajectory from q_0 to the attractor $\mathscr{A}_i$ in whose domain of attraction q_0 has been chosen. Along this part of the path L_0 vanishes, i.e. this part of the path does not contribute to $\Phi(q)$. Near the attractor an infinite amount of time elapses before the trajectory goes on the final point q, possibly passing through small neighborhoods of other singular points. This final part, leading away from the attractor $\mathscr{A}_i$, is necessarily nondeterministic, i.e. $\dot{q}^{v} \neq K^{v}$, $L_0 > 0$, and Φ receives a positive contribution in (7.14.4). The total path selected by the minimum principle (7.14.4) is, of course, nothing but the most probable path of (7.3.1) from q_0 to q in an infinite time interval. Since L_0 vanishes on the first part of the path, from q_0 to $\mathscr{A}_i$, the lower boundary of the integral in (7.14.4) may be replaced by some point $q_0 \in \mathscr{A}_i$. In this manner each attractor $\mathscr{A}_i$ of the system (bounded away from infinity) generates a potential

$$\Phi_i(q) = \min \int_{\mathscr{A}_i}^{q} L_0(q, \dot{q}) \, d\tau + C(\mathscr{A}_i), \tag{7.14.5}$$

where $C(\mathscr{A}_i)$ is an additive constant. Therefore, if the system has no attractor at all (except infinity), a potential Φ cannot be defined for it. If the system has only a single attractor $\mathscr{A}_i$ then $\Phi = \Phi_i(q)$ for all points q inside its domain of attraction, while for all points q outside its domain Φ is not defined. If the system has several attractors $\mathscr{A}_i$ then Φ has to be constructed from the various pieces $\Phi_i(q)$ according to (7.3.2) by taking the minimum of all the $\Phi_i(q)$ at the point q

$$\Phi(q) = \min_{\mathscr{A}_i} \Phi_i(q). \tag{7.14.6}$$

In order to take the minimum it is necessary to determine the relative size of the constants $C(\mathscr{A}_i)$ appearing in (7.14.5). These constants are fixed by the balance of the probability flow between the domains of the various attractors in the steady state. The net effect of this fixing of the $C(\mathscr{A}_i)$ is the equality of $\Phi_i(q_s)$ and $\Phi_j(q_s)$ for attractors $\mathscr{A}_i$ and $\mathscr{A}_j$ with adjacent basins of attraction in the lowest lying saddle q_s on the separatrix separating their basins. For details we refer to Freidlin and Ventzel (1984) or Graham and Tél (1985b). Equations (7.14.5) and (7.14.6) may be simplified further by noting that

$$L_0(q, \dot{q}) = p_\nu \dot{q}^\nu - H(q, p) \tag{7.14.7}$$

and recalling that in order to calculate the minimal action we only need solutions of (7.11.5) and (7.11.6) satisfying $H(q, p) = 0$. Equation (7.11.6) is thereby reduced to

$$\Phi(q) = \min_{\mathscr{A}_i} \left[\min_{\mathscr{A}_i} \int_{\mathscr{A}_i}^q p_\nu^{(0)}(q) \, dq^\nu + C(\mathscr{A}_i) \right]. \tag{7.14.8}$$

In (7.14.8) the invertibility of the matrix $Q^{\nu\mu}$ is no longer needed. All that is needed is a matrix $Q^{\nu\mu}$ ensuring the uniqueness of $P_\infty(q, \eta)$ in (7.2.3). In (7.14.8) we made use of the fact that the functions $p_\nu(q)$ passing through the attractors $\mathscr{A}_i$ are just the separatrices $p_\nu^{(0)}(q)$ we considered earlier in (7.11.8). The evaluation of Φ is now well defined by (7.14.8) even in the general case of wildly oscillating separatrices. The general scheme which emerges is indicated in Figure 7.10. The rule which implements (7.14.8) in the case of an oscillating separatrix is simply the interpolation of the oscillations by Maxwell's rule of equal areas. As a result, a discontinuity of the first and higher derivatives of Φ appears every time that rule has to be used. Physically, this discontinuity results from the existence of two or more most probable paths from the same attractor $\mathscr{A}_i$ to any point on the surface of discontinuity. Figure 7.11 shows the oscillating separatrix of the example of Section 7.12 after interpolating the various branches by Maxwell's rule. Only a small neighborhood of the saddle near $x = 0$ is shown. The same figure also displays the nondifferentiable potential Φ near the saddle. A discontinuity of the first and higher derivatives of Φ appears also in all points in which two or more most probable paths from different attractors $\mathscr{A}_i$ meet.

It is clear that the discontinuity of the first derivatives is a property of Φ, but it does not appear in $P_\infty(q, \eta)$. The way in which the discontinuity is avoided in the probability density is easily seen by considering the example of two most probable paths from $\mathscr{A}_i$ to a point $q = \bar{q}$. In this case $P(q, \eta)$ close to $q = \bar{q}$ can be represented as

$$P(q, \eta) \simeq Z_1(q) \exp\left(-\frac{\Phi_1(q)}{\eta} \right) + Z_2(q) \exp\left(-\frac{\Phi_2(q)}{\eta} \right), \tag{7.14.9}$$

where $\Phi_{1,2}$ and $Z_{1,2}$ correspond to the two branches of the separatrix $p_\nu = p_\nu^{(0)}(q)$, $(\nu = 1, \ldots, n)$ which are connected by Maxwell's rule at $q = \bar{q}$ when taking $\eta \to 0$. For any arbitrarily small but finite value of η the right hand side of (7.14.9) has continuous first derivatives at $q = \bar{q}$ (for continuously differentiable $\Phi_{1,2}$ and $Z_{1,2}$).

In summary, a large part of the structure of equilibrium thermodynamics can be generalized to nonequilibrium systems. The reversibility properties of equilibrium thermodynamics have to be sacrificed, however, in order to

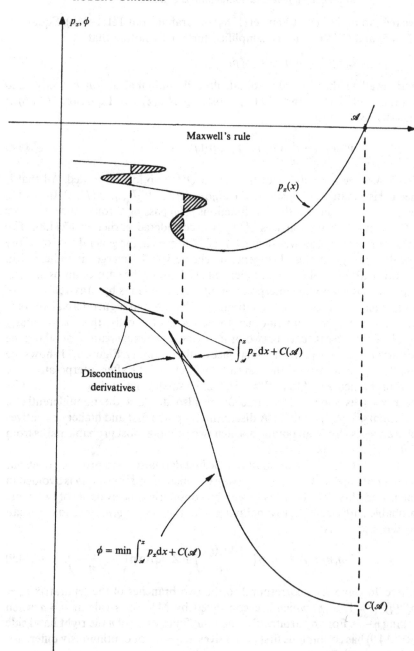

Figure 7.10. Evaluation of $\phi(q)$ from the oscillating 'wild separatrix'. (After Graham and Tél, 1986.)

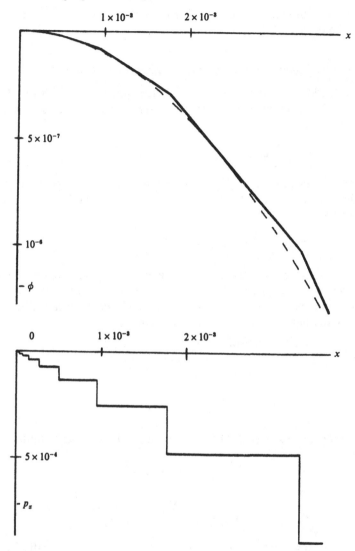

Figure 7.11. Evaluation of $\phi(q)$ for the example of Figure 7.9. (From Graham and Tél, 1985a.)

achieve this generalization. As a consequence the analytical properties of the nonequilibrium potential are less nice than in thermodynamic equilibrium. In general, there may exist surfaces in configuration space where the first derivatives of the potential jumps. Despite these complications, a non-equilibrium potential exists for a very large class of systems (namely those for which (7.3.2) is well defined) and thus provides a common conceptual

261

framework in which systems in thermodynamics equilibrium and far from it may be discussed.

7.15   Application to noise in Josephson junctions

We consider a resistively shunted Josephson junction. The potential difference across the junction is given by the Josephson equation (Josephson, 1962; Solymar, 1972)

$$V = \frac{\hbar}{2e} \dot{\varphi} \tag{7.15.1}$$

where φ is the phase difference of the macroscopic wave function on both sides of the junction and e is the electron charge. The current through the junction is given by

$$I_{\text{ex}} = \frac{V}{R} + C\dot{V} + I_0 \sin\varphi + \delta I(t) \tag{7.15.2}$$

where R and C are the resistance and the capacitance of the junction, respectively, and I_0 denotes the maximum of the current due to tunnelling of Cooper pairs. The current I_{ex} is externally controlled and consists of a systematic part I_s and a thermal noise current

$$I_{\text{ex}} = I_s + \delta I(t), \tag{7.15.3}$$

where

$$\left.\begin{aligned}
\langle \delta I(t) \rangle &= 0, \\
\langle \delta I(t)\delta I(t') \rangle &= \frac{2k_{\text{B}}T}{R} \delta(t - t').
\end{aligned}\right\} \tag{7.15.4}$$

Combining (7.15.1)–(7.15.4) and introducing dimensionless units

$$\left.\begin{aligned}
x &= \varphi, \quad v = \frac{2e}{\hbar} V, \quad f(x) = \sin x \\
F &= \frac{I_{\text{ex}}}{I_0}, \quad \gamma = \left(\frac{\hbar}{2eI_0 R^2 C}\right)^{1/2}, \quad \eta = \frac{2ek_{\text{B}}T}{\hbar I_0}
\end{aligned}\right\} \tag{7.15.5}$$

we obtain the equations of Brownian motion of a mathematical pendulum with constant externally applied torque

$$\left.\begin{aligned}
\dot{x} &= v \\
\dot{v} &= -\gamma v - f(x) + F + \xi(t) \\
\langle \xi(t)\xi(t') \rangle &= 2\eta\gamma\,\delta(t - t').
\end{aligned}\right\} \tag{7.15.6}$$

The configuration space of (7.15.6) is the cylinder

$$\left.\begin{aligned}
0 \leqslant x &< 2\pi \\
-\infty < v &< +\infty.
\end{aligned}\right\} \tag{7.15.7}$$

Equation (7.15.6) now constitutes the model for which we wish to determine the nonequilibrium potential (Ben-Jacob, Bergmann, Matkowsky and Schuss, 1982; Graham and Tél, 1985b). At first glance it might appear that an exact analytical result for Φ can immediately be written down, namely

$$\Phi = \Phi_0(x, v) \equiv \frac{v^2}{2} - \cos x - Fx + \text{const.}, \tag{7.15.8}$$

which is an obvious solution of the Hamilton Jacobi equation (7.3.7). In fact, Φ_0 is just the energy of the 'pendulum'. However, even though Φ_0 satisfies the Hamilton Jacobi equation it is not a 2π-periodic function of the phase variable x and hence does not satisfy (7.3.2). It is therefore not acceptable as a nonequilibrium potential, i.e. for $F \neq 0$ the function (7.15.7) is not the minimum in the sense of (7.14.7) everywhere on the cylindrical configuration space (7.15.7). Rather, Φ_0 must be replaced at least in some parts of this space by a lower lying potential $\bar{\Phi}_1$ still to be determined. When $\bar{\Phi}_1$ is continuously joined with Φ_0 along some curve (also yet to be determined) with discontinuous first and higher derivatives, the two functions $\Phi_0, \bar{\Phi}_1$ together define a periodic potential Φ. In order to find $\bar{\Phi}_1$ and the curves of discontinuity of derivatives the Hamiltonian equations (7.11.5), (7.11.6) and the extremum principle (7.14.5), (7.14.6) must be evaluated. In the present case the Hamiltonian equations at $H = 0$ can be reduced to the form (Graham and Tél, 1985b; Graham, Roekaerts and Tél, 1985)

$$\left.\begin{array}{l} \dot{x} = v \\[2mm] \dot{v} = -\gamma(t)v - f(x) + F, \end{array}\right\} \tag{7.15.9}$$

with

$$\gamma(t) = \gamma \frac{A\exp(-\gamma t) - 1}{A\exp(-\gamma t) + 1}. \tag{7.15.10}$$

In (7.15.10) A is a constant of integration which must be fixed in such a way that an initial point at $t \to -\infty$ in one of the attractors of the deterministic system ((7.15.6) with $\xi = 0$) is connected with a given final point (x, v) at $t = 0$ by a solution of (7.15.9). The minimum principle (7.14.5), (7.14.6) or (7.14.8), in the present case, takes the form

$$\Phi(x, v) = \gamma \min_{\mathscr{A}_i} \left[\min_{\mathscr{A}_i} \int^{(x,v)} \frac{\dot{x}^2 \, dt}{(1 + A\mathrm{e}^{-\gamma t})^2} + C(\mathscr{A}_i) \right], \tag{7.15.11}$$

where $x(t)$ satisfies (7.15.9). The attractors $\mathscr{A}_i$ over which the minimum is taken are: (i) a single fixed point $P_0 = (x_0, v)$ if F is sufficiently small, $0 < F < F_{\mathrm{c}}(\gamma)$, in which case there is no need to take a minimum over different attractors; (ii) the fixed point P_0 coexisting with a limit cycle $v = v_{\mathrm{c}}(x)$, $0 \leqslant x \leqslant 2\pi$ for $F_{\mathrm{c}}(\gamma) < F < 1$; (iii) the limit cycle for $F > 1$. Here $F_{\mathrm{c}}(\gamma)$ is a critical value of the control parameter F, which depends on γ. In the coexistence region $F_{\mathrm{c}}(\gamma) < F < 1$ of the two attractors (the fixed point and the limit cycle) the two constants $C(\mathscr{A}_{1,2})$ have to be adjusted in such a way that Φ is continuous in the saddle point $S = (x_s, 0)$ of the deterministic system ((7.15.5) with $\xi \equiv 0$). For the details of the construction of Φ by using (7.15.9) and (7.15.11) we refer to Graham and Tél (1985b). In Figures 7.12–7.14 we display the results in the three different regimes. If the attracting fixed point P_0 exists $(0 < F < 1)$ the potential Φ is given by (7.15.8) for all q sufficiently close to P_0. Figure 7.12 shows the equipotential curves $\Phi = $ const. on the (x, v)-cylinder $(x_s - 2\pi < x < x_s, \ -\infty < v < +\infty)$ for $-6 < v < 6$ in the case $0 < F < F_c(\gamma)$. The domains where Φ_0 and $\bar{\Phi}_1$ represent the potential are indicated. Along the dashed–dotted line $\Phi_0 = \bar{\Phi}_1$ holds, but with discontinuous first derivatives. Points along this line can be reached at $t = 0$ by two different paths starting at $t = -\infty$ in P_0 and satisfying (7.15.11). The first of these paths proceeds directly from P_0 to the endpoint. This first class of paths dominates in the domain of Φ_0. The second path first proceeds from the fixed point P_0 to the saddle $S = (x_s, 0)$, which is equivalent to $S' = (x_s, -2\pi, 0)$ and then from S' to the endpoint. This second class of paths dominates in the domain $\bar{\Phi}_1$. Along the dashed lines shown in Figure 7.12 both classes of paths continuously merge into each other. Thus, along these lines the two functions Φ_0 and $\bar{\Phi}_1$ are connected with continuous derivatives. As shown in Graham and Tél (1985b) the dashed lines are obtained from the stable manifolds $v = \tilde{v}(x)$ of the saddle S (equivalent to S') by the reflection $v \to -v$, i.e. these lines are the curves $v = -\tilde{v}(x)$. The equipotential line connecting the saddle $S(S')$ with the point $\bar{P}$ is a part of the unstable manifold of S'. Since it is a deterministic trajectory beginning in S', L_0 vanishes there, i.e. Φ is constant. For the case $0 < F < F_c(\gamma)$ shown in Figure 7.12 the potential has a single global minimum in P_0.

If F is increased, the point $\bar{P}$ in Figure 7.12 moves along the upper part of the loop enclosing P_0 to the right towards S until it reaches the point S for $F = F_c(\gamma)$. In this case an equipotential line connects the equivalent saddle points S, S' (homoclinic line), and this line now becomes a new local minimum of the potential, representing a new attractor – the limit cycle which is here borne by a homoclinic bifurcation. Upon increase of F beyond $F_c(\gamma)$ the limit cycle is disconnected from S', S. Equipotential contours for this case are shown in Figure 7.13, for values of Φ below the saddle point value $\Phi(S) = \Phi(S')$. $\Phi(S)$ is normalized to zero, as in Figure 7.12. The dashed line shows an approximate result for the equipotential line $\Phi = \Phi_s$ obtained in Ben-Jacob et al. (1982). For the particular case shown in Figure 7.13 the minimum of Φ at the limit cycle (at $\Phi = -8.8$) is much deeper than the minimum at P_0, indicating that the limit cycle has much higher stability than P_0 in this case. Such a quantitative comparison of the stability of the coexisting attractors is only possible because a potential has been constructed. The general properties of Φ discussed in Section 7.3 ensure that all points of Figure 7.13 above the equipotential line $\Phi = \Phi(S)$ connecting S' and S are attracted by the limit cycle in the deterministic

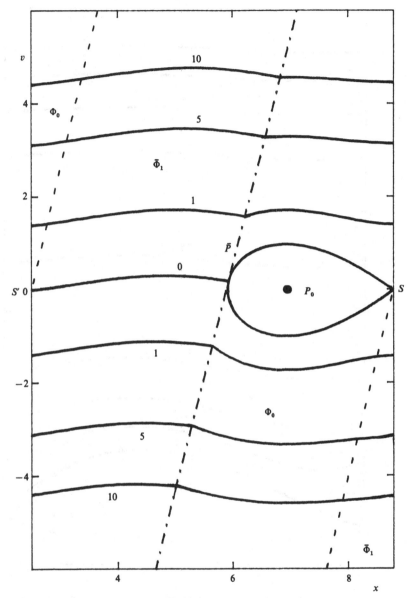

Figure 7.12. Equipotential curves of the potential (7.15.11) for $0 < F < F_c(\gamma)$
($\gamma = 5$, $F = 0.6$). (From Graham and Tél, 1985b.)

system and can reach P_0 only by a fluctuation. Similarly all points within the
closed loop $\Phi = \Phi_s$ surrounding P_0 are attracted by P_0 in the deterministic
system, and can reach the limit cycle only by a fluctuation. As the potential
minimum at P_0 is rather shallow in Figure 7.13 the mean first exit time for the

265

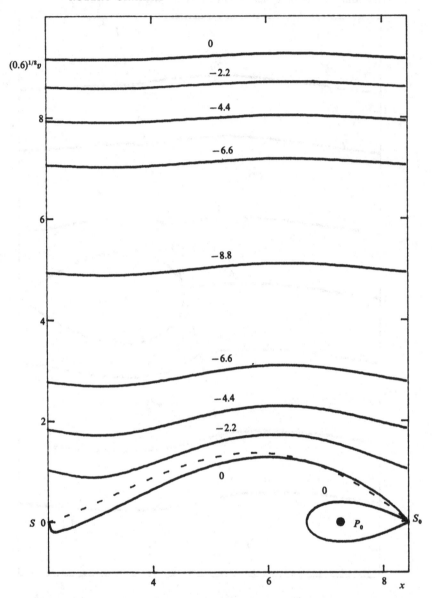

Figure 7.13. Equipotential curves of the potential (7.15.11) for $F_c(\gamma) < F < 1$ ($\gamma = 0.13$, $F = 0.83$, values of $(0.6)\phi$ indicated). (From Graham and Tél, 1985b.)

latter case is much shorter than for the opposite case. For initial points in the domain where $\Phi > \Phi_s$ in principle both attractors can be reached by going downwards in the potential. In this region the domain of the two attractors is therefore not determined by Φ alone but only by Φ together with r^ν as defined in (7.3.10).

266

If F is increased further, the point P_0 in Figure 7.13 moves to the right towards S and reaches S at $F = 1$. The loop $\Phi = 0$ around P_0 has then shrunk to a point coinciding with S, and the attractor P_0 disappears by a saddle-node bifurcation. The attractor P_0 generating the potential Φ_0 in its neighborhood then disappears together with the potential Φ_0. For $F > 1$ only the limit cycle remains, and the potential generated from it via (7.14.5) is automatically periodic. Equipotential contours for this case are shown in Figure 7.14.

In summary, we see that the nonequilibrium steady state of a Josephson junction driven by a constant external current minimizes a macroscopic potential Φ whose form can be exhibited in the qualitatively different operation regimes of the junction. The form of Φ allows us to compare the stability and the mean times of first exit of the coexisting attractors for $F_c(\gamma) < F < 1$ and to identify those parts of their domains of attraction where the potential is below its saddle value.

7.16   Lorenz attractor and noise

The Lorenz model was originally introduced as a (not very realistic) model for thermal convection (Lorenz, 1963; Sparrow, 1982). However, it also arises in quantum optics as the fundamental model of a homogeneously broadened single-mode laser (Haken, 1975). Its theoretical interest derives from the appearance of a strange attractor. Recently, the parameter conditions for the Lorenz model to display a strange attractor seem to have been realized physically for the first time in a far-infrared laser (Klische and Weiss, 1985). We are here, of course, not counting realizations in analogue computers, which are very easy. In the following we consider the Lorenz model perturbed by weak noise, under conditions where the deterministic dynamics is described by the Lorenz attractor (Dörfle and Graham, 1983; Zippelius and Lücke, 1981). The equations of motion of the model read

$$
\left.
\begin{aligned}
\dot{x} &= -\sigma(x - y) + \eta^{1/2}\xi_x(t) \\
\dot{y} &= -y + (r - z)x + (\eta\alpha)^{1/2}\xi_y(t) \\
\dot{z} &= -bz + xy + (\eta\beta)^{1/2}\xi_z(t),
\end{aligned}
\right\}
\tag{7.16.1}
$$

where ξ_x, ξ_y, ξ_z are statistically independent sources of Gaussian white noise, normalized to unit intensity. The parameter η controls the weak noise limit as before. The Lorenz model without noise has three parameters r, σ, b. In numerical examples we shall assign the standard values (Lorenz, 1963) $r = 28$, $\sigma = 10$, $b = 8/3$ to place ourselves in the domain of parameter space where the Lorenz attractor is realized. The parameters α, β of order unity describe the relative noise strengths of the three fluctuating forces. The variables x, y, z are assumed to be real, for simplicity. In applications to the single-mode laser x and y must be complex, if fluctuations are allowed for. We note that stochastic perturbations of the form of the noise terms in (7.16.1) appear automatically in

ROBERT GRAHAM

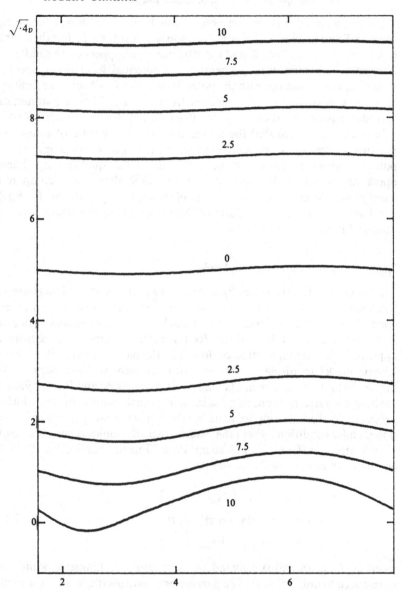

Figure 7.14. Equipotential curves of the potential (7.15.11) for $F > 1$ ($\gamma = 0.16$, $F = 1.25$, values of (0.4) ϕ indicated). (From Graham and Tél, 1985b.)

the quantum theory of the laser, i.e. in the quantized Lorenz model. The necessary generalizations of the following discussion to this case have been given in Graham (1984).

The properties of the Lorenz model without external noise have been the subject of many papers, following Lorenz (1963). A review has been given by Sparrow (1982). Here we are interested in the construction of the steady state probability of the model (7.16.1) along the lines of Section 7.3 (Dörfle and Graham, 1983). This requires that η is sufficiently small

$$\left(\frac{\eta}{\sigma + 1 + b}\right)^{1/2} \ll 1. \tag{7.16.2}$$

As a new problem we have to cope here with the fact that in the present example the attractor is a fractal set with scales of arbitrarily small size. It is clear from the outset that only a finite part of this infinite hierarchy of small scales can be resolved in the steady state distribution for any arbitrarily small but finite value of η. In other words, there are always scales smaller than the scale (7.16.2) set by η, which are relevant for the deterministic Lorenz attractor. In the following we shall choose η in such a way that the main two-leaf structure of the Lorenz attractor is resolved, but the Cantor substructure of each of the two leaves is still swamped by noise. This is achieved by choosing η inside the window

$$e^{-(\sigma + 1 + b)T} \ll \left(\frac{\eta}{\sigma + 1 + b}\right)^{1/2} \ll 1, \tag{7.16.3}$$

where T is a typical round trip time on the attractor, which is of order unity for the parameter values chosen. As the left hand side of (7.16.3) is very small, we can still use asymptotic approximations for $\eta \to 0$.

The steady state probability density of the model (7.16.1) is now sought in the form (7.3.3), where $\Phi(x, y, z)$ satisfies the Hamilton Jacobi equation (7.3.7)

$$-\sigma(x - y)\frac{\partial\Phi}{\partial x} - (y + (z - r)x)\frac{\partial\Phi}{\partial y} - (bz - xy)\frac{\partial\Phi}{\partial z}$$

$$+ \frac{1}{2}\left(\frac{\partial\Phi}{\partial x}\right)^2 + \frac{\alpha}{2}\left(\frac{\partial\Phi}{\partial y}\right)^2 + \frac{\beta}{2}\left(\frac{\partial\Phi}{\partial z}\right)^2 = 0, \tag{7.16.4}$$

and $Z(x, y, z)$ satisfies (7.3.8)

$$\left[-\sigma(x - y) + \frac{\partial\Phi}{\partial x}\right]\frac{\partial Z}{\partial x} + \left[-y - x(z - r) + \alpha\frac{\partial\Phi}{\partial y}\right]\frac{\partial Z}{\partial y}$$

$$+ \left[-bz + xy + \beta\frac{\partial\Phi}{\partial z}\right]\frac{\partial Z}{\partial z}$$

$$+ \left[-\sigma - 1 - b + \frac{1}{2}\left(\frac{\partial^2\Phi}{\partial x^2} + \alpha\frac{\partial^2\Phi}{\partial y^2} + \beta\frac{\partial^2\Phi}{\partial z^2}\right)\right]Z = 0. \tag{7.16.5}$$

As was discussed in Section 7.3 the potential $\Phi(x, y, z)$ must be a minimum and constant on the attractor of the system, which, in the case of the fractal Lorenz

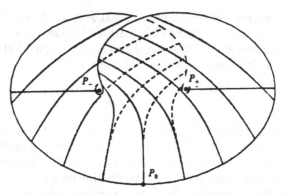

Figure 7.15. The two effective branches of the Lorenz attractor for $\gamma = 28$, $\sigma = 10$, $b = 8/3$ and their boundary shown by contour lines $x = $ const. in the (y, z)-plane with origin in P_0. The two branches are assumed to be joined along the horizontal cuts. (From Dörfle and Graham, 1983.)

attractor, implies that Φ must be a highly complicated function, which at present we cannot determine analytically. However, if η is in the window defined by (7.16.3), the attractor may be approximated by a branched two-dimensional manifold with two leaves (see Figure 7.15), and we may look for an effective potential $\tilde{\Phi}(x, y, z)$ which has minima on the two main branches of the attractor. The applicability of this simplification depends on the fact that the two main branches of the attractor appear on a scale which is much bigger than even the largest scales of the Cantor substructure of the attractor, which is satisfied for the parameter values chosen here. Making this simplification, we can then go one step further and expand $\tilde{\Phi}(x, y, z)$ locally around the two effective branches of the attractor, written as

$$x = f(y, z). \tag{7.16.6}$$

Keeping terms up to second order we put

$$\tilde{\Phi}(x, y, z) = \text{const.} + \frac{1}{2}\frac{(x - f(y, z))^2}{\varphi(x, y)} + O((x - f)^3) \tag{7.16.7}$$

with a yet undetermined function $\varphi(x, y)$. From (7.16.4) we then find, by comparing the coefficients of $(x - f)$,

$$(y + (z - r)f)\frac{\partial f}{\partial y} + (bz - yf)\frac{\partial f}{\partial z} + (y - f)\sigma = 0, \tag{7.16.8}$$

which is the partial differential equation of an invariant surface of the deterministic equations (7.16.1) with $\eta = 0$. Within the finite accuracy permitted by the window of η-values (7.16.3), (7.16.8) can be solved by a two-dimensional branched manifold consisting of the two leaves

$$f(y,z) = \begin{cases} f_+(y,z) & \begin{cases} y > -(b(r-1))^{1/2}, & z > 0 \\ y < -(b(r-1))^{1/2}, & 0 < z < r-1 \end{cases} \\ \\ f_-(y,z) & \begin{cases} y < (b(r-1))^{1/2}, & z > 0 \\ y > (b(r-1))^{1/2}, & 0 < z < r-1. \end{cases} \end{cases} \tag{7.16.9}$$

The two branches are taken to be joined for $z = r - 1$ and $y < -(b(r-1))^{1/2}$ and $y > (b(r-1))^{1/2}$. Obviously this cannot hold for exact invariant manifolds of the deterministic system, hence the Cantor structure, but it holds approximately to a high degree of accuracy on scales of the order of the left hand side of (7.16.2) which is all we need in our approximation. Explicit representations of the two branches f_+, f_- can be obtained by determining the two-dimensional unstable manifolds of the two fixed points $P_\pm = [\pm (b(r-1))^{1/2}, \pm (b(r-1))^{1/2}, r - 1]$ in a sufficiently large neighborhood of $P_\pm$ (Graham and Scholz, 1980).

We also obtain an equation for $\varphi(y,z)$ by putting (7.16.7) into (7.16.4) and comparing the coefficients of $(x - f)^2$. It is easy to check that the $O((x - f)^3)$ terms in (7.16.7) cannot yet contribute to this order. We then find

$$(y - (r - z)f)\frac{\partial \varphi}{\partial y} + (bz - yf)\frac{\partial \varphi}{\partial z} + 2\lambda_\perp(y,z)\varphi$$

$$= -1 - \alpha\left(\frac{\partial f}{\partial y}\right)^2 - \beta\left(\frac{\partial f}{\partial z}\right)^2. \tag{7.16.10}$$

Here we introduced the local rate of attraction $\lambda_\perp(y,z)$ of the attractor (7.16.6), given by

$$\lambda_\perp(y,z) = -\sigma - (r - z)\frac{\partial f}{\partial y} - y\frac{\partial f}{\partial z}, \tag{7.16.11}$$

which is negative, i.e. $\lambda_\perp < 0$. Equation (7.16.10) can be solved by the method of characteristics. The characteristics satisfy

$$\dot{y} = -y + (r - z)f(y,z)$$
$$\dot{z} = -bz + yf(y,z). \tag{7.16.12}$$

Equations (7.16.12) are just the deterministic equations (7.16.1) for $\eta = 0$ restricted to $x = f(y,z)$. This restriction is possible, approximately, since in the present approximation $x = f(y,z)$ is an invariant manifold of (7.16.1) for $\eta = 0$. Integrating (7.16.10) from $t = -\infty$ along one of the characteristics ending at time $t = 0$ in the point (x, y) of the branched manifold we obtain

$$\varphi(y, z) = \int_{-\infty}^0 d\tau (1 + \alpha f_y^2(\tau) + \beta f_z^2(\tau))$$

$$\times \exp\left[2 \int_\tau^0 \lambda_\perp(\tau') d\tau' \right], \tag{7.16.13}$$

where

$$f_y = \frac{\partial f}{\partial y}, \quad f(\tau) = f(y(\tau), z(\tau)). \tag{7.16.14}$$

Equations (7.16.12) actually describe a semiflow which is noninvertible (due to our approximation for f). Therefore, there are actually nondenumerably many different characteristics of the semiflow ending in any given point (x, y), which differ in the way they are continued backward through the cuts in (7.16.9), but due to the exponential damping provided by $\lambda_\perp(\tau)$ it is guaranteed that the differences in $\varphi(y, z)$ among these different characteristics are of the order of the factor on the left hand side of (7.16.3). In fact, since the local attraction rate $\lambda_\perp(y, z)$ in the Lorenz model for the standard parameter values is rather large, we may even safely assume that the relative magnitudes of $f_y(\tau)$, $f_z(\tau)$ and $\lambda_\perp(\tau)$ change negligibly along that part of the characteristics which contributes in (7.16.13), in which case (7.16.13) simplifies to

$$\varphi(y, z) = -\frac{1}{2} \frac{1}{\lambda_\perp(y, z)} (1 + \alpha f_y^2(y, z) + \beta f_z^2(y, z)). \tag{7.16.15}$$

Turning now to the prefactor $Z(x, y, z)$ governed by (7.16.5) we first note that, within the present approximation, it is sufficient to determine Z only for values of x on the attractor, i.e. we have to determine the function

$$\tilde{Z}(y, z) = Z(f(y, z), y, z). \tag{7.16.16}$$

It is useful to relate $\tilde{Z}(y, z)$ to the probability density $\rho_\infty(y, z)$ on the attractor. The latter is obtained from the full probability density $P_\infty(x, y, z)$ by

$$\rho_\infty(y, z) = \int dx \, P_\infty(x, y, z) = \rho_+(y, z) + \rho_-(y, z). \tag{7.16.17}$$

In (7.16.17) we may insert the form (7.3.3) with $\Phi \simeq \tilde{\Phi}$ given by (7.16.7). Evaluating the integral in (7.16.17) in saddle point approximation we then find

$$\rho(y, z) = \begin{cases} \rho_+(y, z) & \begin{cases} y > -(b(r-1))^{1/2}, & z > 0 \\ y < -(b(r-1))^{1/2}, & 0 < z < r - 1 \end{cases} \\ \\ \rho_-(y, z) & \begin{cases} y < (b(r-1))^{1/2}, & z > 0 \\ y > (b(r-1))^{1/2}, & 0 < z < r - 1, \end{cases} \end{cases} \tag{7.16.18}$$

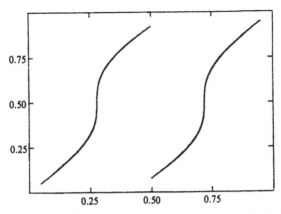

Figure 7.16. Effective one-dimensional return map on the horizontal cuts of Figure 7.15 parametrized from 0 at P_- at $\frac{1}{2}$ at the left border and from $\frac{1}{2}$ at the right border to 1 at P_+. (From Dörfle and Graham, 1983.)

with

$$\rho_\pm(y,z) = (2\pi\eta\varphi_\pm(y,z))^{1/2}\tilde{Z}_\pm(y,z). \qquad (7.16.19)$$

Here, and in (7.16.17), the subscript $\pm$ serves to distinguish the two branches of the attractor defined in (7.16.9). We shall omit this subscript (as in (7.16.10)–(7.16.16)) whenever this distinction has not to be made explicitly. From the equations satisfied by $\varphi(y,z)$ and $Z(x,y,z)$ we obtain the equation for ρ:

$$-(y+(z-r)f)\frac{\partial\rho}{\partial y} - (bz - yf)\frac{\partial\rho}{\partial z} + \lambda_\parallel(y,z)\rho = 0, \qquad (7.16.20)$$

where

$$\lambda_\parallel(y,z) = -(\sigma + 1 + b) - \lambda_\perp(y,z) \qquad (7.16.21)$$

is the local rate of divergence of the semiflow (7.16.12) on the attractor.

Equation (7.16.20) is the local conservation law of probability on the attractor. We only briefly summarize its solution here and refer to Dörfle and Graham (1983) for details. The method makes use of the return map induced by the semiflow (7.16.12) of the curve C, along which the two branches of the attractor are joined (see Figure 7.16). This curve C is parametrized by the y-coordinates of its points (the z-coordinates are constant, $z = r - 1$, and the x-coordinates are given by $x = f(y,z)$). First the invariant probability density $\mu_\infty(y)$ of this return map has to be determined, which is easy if done numerically (see Figure 7.17), but may also be done analytically, at least approximately. A simple geometrical argument then shows that the probability density $\rho(y,z)$ restricted to the curve C is obtained by the ratio of $\mu_\infty(y)$ and the absolute value of the orthogonal component of the velocity of the semiflow crossing the curve C at the point $x = f(y,z)$, $y,z = r - 1$ on C. Thus

$$\rho(y, r-1) = \text{const.} \cdot \frac{\mu_\infty(y)}{|-b(r-1) + yf(y, r-1)|}. \qquad (7.16.22)$$

273

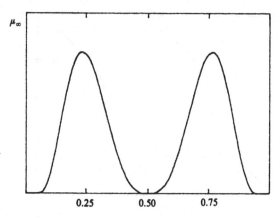

Figure 7.17. Invariant probability density of the return map of Figure 7.16. (From Dörfle and Graham, 1983.)

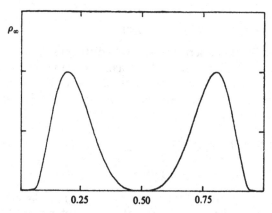

Figure 7.18. Invariant probability density on the two effective branches of Figure 7.15 restricted to the horizontal cuts. (From Dörfle and Graham, 1983.)

This probability density is shown in Figure 7.18. Having determined ρ on the curve C the method of characteristics may be used to obtain the solution for arbitrary (y, z). The characteristics of (7.16.20) are again given by (7.16.12). Thus, we have

$$\rho(y, z) = \rho(\bar{y}(y, z), r - 1) \exp \left[- \int_0^{t(y,z)} \lambda_{\parallel}(\tau) \, d\tau \right]. \qquad (7.16.23)$$

Here the time integral is carried out over the characteristic curve ending in the point (y, z) and starting a time interval $t(y, z)$ earlier in the corresponding point $(\bar{y}(y, z), r - 1)$ on the curve C. The result (7.16.23) is shown in Figure 7.19 for $\rho_+(y, z)$ and in Figure 7.20 for

$$\rho_{\infty}(y, z) = \rho_+(y, z) + \rho_-(y, z). \qquad (7.16.24)$$

274

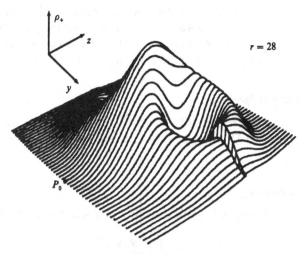

Figure 7.19. Invariant probability density on one of the two effective branches of Figure 7.15. (From Dörfle and Graham, 1983.)

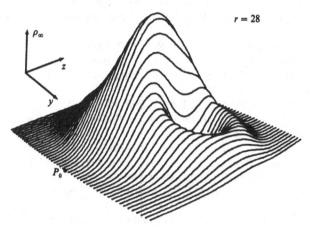

Figure 7.20. Invariant probability density summed over the two effective branches of Figure 7.15. (From Dörfle and Graham, 1983.)

In summary, the method of analysis we have described nicely separates the deterministic and the stochastic aspects of the probability density (7.3.3) and its constituents Φ ($\tilde{\Phi}$, in the present case) and Z (or $\tilde{Z}$). The noise intensities are found to enter explicitly only the curvature of $\tilde{\Phi}$ in its minimum $x = f$. Implicitly, the noise is also important for the neglect of the Cantor substructure, but only in as much as it is important that η lies within the window (7.16.3). The other results we have shown (i.e. the branched manifold $x = f$ and the probability density ρ on it) refer to the deterministic system only. One may conclude from this that the probability density induced by the

275

deterministic chaotic behavior in the Lorenz model is quite robust under small random fluctuations. Rigorous proofs of such a stability of chaos under random perturbations for a restricted class of systems (Axiom A systems not including the Lorenz model) have been given by Kifer (1974).

Acknowledgements

The author has enjoyed fruitful collaborations on various topics treated in this chapter with Michael Dörfle, Axel Schenzle, and, in particular, with Tamas Tél.

References

Ben-Jacob, E., Bergmann, D. J., Matkowsky, B. J. and Schuss, Z. 1982. *Phys. Rev. A* **26**, 2805.
Brand, H. and Steinberg, V. 1983. *Phys. Lett.* **93A**, 333.
Cohen, J. K. and Lewis, R. M. 1967. *J. Inst. Math. Appl.* **3**, 266.
Cross, M. C. 1982. *Phys. Rev. A* **25**, 1065.
DeGiorgio, V. and Scully, M. O. 1970. *Phys. Rev. A* **2**, 1170.
De Groot, S. R. and Mazur, P. 1962. *Nonequilibrium Thermodynamics.* Amsterdam: North-Holland.
Dörfle, M. and Graham, R. 1983. *Phys. Rev. A* **27**, 1096.
Ebeling, W. and Schimansky-Geier, L. 1985. *Fluid Dynam. Trans.* **12**, 7.
Ebeling, W., Herzel, H., Richert, W. and Schimansky-Geier, L. 1986. *Z. Angew. Math. Mech. (ZAMM)* **66**, 141.
Fraikin, A. and Lemarchand, H. L. 1985. *J. Stat. Phys.* **41**, 531.
Freidlin, M. I. and Ventzel, A. D. 1984. *Random Perturbations of Dynamical Systems.* Berlin: Springer.
Gardiner, C. 1983. *Handbook of Stochastic Methods.* Berlin: Springer.
Grabert, H., Graham, R. and Green, M. 1980. *Phys. Rev. A* **21**, 2136.
Graham, R. 1973a. In *Quantum Statistics in Optics and Solid-State Physics.* (G. Höhler, ed.), Springer Tracts in Modern Physics 66, pp. 1–97. Berlin: Springer.
Graham, R. 1973b. In *Coherence and Quantum Optics* (L. Mandel and E. Wolf, eds.). New York: Plenum.
Graham, R. 1973c. *Phys. Rev. Lett.* **31**, 1479.
Graham, R. 1974. *Phys. Rev. A* **10**, 1762.
Graham, R. 1975. In *Fluctuations, Instabilities and Phase Transitions* (T. Riste, ed.). New York: Plenum.
Graham, R. 1977. *Z. Phys. B* **26**, 281.
Graham, R. 1980. *Z. Phys. B* **40**, 149.
Graham, R. 1981. In *Stochastic Nonlinear Systems* (L. Arnold and R. Lefever, eds.). Berlin: Springer.
Graham, R. 1984. *Phys. Rev. Lett.* **53**, 2020.
Graham, R. 1986. *Europhys. Lett.* **2**, 901.
Graham, R. 1987a. In *Fluctuations and Stochastic Phenomena in Condensed Matter* (L. Garrido, ed.), Lecture Notes in Physics, vol. 268. Berlin: Springer.

Graham, R. 1987b. In *Proceedings of the 1986 Varenna School* (G. Caglioti, H. Haken and L. A. Lugiato, eds.). New York: Academic Press.
Graham, R. and Haken, H. 1970. *Z. Phys.* **237**, 31.
Graham, R. and Haken, H. 1971a. *Z. Phys.* **243**, 289.
Graham, R. and Haken, H. 1971b. *Z. Phys.* **245**, 141.
Graham, R. and Schenzle, A. 1983. *Phys. Rev. A* **27**, 1096.
Graham, R. and Scholz, H. J. 1980. *Phys. Rev. A* **22**, 1198.
Graham, R. and Tél, T. 1984a. *Phys. Rev. Lett.* **52**, 9.
Graham, R. and Tél, T. 1984b. *J. Stat. Phys.* **35**, 729.
Graham, R. and Tél, T. 1985a. *Phys. Rev. A* **31**, 1109.
Graham, R. and Tél, T. 1985b. *Phys. Rev. A* **33**, 1322.
Graham, R. and Tél, T. 1986. In *Stochastic Processes – Mathematics and Physics* (S. Albeverio, P. Blanchard and L. Streit, eds.). Lecture Notes in Physics 1158. Berlin: Springer.
Graham, R. and Tél, T. 1987. *Phys. Rev. A* **35**, 1328.
Graham, R., Roekaerts, D. and Tél, T. 1985. *Phys. Rev. A* **31**, 3364.
Green, M. S. 1952. *J. Chem. Phys.* **20**, 1281.
Guckenheimer, J. and Holmes, P. 1983. *Nonlinear Oscillations, Dynamical Systems, and Bifurcations of Vector Fields.* Berlin: Springer.
Haken, H. 1964. *Phys. Rev. Lett.* **13**, 326.
Haken, H. 1970. In *Encyclopedia of Physics* (S. Flügge, ed.), vol. XXV/2c. Berlin: Springer.
Haken, H. 1973. *Phys. Lett A.* **46**, 193.
Haken, H. 1975. *Rev. Mod. Phys.* **47**, 67.
Haken, H. 1977. *Synergetics.* Berlin: Springer.
Haken, H. 1983. *Advanced Synergetics.* Berlin: Springer.
Hänggi, P. 1986. *J. Stat. Phys.* **42**, 105.
Hänggi, P. and Thomas, H. 1982. *Phys. Rep.* **88**, 207.
Hietarinta, J. 1985. *J. Math. Phys.* **26**, 1970.
Jauslin, H. R. 1986. *Stat. Phys.* **42**, 573.
Jona-Lasinio, G. 1985. In *Turbulence and Predictability in Geophysical Fluid Dynamics and Climate Dynamics* (M. Ghil, R. Benzi and G. Parisi, eds.). Amsterdam: North-Holland.
Josephson, B. D. 1962. *Phys. Lett.* **1**, 251.
Kifer, I. Y. 1974. *Math. USSR Izvestija* **8**, 1083.
Kitahara, K. 1975. *Adv. Chem. Phys.* **29**, 85.
Klische, W. and Weiss, C. O. 1985. *Phys. Rev. A* **31**, 4049.
Knobloch, E. and Wiesenfeld, K. A. 1983. *J. Stat. Phys.* **33**, 611.
Kubo, R., Matsuo, K. and Kitahara, K. 1973. *J. Stat. Phys.* **9**, 51.
Landau, L. D. and Lifshitz, E. M. 1958. *Statistical Physics.* Oxford: Pergamon.
Langouche, F., Roekaerts, D. and Tirapegui, E. 1982. *Functional Integration and Semiclassical Expansion.* Dordrecht: Reidel.
Lebowitz, J. L. and Bergmann, P. G. 1957. *Ann. Phys. (N. Y.)* **1**, 1.
Lemarchand, H. 1984. *Bull. Acad. R. Belg.* **70**, 40.
Lemarchand, H. and Nicolis G. 1984. *J. Stat. Phys.* **31**, 609.
Lorenz, E. N. 1963. *J. Atmos. Sci.* **20**, 130.
Ludwig, D. 1974. *Stochastic Population Theories.* Berlin: Springer.
Ludwig, D. 1975. *SIAM Rev.* **17**, 4.

Newel, A. and Whitehead, J. 1969. *J. Fluid Mech.* **38**, 279.

Nicolis, G. and Prigogine, I. 1977. *Self-Organization in Non-Equilibrium Systems.* New York: Wiley.

Rehberg, I. and Ahlers, G. 1985. *Phys. Rev. Lett.* **55**, 500.

Risken, H. 1965. *Z. Phys.* **186**, 85.

Risken, H. 1983. *The Fokker-Planck Equation.* Berlin: Springer.

Roekaerts, D. and Schwartz, F. 1987. Preprint.

Schenzle, A. 1981. In *Proceedings of the International Conference Lasers '81* (C. B. Collins, ed.). McLean: STS Press.

Schenzle, A. and Brand, H. 1978. *Opt. Commun.* **27**, 485.

Schenzle, A. and Tél, T. 1985. *Phys. Rev. A* **32**, 596.

Schlüter, A., Lortz, D. and Busse, F. H. 1965. *J. Fluid Mech.* **23**, 129.

Schuss, Z. 1980. *Theory and Applications of Stochastic Differential Equations.* New York: John Wiley.

Segel, L. A. 1969. *J. Fluid Mech.* **38**, 203.

Solymar, C. P. L. 1972. *Superconductive Tunnelling and Applications.* London: Chapman and Hall.

Sparrow, C. T. 1982. *The Lorenz Equations: Bifurcations, Chaos and Strange Attractors.* Berlin: Springer.

Stratonovich, R. L. 1963. *Topics in The Theory of Random Noise*, vol. I. New York: Gordon and Breach.

Stratonovich, R. L. 1967. *Topics in The Theory of Random Noise*, vol. II. New York: Gordon and Breach.

Sulpice, E., Lemarchand, A. and Lemarchand, H. 1987. *Phys. Lett. A* **121**, 67.

Swift, J. and Hohenberg, P. C. 1977. *Phys. Rev. A* **15**, 319.

Talkner, P. and Hänggi, P. 1984. *Phys. Rev. A* **29**, 768.

Tél, T. 1988. *J. Stat. Phys.* **50**, 897.

Tomita, K. and Tomita, H. 1974. *Progr. Theor. Phys.* **51**, 1731.

Van Kampen, N. 1981. *Stochastic Processes in Physics and Chemistry.* Amsterdam: North-Holland.

Ventzel, A. D. and Freidlin, M. I. 1970. *Usp. Mat. Nauk.* **25**, 3.

Wunderlin, A. and Haken, H. 1981. *Z. Phys. B* **44**, 135.

Zippelius, A. and Lücke, M. 1981. *J. Stat. Phys.* **24**, 345.

8 Transition phenomena in multidimensional systems – models of evolution

W. EBELING and L. SCHIMANSKY-GEIER

8.1 Introduction

In this chapter we investigate the influence of noise in systems where many degrees of freedom are participating. The phenomenon we want to investigate is the possibility to leave and to enter attractor regions in bi- and multi-stable systems. It implies the ability to cross through separatrices (to jump over potential barriers) to alternative attractors of the systems after a certain time.

Since the pioneering work of Kramers this phenomenon has been widely investigated in one- and two-dimensional bistable systems (for reviews see Hänggi, 1986, and Weiss, 1986); several papers also appeared on multistable situations (see, e.g., Büttiker and Landauer, 1982). Stochastic transitions in multistable multidimensional systems are of importance for several physical and nonphysical problems, for example for nucleation induced phase transitions, for the molecular dynamics and the kinetics of chemical reactions, for the storage of information in electronic circuits, for learning procedures in networks of bistable elements, for the search of extrema in combinatorial optimization problems, etc. All those problems possess in general many interacting degrees of freedom, have a large number of stable states, configurations, etc.

Though in all the problems mentioned above stochastic transitions between stable states occur, their dynamics may be quite different. In some situations, for example chemical catalysis, it is desirable that the transition takes place very quickly. In other situations transitions should be avoided; for example, computer memory becomes meaningless if transitions occur on time scales which are comparable with the time of calculations. Thus one is often interested in manipulating (controlling) the transition time from outside the system. This problem plays a special role in multidimensional systems. In Sections 8.2 and 8.3 we use up to four degrees of freedom to produce more complicated noise than Gaussian white noise. In this way we are able to drive bistable systems by an Ornstein–Uhlenbeck process or by stochastic oscillators, thereby introducing, as additional control parameters, the correlation time and the oscillator frequency, respectively. Obviously that is the situation we often meet in physical problems. For example, the activation of the substrate inside an enzyme cannot be described with Gaussian white noise only. A more realistic model has to include the several experimentally measured frequencies of intramolecular motions of the enzyme which influence the transition from the substrate to the product. In Section 8.4 we consider networks of bistable elements. Some simple cases of interactions between the elements are investigated and we calculate mean transition times between the stable states. Section 8.5 is devoted to the study of a special class of evolution processes which are connected with transition phenomena in high-dimensional spaces (phenotypic spaces).

8.2 Bistable systems influenced by exponentially correlated Gaussian noise

We start by considering systems described by two stochastic differential equations

$$\dot{x} = f(x) + y(t), \tag{8.2.1}$$

$$\dot{y} = -\frac{1}{\tau}y + \frac{(2D)^{1/2}}{\tau}\xi(t), \tag{8.2.2}$$

$$\langle \xi(t) \rangle = 0; \quad \langle \xi(t)\xi(t') \rangle = \delta(t - t'). \tag{8.2.3}$$

Then $y(t)$ stands for a Gaussian noise whose correlation function is

$$\langle y(t)y(t') \rangle = \frac{D}{\tau}\exp\{-|t - t'|/\tau\}. \tag{8.2.4}$$

Such noise models physical systems in a more realistic way than with Gaussian white noise (see Horsthemke and Lefever, 1984). Due to the cut-off at $\omega_c \sim \tau^{-1}$ in its spectrum (Landau and Lifshits, 1971)

$$S_{yy}(\omega) = \frac{D}{\pi(1 + \tau^2\omega^2)} \tag{8.2.5}$$

only a finite value of frequencies will be excited.

Current theories often use (8.2.1) and (8.2.4) as the starting point of the derivation of a kinetic equation for the probability distribution density $P_t(x, t)$. One is then interested in its stationary solution (Fox, 1983; Fronzoni et al., 1986; Hänggi, Mroczkowski, Moss and McClintock, 1985; Lindenberg and West, 1983; Sancho, San Miguel, Katz and Gunton, 1982; Schenzle and Tél, 1985). These approaches show good results for one-dimensional scalar variables $x(t)$; the resulting kinetic equation (which is usually approximated by Fokker–Planck equation (FPE) with an effective diffusion coefficient) can be easily solved.

We follow here an alternative approach that involves enlarging the

280

dimensionality of the state space and introducing the auxiliary equation (8.2.2). In this way we arrive at two-dimensional stochastic processes. Several methods for investigating problems of this type have been presented. Solutions of the corresponding two-dimensional Fokker–Planck equation were given in terms of infinite matrix continued fractions by Jung and Risken (1985). Phase portraits of the stationary probability distribution density (SPDD) resulting from modelling (8.2.1) and (8.2.4) by electronic circuits were shown by Moss and McClintock (1985). Remarkable effects due to finite relaxation times τ are skewed probability densities (cf. Vogel *et al.*, 1987) and an increase in the mean transition times between the stable states of a bistable system for small τ.

In the following we shall give an approximation formula of the SPDD which reflects these properties and solves the FPE for small τ (Malchow and Schimansky-Geier, 1985). The stationary FPE reads (Stratonovich, 1963)

$$LP^0(x, y) = -\frac{\partial}{\partial x}[f(x) + y]P^0 - \frac{\partial}{\partial y}\left\{-\frac{1}{\tau}y - \frac{D}{\tau^2}\frac{\partial}{\partial y}\right\}P^0 = 0.$$

(8.2.6)

Introducing the potential (Graham, 1973)

$$\Phi(x, y) = -D \ln P^0(x, y) \qquad (8.2.7)$$

it follows that

$$F\Phi = (f + y)\frac{\partial\Phi}{\partial x} + \frac{1}{\tau}\frac{\partial\Phi}{\partial y}\left(-y + \frac{1}{\tau}\frac{\partial\Phi}{\partial y}\right)$$

$$- D\left(\frac{\partial f}{\partial x} + \frac{1}{\tau^2}\frac{\partial^2\Phi}{\partial y^2}\right) = 0, \qquad (8.2.8)$$

where F is a nonlinear operator.

Equation (8.2.8) will be solved by a series in small τ. After successive iteration with increasing powers of τ we get

$$\Phi = \frac{\tau}{2}(1 - \tau f'(x))(y + f(x))^2 - \int f(1 - f'\tau + \tfrac{1}{2}\tau^2 f'')\,dx$$

$$+ \tfrac{1}{2}f''(x)\tau^3 y^2(\tfrac{1}{2}f + \tfrac{1}{3}y) + \tfrac{3}{2}D\tau f'(x). \qquad (8.2.9)$$

Introducing this ansatz into (8.2.6) one easily confirms

$$LP^0 = O(\tau^3, D\tau^2)P^0.$$

This proves that (8.2.9) is an approximate solution valid for small correlation times τ.

Further on we neglect the item with $O(\tau^3)$. We mentioned it only since it was necessary for the calculation of the term with $O(D\tau)$.

The probability density for a symmetric bistable system

$$f(x) = ax - bx^3 \qquad (8.2.10)$$

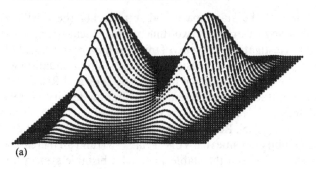

(a)

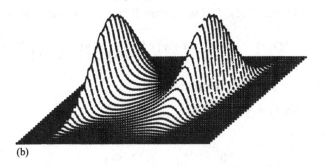

(b)

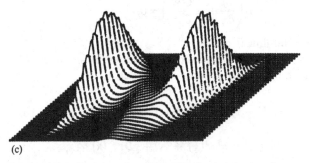

(c)

Figure 8.1. The stationary probability distribution density of a symmetric
bistable system: $a = 1$, $b = 1$, $D = 0.144$. (a) $\tau = 0.1$; (b) $\tau = 0.5$; (c) $\tau = 0.8$.

is shown in Figure 8.1(a–c) for various values of τ. Qualitatively we find good
agreement with the experimentally observed skew phase portraits of Moss and
McClintock (1985) and with the calculations of Jung and Risken (1985) based
on continued matrix fraction. Introducing the reduced PDD

$$P_\tau(x) = \int_{-\infty}^{\infty} \mathrm{d}y \, P^0(x, y)$$

282

the result of Jung and Hänggi (1987) follows (here for small τ):

$$P_\tau(x) \sim |1 - \tau f'(x)| \exp\left\{\frac{1}{D} \int f(1 - f'\tau)\,dx\right\} \qquad (8.2.11)$$

Further let us estimate the mean transition time between the maximal states of probability by considering the concept of the mean first passage time through an unstable state (Gardiner, 1983; Landauer and Swanson, 1961; Langer, 1969; Matkowsky and Schuss, 1977; Van den Broeck and Hänggi, 1984).

In the two-dimensional case considered above, which is a problem without detailed balance (Graham, 1973), we use the approach developed by Talkner, Hänggi and Ryter (Ryter, 1985; Talkner and Hänggi, 1984; Talkner and Ryter, 1983). The mean first passage time to leave Ω through a boundary $\partial\Omega$ in terms of the SPDD yields

$$T = \frac{-\displaystyle\int_\Omega P^0(x)\,dx}{\displaystyle\int_{\partial\Omega} dS\,\hat{D}\,\nabla\tilde{f}\,P^0}, \qquad (8.2.12)$$

where $\tilde{f}$ is a special solution of the Pontryagin equation

$$L^+\tilde{f} = (f + y)\frac{\partial\tilde{f}}{\partial x} - \frac{1}{\tau}y\frac{\partial\tilde{f}}{\partial y} + D\frac{\partial^2\tilde{f}}{\partial y^2} = T^{-1} \simeq 0,$$

with $\tilde{f} = 0$ on $\partial\Omega$ and $\tilde{f} = 1$ on the inner boundary of a supposed small $\Delta\Omega$. Use of the abovementioned approach for the valuation of the leading contributions for low noise intensity requires the definition of the boundary of attraction $\partial\Omega$. It can be given approximately by the tangent of the separatrix in the saddle point $(x_s, y_s = 0)$ (we further assume $x_s = 0$).

$$y + \left[f'(x_s) + \frac{1}{\tau}\right]x = 0. \qquad (8.2.13)$$

Application of (8.2.11) and the calculation of the integrals therein by the method of steepest descent gives

$$T = T_0\frac{(1 - \tau f'(x_s))}{|1 - \tau f'(x_A)|}\exp\{\tfrac{3}{2}\tau[f'(x_s) - f'(x_A)]\}, \qquad (8.2.14)$$

where x_A is the attractor state which is being left and T_0 is the transition time for the white noise case. If we introduce by definition the effective potential barrier $\Delta\Phi_{eff}$

$$T \sim \exp\{\Delta\Phi_{eff}/D\}$$

then it follows that:

$$\Delta\Phi_{eff} = \Delta\Phi_{\tau=0}t_0 + D\tau\{f'(x_s) - \tfrac{1}{2}f'(x_A)\}. \qquad (8.2.15)$$

283

We see that the leading contribution in $\Delta\Phi_{\text{eff}}$ is not altered in comparison with the white noise case. For the bistable system (8.2.10) it becomes

$$\Delta\Phi_{\text{eff}} = \frac{1}{4}\frac{a^2}{b} + \tfrac{3}{2}aD\tau + O(\tau^2). \tag{8.2.16}$$

This expression is consistent with the measured data for small τ (Fronzoni et al., 1986; Hänggi et al., 1985; see for an analytical calculation: Hänggi, Marchesoni and Grigolini, 1984).

So far we have studied the so-called overdamped case. Let us include now inertia. We consider the motion of a particle in an external force-field $f(x)$ (L. Schimansky-Geier, 1988):

$$\begin{aligned} \dot{x} &= v \\ \dot{v} &= -\gamma v + f(x) + y(t), \end{aligned} \tag{8.2.17}$$

where $y(t)$ is defined by (8.2.2). Again we enlarge the dimensionality by introducing (8.2.3) from which we get immediately the stationary FPE for $P^0(x,v,y)$ depending now on three variables:

$$LP^0 = -v\frac{\partial P^0}{\partial x} - \frac{\partial}{\partial v}(-\gamma v + f(x) + y)P^0$$

$$+ \frac{1}{\tau}\frac{\partial}{\partial y}\left(y - \frac{D}{\tau}\frac{\partial}{\partial y}\right)P^0 = 0. \tag{8.2.18}$$

Let the potential again be a series in τ. A successive iteration repeated three times in increasing powers of τ yields

$$\Phi = -\gamma \int f\left(1 - f'\frac{\tau^2}{1+\gamma\tau} + \frac{\tau^2}{2\gamma^2}f''\zeta\right)dx$$

$$+ \tfrac{1}{2}\gamma(1 + \tau(\gamma - \tau f'(x)))v^2 - \frac{1}{2}\frac{f''}{\gamma}\tau^2 \int v(\gamma v - f)\,dv$$

$$+ \tfrac{1}{2}\tau(1 + \gamma\tau)\left\{y - \gamma v + \frac{\gamma\tau f(x)}{1+\gamma\tau}\right\}^2$$

$$+ \tfrac{1}{2}\tau^3\frac{f''}{\gamma}\int v(\gamma v - f)\,dy. \tag{8.2.19}$$

That potential obeys the FPE except for a small term

$$LP^0 = [\tau^2\{(f'''f + f''f')v^3/\gamma - \tfrac{1}{6}f'''v^4\} + O(\tau^3)]P^0.$$

The solution (8.2.19) is valid for small τ and locally in the v-direction.

Let us discuss (8.2.19). For $\tau \to 0$ it follows the well known Boltzmann distribution ($D = \gamma kT$). The former result (8.2.9) is obtained in the overdamped case $\gamma \to \infty$ by simultaneous rescaling $f \to f/\gamma$; $y \to y/\gamma$; $D \to D/\gamma^2$. The SPDD is drawn in Figure 8.2 for different τ. Moss, Hänggi, Mannella and

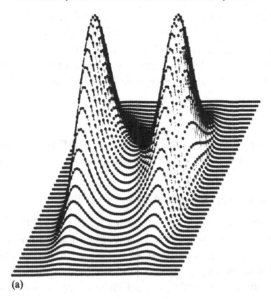

(a)

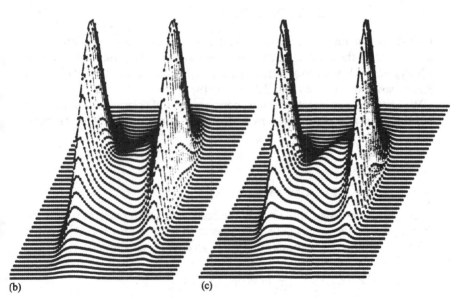

(b)   (c)

Figure 8.2. The stationary probability distribution density of the bistable oscillator. $P^0(x, v, y = 0)$; $a = 1$, $b = 1$, $\gamma = 1$, $D = 0.144$. (a) $\tau = 0.1$; (b) $\tau = 0.8$; (c) $\tau = 1.2$.

285

McClintock (1986) obtained phase portraits from measurements of electronic circuits. Again we find a good agreement with the skew portraits. We also prove the occurrence of the third maximum at the saddle point for larger τ as was detected by Moss et al. (1986). It originates from a term $\sim \gamma v^2 f'(x)\tau^2$. A first estimate of the mean first passage time (8.2.12) gives

$$T_{\gamma,\tau} = T_{\gamma,0} \frac{[(1 + \gamma\tau)\{1 + \tau(\gamma - \tau f'(x_s))\}]^{1/2}}{|1 + \tau(\gamma - f'(x_A))|} \exp\{O(\tau f'(x_s))\} \quad (8.2.20)$$

Again the main contribution in an effective potential difference is equal to the $\Delta\Phi$ of the white noise case. Effects of the finiteness of the noise occur with $O(D\tau)$.

8.3   Bistable systems influenced by noisy oscillators

The present section is devoted to a coupling of the bistable system with oscillating degrees of freedom. This is a common situation in reacting molecular systems. As usual only parts of the molecule are involved in the reaction, and other ones may carry out vibrations, librations, etc. It is true especially for large macromolecules, say for enzymes catalysing reactions. In order to model these effects in a simple way we study the coupling of a harmonic oscillator in the bistable situation. It is a straight generalization of the former section. We consider harmonic noise with the spectrum (Landau and Lifshits, 1971)

$$S_{yy}(\omega) = \frac{1}{\pi} \frac{\Omega_0^4 \varepsilon}{(\omega^2 - \Omega_0^2)^2 + \omega^2 \Gamma^2}. \quad (8.3.1)$$

It cuts off not only at a certain frequency but the intensity is differently distributed on the allowed frequencies. The harmonic oscillator acts as a filter which gives preference to a special frequency. (For a review of several filters see Kuznetsov, Stratonovich and Tikhonov, 1965.)

We continue our investigation by consideration of enlarged state spaces. A bistable flow $f(x)$ perturbated by a stochastic harmonic oscillator satisfies

$$\left.\begin{aligned} \dot{x} &= f(x) + y \\ \dot{y} &= s \\ \dot{s} &= -\Gamma s - \Omega_0^2 y + \Omega_0^2 (2D)^{1/2} \xi(t), \end{aligned}\right\} \quad (8.3.2)$$

where $\xi(t)$ is Gaussian white noise. We scale with Ω_0^2 to perform simply the limit $\Gamma \to \infty$ to the former case of the spectrum (8.2.5) with $\tau = \Gamma/\Omega_0^2 = \text{const.}$
The corresponding stationary FPE for $P(x, y, s)$ results immediately:

$$LP^0(x, y, s) = -\frac{\partial}{\partial x}(f(x) + y)P^0 - s\frac{\partial}{\partial y}P^0$$

$$-\frac{\partial}{\partial s}\left(-\Gamma s - \Omega_0^2 y + \Omega_0^4 D\frac{\partial}{\partial s}\right)P^0 = 0. \quad (8.3.3)$$

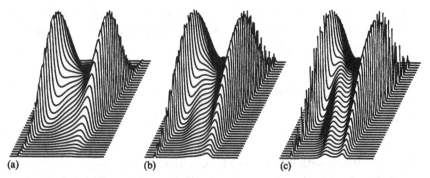

Figure 8.3. The stationary probability distribution density of a bistable symmetric system driven by harmonic noise $P(x, y, s = 0)$: $a = 1$, $b = 1$, $D = 0.250$, $\tau = 0.5$. (a) $\Omega^2 = 4$; (b) $\Omega^2 = 2$; (c) $\Omega^2 = 1$.

Let us look for solutions for small $\tau = \Gamma/\Omega_0^2$ and high frequencies Ω_0^2 if $\Omega_0^2\tau > f'(x_s)$ or $\Gamma > f'(x_s)$. The potential we have found is

$$\Phi = -\int f\,dx + \tfrac{1}{2}f^2\left(\tau - \frac{f'}{\Omega_0^2}\right) - \frac{1}{\Omega_0^2}\int f^2 f''\,dx$$

$$+ \tfrac{1}{2}\tau\left(1 - f'\left(\tau - \frac{f'}{\Omega_0^2}\right)\right)(y + f)^2$$

$$+ \frac{1}{2\Omega_0^2}\left\{\tau - \frac{f'}{\Omega_0^2}\right\}\{s + f'(f + y)\}^2$$

$$- \frac{1}{\Omega_0^2}\{f(s + f'y)\} - \frac{1}{\Omega_0^4}ff''s(y + f). \tag{8.3.4}$$

One easily confirms that this solution obeys the FPE (8.3.3) except for a small term

$$LP^0 = O(\Omega_0^4, \tau^2)P^0.$$

Unfortunately experimental data are not yet available so we cannot compare the PDD with measurements. Its shape for several frequencies Ω_0 and damping coefficients becomes more and more skew in comparison with (8.2.9) (see Figure 8.3). In this approximation, $\Delta\Phi$ again does not depend on the noise parameters. Nevertheless one can state that possible changes of the mean transition times occur with $O(D\tau)$. Application of (8.2.11) yields, at first inspection,

$$T_{\tau,\Omega_0^{-2}} = T_{0,0}\frac{\{[1 - f'(x_s)(\tau - f'(x_s)/\Omega_0^2)][\tau - f'(x_s)/\Omega_0^2]\}^{1/2}}{\{[1 - f'(x_A)(\tau - f'(x_A)/\Omega_0^2)][\tau - f'(x_A)/\Omega_0^2]\}^{1/2}}$$

$$\times \frac{\exp\{O(\tau f'(x_s))\}}{[1 - f'(x_A)\{\tau - 1/[\Omega_0^2(\tau - f'(x_A)/\Omega_0^2)]\}]^{1/2}}.$$

287

Let us include now the influence of inertia; we consider the motion in an external field with finite friction but now driven by an oscillatory noise. The stochastic differential equations become

$$
\left.
\begin{aligned}
\dot{x} &= v \\
\dot{v} &= -\gamma v + f(x) + y \\
\dot{y} &= s \\
\dot{s} &= -\Gamma s - \Omega_0^2 y - \Omega_0^2 (2D)\xi(t).
\end{aligned}
\right\}
\tag{8.3.5}
$$

Related problems without noise and for low friction γ were widely investigated and are known as Holmes-oscillators (Holmes, 1979). Particularly close attention has been paid to resonance situations, for which complicated trajectories have been observed. Similar stochastic problems can be found in papers by Bulsara, Lindenberg and Shuler (1982), Carmeli and Nitzan (1985), de Pasquale, Racz, San Miguel and Tartaglia (1984), Iyengar (1986), Klimontovich (1983), San Miguel (1984).

Since $\xi(t)$ is the Gaussian white noise the FPE can readily be obtained:

$$
LP^0(x, v, y, s) = -v\frac{\partial}{\partial x}P^0 - \frac{\partial}{\partial v}(-\gamma v + f(x) + y)P^0 - s\frac{\partial}{\partial y}P^0
$$

$$
- \frac{\partial}{\partial s}\left(-\Gamma s - \Omega_0^2 y + \Omega_0^4 D\frac{\partial}{\partial s}\right)P^0 = 0. \tag{8.3.6}
$$

Again one can find a solution of the FPE which is valid for high frequencies Ω_0^2 and locally in the v-direction

$$
\Phi = -\gamma \int f \, dx + \tfrac{1}{2}f^2 \frac{\Gamma + \gamma}{\Omega_0^2}\left\{1 - \frac{(\Gamma + \gamma)\Omega_0^2}{(\Gamma + \gamma)\Gamma\gamma + \Gamma\Omega_0^2 - \gamma f'}\right\}
$$

$$
- \frac{1}{2\gamma}\int f^2 f'' \, dx + \tfrac{1}{2}\gamma v^2\left\{1 + \frac{2f'}{\Omega_0^2} + \frac{\gamma(\Gamma + 2\gamma)}{\Omega_0^2}\right.
$$

$$
\left. - f'\left(\frac{\Gamma^2}{\Omega_0^4} - \frac{2\gamma^2}{\Omega_0^4}\right) + \frac{f'^2}{\Omega_0^4} + \frac{\gamma^3(\Gamma + \gamma)}{\Omega_0^4}\right\}
$$

$$
- \frac{1}{2}\frac{f''}{\gamma\Omega_0^2}\int v(\gamma v - f)\,dv + \frac{1}{2}\left\{\frac{\Gamma}{\Omega_0^2} - \frac{f'\gamma}{\Omega_0^4} + \frac{\Gamma\gamma(\Gamma + \gamma)}{\Omega_0^4}\right\}
$$

$$
\times \left\{y - \gamma v + f\left(1 - \frac{(\Gamma + \gamma)\Omega_0^2}{(\Gamma + \gamma)\Gamma\gamma + \Gamma\Omega_0^2 - \gamma f'}\right)\right\}^2
$$

$$
+ \frac{1}{2\Omega_0^4}(\Gamma + \gamma)\left\{s - \gamma(y + f - \gamma v) + \gamma\frac{f' + \Omega_0^2}{\Gamma + \gamma}v\right\}^2
$$

$$
- \frac{1}{2\Omega_0^2}\frac{f''}{\gamma}\int v(\gamma v - f)\,ds. \tag{8.3.7}
$$

Inserting this potential into the FPE yields

$$LP^0 = \left\{\frac{1}{6\gamma\Omega_0^2} f'''v^4 - \frac{1}{4\gamma\Omega_0^2}[f'''f + f''f']v^3 + O(\Omega_0^{-4})\right\}P^0.$$

(8.3.8)

Using (8.2.12) and the PDD given above we get the following estimate:

$$T_{\Gamma,\Omega_{0,\gamma}^2} \simeq T_{0,0,\gamma}\left(\frac{1 - f'(x_s)\tau_\gamma/(\Omega_0^2\tau_\gamma) + \gamma(\tau_\Gamma + \tau_\gamma)\tau_\Gamma}{1 - f'(x_A)\tau_\gamma/(\Omega_0^2\tau_\Gamma) + \gamma(\tau_\Gamma + \tau_\gamma)\tau_\Gamma}\right)^{1/2}$$

$$\times \left(\frac{(1 + f'(x_s)/\Omega_0^2 + \gamma\tau_\gamma)^2 + \gamma\tau_\Gamma(1 + \tau_\gamma\gamma) - f'(x_s)\tau_\Gamma^2}{(1 + f'(x_A)/\Omega_0^2 + \gamma\tau_\gamma)^2 + \gamma\tau_\Gamma(1 + \tau_\gamma\gamma) - f'(x_A)\tau_\Gamma^2}\right)^{1/2}$$

$$\times \frac{\exp\{O(f'(x_s)\tau,\Omega_0^{-2})\}}{\{\gamma - f'(x_A)(\tau_\Gamma + \tau_\gamma)(1 - (\tau_\Gamma + \tau_\gamma)/[\gamma\tau_\Gamma(\tau_\gamma + \tau_\Gamma) + \tau_\Gamma - \tau_\gamma f'(x_A)/\Omega_0^2])\}^{1/2}},$$

(8.3.9)

where $\tau_\Gamma = \Gamma/\Omega_0^2, \tau_\gamma = \gamma/\Omega_0^2$. $T_{0,0,\gamma}$ is the white noise case which can be reached by two steps: (i) $\Gamma \to \infty$, $\Omega_0^2 \to \infty$, $\tau_\Gamma = \Gamma/\Omega_0^2 = $ const. (which yields (8.2.19) or (8.2.20), respectively); (ii) $\tau_\Gamma \to 0$. The first correction factor arising from harmonic noise here and in the former overdamped case is always smaller than one due to

$$f'(x_s) > 0; \quad f'(x_A) < 0,$$

$$\frac{1 - (f'(x_s)\tau_\gamma/(\Omega_0^2\tau_\Gamma)) + \gamma(\tau_\Gamma + \tau_\gamma)\tau_\Gamma}{1 - (f'(x_A)\tau_\gamma/(\Omega_0^2\tau_\Gamma)) + \gamma(\tau_\Gamma + \tau_\gamma)\tau_\Gamma} < 1$$

or, respectively,

$$\frac{\tau - f'(x_s)/\Omega_0^2}{\tau - f'(x_A)/\Omega_0^2} < 1.$$

It can be seen that this expresses a tendency to decrease the transition times. The physical explanation of this effect is connected with resonance phenomena. The factor gets smaller as it approaches the resonance condition, which is beyond the range of validity of our formulae. Also in the last exponential factor in (8.3.9) there appear unknown terms of order $O(D\tau,\Omega_0^{-2})$. However, the present authors believe that the dominant effect of harmonic noise is the decrease in transition time in the resonance region.

So far we have considered an additive influence of an independent oscillator on the bistable system. A new situation is reached if the bistable system interacts with the oscillator by a potential. For simplicity we consider an interaction by a potential which saves the harmonic character of the oscillator (Ebeling and Romanovsky, 1985; Brettschneider, 1986, diploma thesis):

$$W(x, y) = -\psi(x)y - \tfrac{1}{2}\eta(x)y^2.$$

(8.3.10)

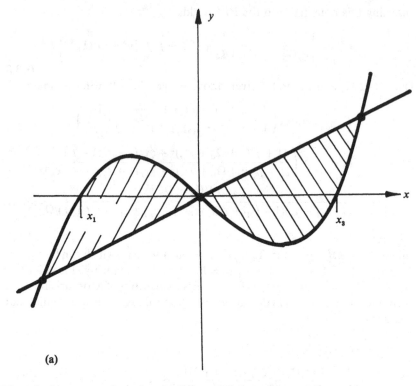

(a)

The whole system of stochastic differential equations can be written

$$
\left.
\begin{aligned}
\dot{x} &= v \\
\dot{v} &= -\gamma v + f(x) - \frac{\partial W}{\partial x} + (2\varepsilon_1)^{1/2}\xi_1(t) \\
\dot{y} &= s \\
\dot{s} &= -\Gamma s - \Omega^2 y - \frac{\partial W}{\partial y} + (2\varepsilon_2)\xi_2(t).
\end{aligned}
\right\}
\tag{8.3.11}
$$

$\xi_1(t)$ and $\xi_2(t)$ are independent white noise sources. In thermal equilibrium we have $kT = \varepsilon_1/\gamma = \varepsilon_2/\Gamma$ and the Maxwell–Boltzmann distribution solves the stationary FPE:

$$
P^0(x, v, y, s) \sim \exp\left\{-\frac{1}{kT}\left[\frac{v^2}{2} + \frac{s^2}{2} - \int^x f(x')\,\mathrm{d}x'\right.\right.
$$

$$
\left.\left. + \tfrac{1}{2}\Omega^2 y^2 + W(x_1 y)\right]\right\}.
\tag{8.3.12}
$$

Let us consider the effective activation barrier $\Delta\Phi_{\mathrm{eff}}$. It follows from the

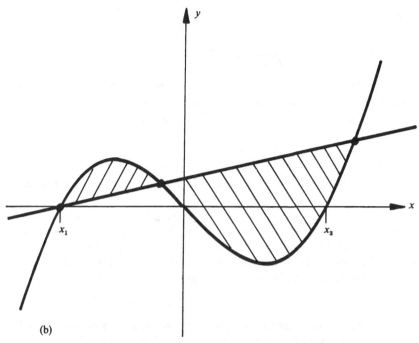

Figure 8.4. Geometrical solution of the stationary states of (8.3.11): $\eta = 0$, (a) $\psi = \psi^0 x$; (b) $\psi = \psi^0(x - x_1)$. Shaded regions correspond to the value of the effective potential barriers.

reduced PDD of the bistable oscillator $(\Omega^2 > \eta(x_A))$

$$P^0(x) = \int dv\, dy\, ds\, P^0(x, v, y, s) \sim \frac{1}{(\Omega^2 - \eta(x))^{1/2}}$$

$$\times \exp\left\{\frac{1}{kT}\left[\int^x f(x')\,dx' + \frac{1}{2}\frac{\psi^2(x)}{\Omega^2 - \eta(x)}\right]\right\}. \qquad (8.3.13)$$

Additionally to $\Delta\Phi_{\Omega\to\infty}$ which is the potential barrier of the decoupled case, a static contribution results in $\Delta\Phi_{\text{eff}}$ from the first item in the interaction potential (8.3.10):

$$\Delta\Phi_{\text{eff}} = \Delta\Phi_{\Omega\to\infty} + \Delta\Phi_{\text{static}} + O(kT).$$

Its sign can be different and depends on the shape of $\psi(x)$. This function defines the rest position of the interacting oscillator

$$y^0 = \frac{\psi(x)}{\Omega^2 - \eta(x)}. \qquad (8.3.14)$$

Two limit situations will be of interest. First let $\psi = \psi^0 x$. (For simplicity we set $\eta(x) = 0$.) It results in an increase of $\Delta\Phi_{\text{eff}}$ as can be seen from Figure 8.4(a).

Explicitly it follows

$$\Delta\Phi_{\text{static}} = \frac{1}{2b}\frac{\psi_0^2}{\Omega^2}\left(a + \frac{\psi_0^2}{2\Omega^2}\right). \tag{8.3.15}$$

In the second case, let ψ vanish in one of the stable points, $\psi = \psi^0(x - x_1)$. Here transitions outgoing from x_1 to the other stable state will be favored (see Figure 8.4b). Transitions in the other direction become less probable:

$$\Delta\Phi_{\text{static}} \simeq -\frac{1}{2\Omega^2}\frac{\psi_0^2}{b}\frac{a}{b} + O\left(\frac{\psi_0^4}{\Omega^4}\right). \tag{8.3.16}$$

We remark that both effects in the considered cases increase with decreasing Ω.

A further change of the effective potential barrier will result if we assume a nonequilibrium situation. Thus let, e.g., the harmonic oscillator possess an energy E different from kT. Either it will be pumped with a different temperature or the oscillator has obtained an initial energy which is not yet dissipated. Approximately we get for small Γ

$$E(t) \simeq E(t = 0)\mathrm{e}^{-\Gamma t} + kT_2(1 - \mathrm{e}^{-\Gamma t}) \simeq \text{const.}$$

Then due to the second item in (8.3.10) a permanent energy flow will attack the dynamics of the bistable oscillator. For a first valuation we will average over fast oscillations and write

$$\left.\begin{array}{l} y \to \langle y \rangle = \dfrac{\psi(x)}{\Omega^2 - \eta(x)} \\[4mm] y^2 \to \langle y^2 \rangle = \langle y \rangle^2 + \dfrac{E}{\Omega^2 - \eta(x)}. \end{array}\right\} \tag{8.3.17}$$

Neglecting the static interaction ($\psi = 0$) the dynamics of the first oscillator is modified and reads

$$\ddot{x} + \gamma\dot{x} = f(x) + \tfrac{1}{2}\eta'(x)\frac{E}{\Omega^2 - \eta(x)}. \tag{8.3.18}$$

Again the effective potential difference obtains a second contribution resulting now from the transferred energy

$$\left.\begin{array}{l} \Delta\Phi_{\text{eff}} = \Delta\Phi_{\Omega\to\infty} + \Delta\Phi_{\text{dyn}} \\[2mm] \Delta\Phi_{\text{dyn}} \sim E. \end{array}\right\} \tag{8.3.19}$$

For $\eta(x) = \eta^0 x$, ($\eta^0 > 0$) the change of $\Delta\Phi$ can be seen from the geometrical construction of the steady states of (8.3.18) in Figure 8.5. $\eta^0 > 0$ favors transitions $x_1 \to x_3$, in the opposite case ($\eta^0 < 0$) the state x_1 will be stabilized.

The ability of coupled systems to assume quite new behavior can be

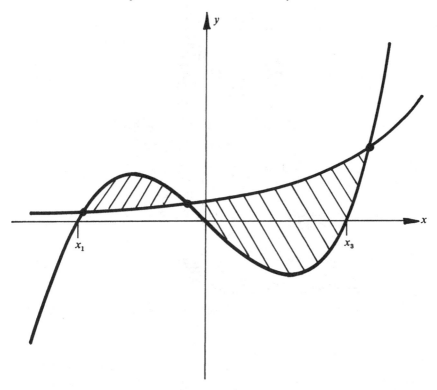

Figure 8.5. Geometrical solution of the stationary states of (8.3.11): $\psi = 0$, $\eta = \eta^0 x$, $\eta^0 > 0$. Shaded regions correspond to the value of the effective potential barriers.

demonstrated by the next simple model. Let the second oscillator perform in the decoupled case large oscillations in the degenerated potential

$$V(y) = \tfrac{1}{4} q_2 y^4.$$

The interaction we define by (8.3.20) with

$$\psi(x); \quad \eta(x) = -(x - x_1)(x - x_3), \tag{8.3.20}$$

where x_1 and x_3 are the stable states of the bistable dynamics $f(x) = -q_1 x(x - x_1)(x - x_3)$; sign. $(x_1 x_3) < 0$. In this case the Maxwell–Boltzmann distribution with the potential

$$\Phi = q_1 \int x(x - x_1)(x - x_3) \, dx$$

$$+ \tfrac{1}{2} y^2 (x - x_1)(x - x_3) + \tfrac{1}{4} q_2 y^4 \tag{8.3.21}$$

is similar to the crater-distribution of limit cycles (see Figure 8.6) (Ebeling and

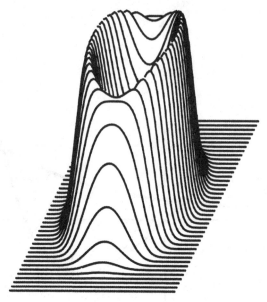

Figure 8.6. The stationary probability distribution with potential energy (8.3.21).

Klimontovich, 1984). The unfavorable state $x = 0$, $y = 0$ will never be reached since a saddle line connecting $(x_1, y = 0)$ and $(x_3, y = 0)$ has formed around it. Its location is defined by the ratio of q_1/q_2 and the stochastic transitions will proceed on it to leave the metastable state on much shorter time scales.

8.4 High-dimensional systems

So far we have restricted our discussion to systems with low dimension, $d \leqslant 4$. However, many important transition processes occur in systems with a rather large number of dimensions, $d \gg 1$. In other words in these transition processes many degrees of freedom are participating. To this class of problems belong, e.g., catalytic processes in complicated molecules and especially in enzymes (Brettschneider and Schimansky-Geier 1986; Ebeling and Romanovsky, 1985) one can show that the high catalytic activity of special systems as enzymes, zeolites, etc., are connected with transitions between potential minima which are accompanied by the excitation of many degrees of freedom. Further we mention transition phenomena in networks of bistable reactions or circuits. The investigation of networks made by couplings of bistable systems is a very recent field of research (see, e.g., Denker, 1986; Hoffmann *et al.*, 1986; Hopfield and Tank, 1985). In this field problems again arise which are connected with stochastic processes in high-dimensional spaces. Finally we mention problems in combinatorial optimization as, e.g., the travelling salesman problem where a stochastic search of good minima in high-dimensional spaces is the basic strategy (Ebeling and Engel, 1986; Kirkpatrick, Gelatt and Vecchi, 1983). We start the theoretical analysis with a problem of gradient type. Let

$$U(x_1, x_2, \ldots, x_d)$$

be a potential function defined over a high-dimensional space. We assume that U is a complicated function having many minima and maxima.

Let us study a stochastic process described by the Langevin equation

$$\left.\begin{aligned} \dot{x}_i(t) &= -\frac{\partial U}{\partial x_i} + (2D)^{1/2}\xi_i(t) \\[2mm] \langle \xi_i(t+\tau)\xi_j(t)\rangle &= \delta_{ij}\delta(\tau). \end{aligned}\right\} \tag{8.4.1}$$

The corresponding Fokker–Planck equation reads

$$\frac{\partial}{\partial t}P(x_1,\ldots,x_d) = \sum_{i=1}^{d}\frac{\partial}{\partial x_i}\left[\frac{\partial U}{\partial x_i}P + D\frac{\partial P}{\partial x_i}\right]. \tag{8.4.2}$$

As is well known, the stationary solution is (Gardiner, 1983)

$$P^0(x_1,\ldots,x_d) = N\exp\left\{-\frac{1}{D}U(x_1,\ldots,x_d)\right\}. \tag{8.4.3}$$

In order to get time-dependent solutions we define the Schrödinger eigenvalue problem (Van Kampen, 1981; Risken, 1984)

$$D\sum_i\frac{\partial^2}{\partial x_i^2}\psi_n(x_1,\ldots,x_d) + [\varepsilon_n - V(x_1,\ldots,x_d)]\psi_n = 0 \tag{8.4.4}$$

with

$$V(x_1,\ldots,x_d) = \frac{1}{4D}\left(\sum_i\frac{\partial U}{\partial x_i}\right)^2 - \tfrac{1}{2}\sum_i\frac{\partial^2}{\partial x_i^2}U.$$

Then the solution of the Fokker–Planck equation is

$$P(x_1,\ldots,x_d;t) = N\exp\left\{-\frac{U}{2D}\right\}\sum_k b_k\psi_k\exp\{-\varepsilon_k t\}. \tag{8.4.5}$$

The lowest eigenvalue and the corresponding eigenfunction are

$$\varepsilon_0 = 0, \quad \psi_0(x_1,\ldots,x_d) = \text{const.}\cdot\exp\left\{-\frac{U}{2D}\right\}.$$

The relaxation time to the ground state is given by the gap to the first nonzero eigenvalue

$$\tau_{\text{rel}} \sim \varepsilon_1^{-1}.$$

Let us assume now that there are many minima of nearly equal height. Then it will take a long time to find the absolute lowest minimum which is, moreover,

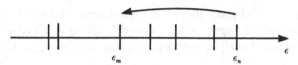

Figure 8.7. The spectrum of eigenvalues.

of no practical importance. For any practical application it will be sufficient to find a good minimum which is near to the absolute minimum. We shall assume that the spectrum is discrete at least in the lower part and that each deep minimum is connected with a localized bound state. Then the evolution of the probability is connected with a hopping down the ladder of eigenvalues (Figure 8.7). Let us assume for a moment that at a certain time only the eigenfunction ψ_m is populated and then we are interested in the transition to the eigenfunction ψ_n. The corresponding time is approximately

$$\tau_{m,n} \sim \frac{1}{\delta\varepsilon_{m,n}}\left[\ln\frac{b_n}{b_m} + 1\right]. \qquad (8.4.6)$$

Here

$$\delta\varepsilon_{m,n} = \varepsilon_m - \varepsilon_n$$

is the gap between the eigenvalues which may be approximated by

$$\delta\varepsilon_{m,n} \sim \exp\left\{-\frac{\Delta U_{m,n}}{D}\right\}, \qquad (8.4.7)$$

where $\Delta U_{m,n}$ is the height of the threshold which leads from minimum m to minimum n. Therefore the population of the new state ψ_n with $b_n \simeq b_m$ needs the time

$$\tau_{m,n} \sim \exp\left\{\frac{\Delta U_{m,n}}{D}\right\}. \qquad (8.4.8)$$

The similarity to the formulae for the low-dimensional case is evident. An exact study of transition phenomena in high-dimensional systems depends strongly on the available knowledge of the corresponding Schrödinger problem. We mention that from the research on disordered solid materials a body of results on the Schrödinger equation for random potentials is now available (Lifshifts, Gredeskul and Pastur, 1982).

Let us now consider more specific problems of the type discussed above. First we mention reaction–diffusion problems which are connected with potentials of the type

$$U(x_1,\ldots,x_d) = \sum_{i=1}^{d} U(x_i) + \frac{\tilde{D}}{2}\sum_{i,j}' (x_i - x_j)^2. \qquad (8.4.9)$$

By means of this potential we may model the diffusion between cells numbered from 1 to d, where $\tilde{D}$ is the diffusion constant. The $\sum'$ is to be extended only over nearest neighbor cells. $U(x_i)$ is the potential of the reaction function; we may assume, e.g., that $U(x_i)$ has two minima corresponding to bistable reactions. The stationary solution of the stochastic reaction–diffusion problem is

$$P^0(x_1, \ldots, x_d) = N \exp \left\{ -\frac{1}{D} \left[\sum_i U(x_i) + \frac{\tilde{D}}{2} \sum_{i,j}{}' (x_i, x_j)^2 \right] \right\}. \quad (8.4.10)$$

The number and location of the extrema of this distribution may easily be studied as may also its bifurcation behavior for different parameter values (Ebeling and Malchow, 1979; Malchow and Schimansky-Geier, 1985; Malchow, Ebeling, Feistel and Schimansky-Geier, 1983). The most important transition process in bistable reaction–diffusion systems is nucleation, i.e. when the system switches from a stable state where all cells are in the same minimum $x^{(1)}$ of $U(x)$:

$$x_1 = x_2 = \cdots = x_d = x^{(1)}$$

to another stable state where all cells are in the other minimum (the new chemical phase):

$$x_1 = x_2 = \cdots = x_d = x^{(3)}.$$

The transition typically goes through intermediate states where groups of cells form nuclei of the new phase (Schimansky-Geier and Ebeling, 1983).

We meet similar but more complicated problems in the study of electrical networks (Hopfield and Tank, 1985). Here one obtains a set of equations of the type

$$\dot{u}_i = I_i - u_i/\tau_i + \sum_{j=1}^{d} T'_{ij} A(u_j), \quad (8.4.11)$$

where u_i is the voltage in the circuit i; further, $A(u)$ is a sigmoid function and T_{ij} a symmetric connection matrix. Introducing the output voltages of the circuit i as new variables

$$x_i = A(u_i)$$

we get

$$\dot{x}_i = \frac{1}{A'(u_i)} \left[I_i - \frac{1}{\tau_i} A^{-1}(x_i) + \sum_j T'_{ij} x_j \right]. \quad (8.4.12)$$

In some approximations we may write this in the form

$$\dot{x}_i = -\frac{\partial}{\partial x_i} U(x_1, \ldots, x_d), \quad (8.4.13)$$

where

$$U = \sum_i U_i(x_i) + \sum_{i,j} T_{ij} x_i x_j. \quad (8.4.14)$$

297

Here $U_i(x_i)$ is a bistable potential with

$$\frac{\partial U_i(x_i)}{\partial x_i} = \frac{1}{\langle A' \rangle}\left\{ -I_i + \frac{1}{\tau_i}A^{-1}(x_i)\right\},$$ (8.4.15)

and

$$T_{ij} = T'_{ij}/\langle A' \rangle,$$

where $\langle A' \rangle$ is some average of the derivative. On including white noise into (8.4.13), we get exactly an equation of the general type (8.4.1) with the stationary solution

$$P^0(x_1,\ldots,x_d) = N\exp\left\{ -\frac{1}{D}\left[\sum_i U_i(x_i) + \sum_{i,j} T_{ij}x_ix_j\right]\right\}.$$ (8.4.16)

The electrical networks described above can solve, by means of their stochastic dynamics, complicated optimization problems as well as recognition problems if the 'synaptic matrix' T_{ij} is chosen in an appropriate way. Typically each element may have many excitatory connections $T_{ij} > 0$ as well as many inhibitory connections $T_{ij} < 0$. In other words, the synaptic matrix resembles in many respects a spin glass coupling matrix. From this it follows that the potential of the problem $U(x_1,\ldots,x_d)$ may have an extremely complicated (frustrated) character. We will not go into details of the very interesting class of optimization and recognition problems here. Let us just mention that a Fokker–Planck equation models a search by Boltzmann strategy and that it is useful in many cases to decrease the noise intensity in time (strategy of simulated annealing).

Finally let us still remark that sometimes the system we want to describe consists of elements showing inertia, e.g. Hamiltonian system of the type

$$\left.\begin{aligned} \dot{x}_i &= \frac{\partial H}{\partial p_i} \\[2mm] \dot{p}_i &= -\frac{\partial H}{\partial x_i} - \gamma_i\frac{\partial H}{\partial p_i} + (2D\gamma_i)^{1/2}\xi_i(t). \end{aligned}\right\}$$ (8.4.17)

The corresponding Fokker–Planck equation possesses the stationary solution

$$P^0(x_1,\ldots,x_d;p_1,\ldots,p_d)$$

$$= N\exp\left\{ -\frac{1}{D}H(x_1,\ldots,x_d;p_1,\ldots,p_d)\right\}.$$ (8.4.18)

Our knowledge about the relaxation of such systems to equilibrium is very limited; we know, however, that the Boltzmann factor

$$\exp\left\{ -\frac{\Delta U}{D}\right\}$$

gives an estimate of the transition time between certain minima which are separated by a potential barrier of height ΔU.

Summarizing, we may state that transition phenomena in high-dimensional spaces are of much importance for the understanding of a large class of real systems. Further we have learned that the simulation of stochastic processes in high-dimensional systems are an important tool for the effective solution of an interesting class of optimization and recognition problems. Other possible applications are connected with a new direction in the design of special-purpose computers, the so-called network-machine. Especially we mention the Boltzmann-machine, which proved to be very successful in simulating learning behavior (Ackley, Hinton and Sejnowski, 1985; Sejnowski, Kienker and Hinton, 1986).

8.5   Fisher–Eigen models of evolution

In this section we want to discuss a special evolution model which is closely related to the problems discussed in the previous section. As is well known one of the basic strategies of all higher evolution processes is the Darwinian strategy, which includes self-reproduction, mutation and selection and optimizes the fitness of the species. In a very simple model, developed in the 1930s by Wright, fitness is modelled by a kind of potential

$$E(x_1, \ldots, x_d),$$

which forms a complicated landscape over the space of the phenotypic properties $x_1, \ldots, x_d$. We may imagine e.g. that x_1 is a number which defines the weight of the species, x_2 the height, etc. In order to define a species we will need at least 10^3 characteristics, i.e. the dimension d will be very high. Further, let us assume that the value function $E(x_1, \ldots, x_d)$ has a quite complex structure, i.e. certain combinations $x_1, \ldots, x_d$ have a high value and others nearby a low value. Typically this function will have a large number of extrema. A simple model for the structure of E is

$$E(x_1, \ldots, x_d) = \sum_{i=1}^{d} E_i(x_i) + \sum_{i,j} T_{ij} x_i x_j. \tag{8.5.1}$$

Here the E_i denote independent contributions to the valuation. The second term gives correlations or anticorrelations of the individual properties corresponding to $T_{ij} > 0$ or $T_{ij} < 0$, respectively. Let us assume for definiteness that the matrix T_{ij} has a spin-glass character. Then the function E which models many conflicting goals will have a large number of extrema. Now following Fisher and Eigen let us assume that the dynamics of the system which determines the population density $n(x_1, \ldots, x_d t)$ is given by

$$\frac{\partial}{\partial t} n(x_1, \ldots, x_d; t) = n \left\{ E(x_1, \ldots, x_d) - \langle E \rangle \right\}$$

$$+ D \sum_i \frac{\partial^2 n}{\partial x_i^2}. \tag{8.5.2}$$

With the analogy $E \sim (-U)$, (8.5.2) reminds us in some respects of the Fokker–Planck equation (8.4.2). At first we see that (8.5.2) with

$$\langle E \rangle = \frac{1}{N} \int dx_1, \ldots, dx_d \, En \tag{8.5.3}$$

conserves the integral

$$N = \int dx_1 \ldots dx_d n.$$

Further the density converges with increasing time to distributions localized around the maxima of the value function. This recalls the behavior of the probability density whose distributions are concentrated around the minima of $U(x_1, \ldots, x_d)$.

Like the Fokker–Planck equation, the Fisher–Eigen equation is related to a Schrödinger eigenvalue problem

$$D \sum_i \frac{\partial^2}{\partial x_i^2} \psi_n(x_1, \ldots, x_d) + [\varepsilon_n - V(x_1, \ldots, x_d)] \psi_n = 0. \tag{8.5.4}$$

This time the potential is given by

$$V(x_1, \ldots, x_d) = -E(x_1, \ldots, x_d).$$

By means of the eigenvalues ε_k and the eigenfunctions ψ_k the complete solution is (Ebeling et al., 1985)

$$n(x_1, \ldots, x_d; t) = \frac{N \sum c_k \exp\{-\varepsilon_k t\} \psi_k}{\sum c_k \exp\{-\varepsilon_k t\}}. \tag{8.5.5}$$

If ε_0 is the lowest eigenvalue then the target distribution which corresponds to the Boltzmann function in the Fokker–Planck case reads

$$n_0(x_1, \ldots, x_d) = N\psi_0(x_1, \ldots, x_d).$$

This eigenfunction will be located around the highest peak of $E(x_1, \ldots, x_d)$. In other words (8.5.2) defines an extremum-finding process. In the limit of long time the relaxation has the characteristic time

$$\tau_{rel} \sim (\varepsilon_1 - \varepsilon_0)^{-1} \tag{8.5.6}$$

and follows the dynamics

$$n(x_1, \ldots, x_d; t) = N \left\{ \psi_0 + \frac{c_1}{c_0} \psi_1 \exp\left(-\frac{t}{t_{rel}}\right) + \cdots \right\}. \tag{8.5.7}$$

Above we have studied the target distribution and the relaxation to it. In real evolution processes the target is never reached due to the fact that the number of maxima of $E(x_1, \ldots, x_d)$ is extremely large. Further it is of importance to know that evolution is usually a problem of frustrated type i.e. there are many maxima of nearly equal height (Figure 8.8). Therefore evolution processes tend

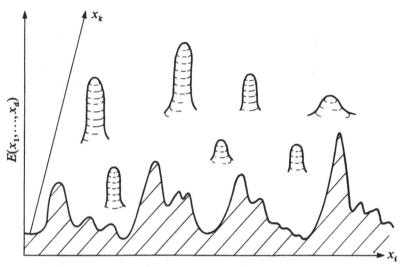

Figure 8.8. Stochastic landscape of an evolutionary process.

to reach only sufficiently good maxima and the hopping between maxima is of fundamental importance. In order to characterize the transient behavior we study now the mean first passage time for the transition from one maximum to another. The real evolution is a long sequence of such transitions. On the scale of eigenvalues this corresponds to a process of falling down the spectrum (Figure 8.7). The efficiency· of the evolutionary search is therefore mainly determined by its ability to leave a local maximum in favor of another higher one, or to leave an eigenvalue in favor of another lower one. Let us denote the maximum where we start by $x^{(m)}$ and the next new maximum by $x^{(n)}$. Further we assume that the barrier ΔE to be crossed is sufficiently high

$$\Delta E \gg D,$$

where

$$\Delta E = E(x^{(m)}) - E(x^{(mn)}).$$

Here $x^{(mn)}$ is the 'best' saddle between $x^{(m)}$ and $x^{(n)}$. A reasonable approximation for the transition time is

$$\tau_{m,m} \sim \frac{1}{\delta\varepsilon_{m,n}} \left[\ln\frac{c_n}{c_m} + 1 \right]. \tag{8.5.8}$$

Two cases must be distinguished: (i) the maxima are of nearly equal height

$$|\delta E_{m,n}| \ll I_{m,n}; \quad \delta E_{m,n} = E(x^{(n)}) - E(x^{(m)}).$$

Here $I_{m,n}$ is the so-called overlap integral of the wave functions, which is estimated by (Ebeling and Engel, 1986):

$$I_{m,n} \sim \exp\left\{ -\left(\frac{\Delta E}{D}\right)^{1/2} |x^{(n)} - x^{(m)}| \right\}.$$

301

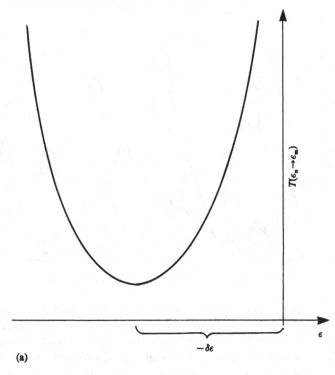

(a)

With $c_n \simeq c_m$ we get the estimate

$$\delta\varepsilon_{m,n} \sim I_{m,n}, \quad \tau_{m,n} \sim |\delta\varepsilon_{m,n}|^{-1} = I_{m,n}^{-1}.$$

(ii) The second maximum is considerably higher

$$|\delta E_{m,n}| \gg I_{m,n}.$$

Then the estimate is

$$\delta\varepsilon_{m,n} \sim (E(x^{(n)}) - E(x^{(m)}))$$

$$c_n/c_m \sim I_{m,n}.$$

Therefore we get

$$\tau_{m,n} \sim \left(\frac{\Delta E}{D}\right)^{1/2} \frac{|x^{(n)} - x^{(m)}|}{E(x^{(n)}) - E(x^{(m)})}. \tag{8.5.9}$$

We see that in the evolution strategy the transition time is mainly determined by the distance between the maxima and by the difference of the values of $E(x)$. On the other hand for thermodynamic processes (Fokker–Planck equations) the transition times depend mainly on the threshold values. Consequently evolution strategies are expected to be advantageous if the extrema of the objective function are separated by narrow high barriers or valleys, respec-

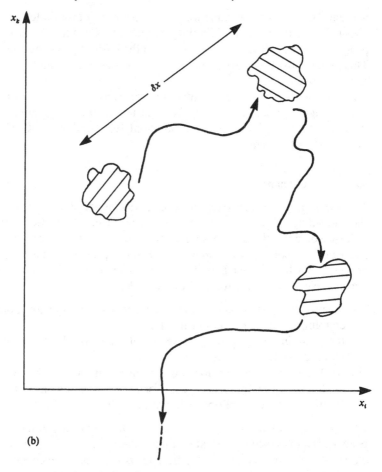

(b)

Figure 8.9. Quantized motion in evolutionary landscape. (a) There exist a minimum of the mean transition times for finite $\delta\varepsilon$. (b) The hopping in the considered phenotypic space processes by jumps over finite distances δx.

tively. For the solution of practical optimization problems one expects that mixtures of both strategies are most appropriate (T. Boseniuk, W. Ebeling and A. Engel, manuscript submitted to *Phys. Lett.*).

Rather than calculating the time for the passage between two definite maxima we may ask instead what is the mean time for jumping from an eigenvalue ε_m to an eigenvalue ε_n. Another question concerns the mean distance of jumps in a stochastic landscape. Under appropriate assumptions (Gaussian distributions) we get the typical distributions shown in Figures 8.9(a), (b). This shows that the hopping processes are quantized with respect to time, improvement and the distance of jumps (Ebeling *et al.*, 1985; Ebeling and Peschel, 1985). On average one gets a finite value for the jumps as well as a finite value for the mean time between jumps. Evolution appears as a punctuated process as proposed already by Gould and others on the basis of palaeontological data. This might show the heuristic value of abstract evolution models of the type shown above.

Let us mention that the 'quantization' of transitions discussed above for Fisher–Eigen equations is also to be expected for thermodynamic processes (Fokker–Planck equations) if the potential surface shows a Gaussian-like distribution of extrema.

8.6   Conclusion

Transition processes in multidimensional systems are of some importance for the modelling of many processes like, e.g., chemical reactions under the influence of molecular degrees of freedom or diffusional degrees of freedom, of strategies for solving frustrated optimization problems and of evolution processes. We have shown here that in many-dimensional transition processes appear several specific features. For example:

(i) the possibility of enhancing or moderating elementary transitions by the coupling to other degrees of freedom;
(ii) the possibility to go through cascades of transitions leading to more and more 'optimal' solutions;
(iii) stochastic distributions for the distance and time of hopping which might be of importance, e.g. for understanding aspects of real evolution or technical optimization procedures.

Let us stress finally that the treatment of multidimensional transition processes is connected with great mathematical difficulties. Therefore in nearly all cases one is forced to use drastic approximations or even to restrict to estimates only.

References

Ackley, D. H., Hinton, G. E. and Sejnowski, T. J. 1985. *Cogn. Sci.* **9**, 147–60.
Boseniuk, T., Ebeling, W. and Engel, A. 1987. *Phys. Lett. A* **125**, 307–10.
Brettschneider, R. and Schimansky-Geier, L. 1986. *Wiss. Zeitschr. der Humb. Univ. Berlin, Math.-Nat. R.* **35**, (5), 450–7.
Bulsara, A. R., Lindenberg, K. and Shuler, K. E. 1982. *J. Stat. Phys.* **27**, 787–808.
Büttiker, M. and Landauer, R. 1982. In *Nonlinear Phenomena at Phase Transitions and Instabilities* (T. Riste, ed.), pp. 111–43. New York: Plenum Press.
Carmeli, B. and Nitzan, A. 1985. *Phys. Rev. A* **32**, 2439–54.
Denker, J. S. 1986. *Physica* **22D**, 216–32.
de Pasquale, F., Racz, Z., San Miguel, M. and Tartaglia, P. 1984. *Phys. Rev. A* **30**, 5228–38.
Ebeling, W. and Engel, A. 1986. *System Anal. Model. Simulation* **3**, 377–85.

Ebeling, W., Engel, A., Esser, B. and Feistel, R. 1985. *J. Stat. Phys.* **37**, 369–85.
Ebeling, W. and Klimontovich, Y. L. 1984. *Selforganization and Turbulence in Liquids*. Leibzig: Teubner.
Ebeling, W. and Malchow, H. 1979. *Ann. der Phys.* **36**, 121–31.
Ebeling, W. and Peschel, M., eds. 1985. *Lotka–Volterre Approach to Cooperation and Competition in Dynamic Systems*. Berlin: Akademie.
Ebeling, W. and Romanovsky, Y. M. 1985. *Z. Phys. Chemie* **266**, 836–43.
Fox, R. F. 1983. *Phys. Rev. A* **94**, 281–8.
Fronzoni, L., Grigolini, P., Hänggi, P., Moss, F., Mannella, R. and McClintock, P. V. E. 1986. *Phys. Rev. A* **33**, 3320–7.
Gardiner, C. W. 1983. *Handbook of Stochastic Methods*. Berlin: Springer.
Graham, R. 1973. In *Quantum Statistics in Optics and Solid-State Physics* (G. Höhler, ed.), pp. 1–97. Springer Tracts in Modern Physics 66. Berlin: Springer.
Hänggi, P. 1986. *J. Stat. Phys.* **42**, 105–48.
Hänggi, P., Marchesoni, F. and Grigolini, P. 1984. *Z. Phys. B* **56**, 333.
Hänggi, P., Mroczkowski, T. J., Moss, F. and McClintock, P. V. E. 1985. *Phys. Rev. A* **32**, 695–8.
Hoffman, G. W., Benson, M. W., Bree, G. M. and Kinahan, P. E. 1986. *Physica* **22D**, 233–46.
Holmes, P. 1979. *Phil. Trans. Roy. Soc. London* **292**, 419–30.
Hopfield, J. J. and Tank, D. W. 1985. *Biol. Cybern.* **52**, 141–52.
Horsthemke, W. and Lefever, R. 1984. *Noise Induced Transitions*. Berlin: Springer.
Iyengar, R. N. 1986. *J. Stat. Phys.* **44**, 907–20.
Jung, P. and Hänggi, P. 1987. *Phys. Rev. A.* **35**, 4464.
Jung, P. and Risken, H. 1985. *Z. Phys. B* **61**, 367–79.
Kirkpatrick, S., Gelatt, C. D. and Vecchi, M. P. 1983. *Science* **220**, 671–9.
Klimontovich, Y. L. 1983. *The Kinetic Theory of Electromagnetic Processes*, pp. 338–42. Berlin: Springer.
Kuznetsov, P. L., Stratonovich, R. L. and Tikhonov, V. I., eds. 1965. *Non-Linear Transformation of Stochastic Processes*. New York: Pergamon.
Landau, L. D. and Lifshits, E. M. 1971. *Statistische Physik*. Berlin: Akademie.
Landauer, R. and Swanson, J. A. 1961. *Phys. Rev.* **121**, 1668–75.
Landauer, R. 1962. *J. Appl. Phys.* **33**, 2209–16.
Langer, J. S. 1969. *Ann. Phys.* **54**, 258–90.
Lifshits, I. M., Gredeskul, S. A. and Pastur, M. A. 1982. *Introduction to the Theory of Disordered Systems*. Moscow: Nauka.
Lindenberg, K. and West, B. J. 1983. *Physica* **119A**, 485–95.
Malchow, H., Ebeling, W., Feistel, R. and Schimansky-Geier, L. 1983. *Ann. der Phys.* **40**, 151–60.
Malchow, H. and Schimansky-Geier, L. 1985. *Noise and Diffusion in Nonequilibrium Bistable Systems*. Leipzig: Teubner.
Matkowsky, B. J. and Schuss, Z. 1977. *SIAM J. Appl. Math.* **33**, 365–82.
Moss, F., Hänggi, P., Mannella, R. and McClintock, P. V. E. 1986. *Phys. Rev. A* **33**, 4459–61.
Moss, F. and McClintock, P. V. E. 1985. *Z. Phys. B* **61**, 381–6.
Pontryagin, L. S., Andronov, A. A. and Vitt, A. A. 1933. *Zh. Exp. Teor. Fiz.* **3**, 165–80. See the Appendix to this volume for a translation.
Risken, H. 1984. *The Fokker–Planck Equation*. Berlin: Springer.

Ryter, D. 1985. *Physica* **130A**, 205–19.

Sancho, J. M., San Miguel, M., Katz, S. L. and Gunton, J. D. 1982. *Phys. Rev. A* **26**, 1589–609.

San Miguel, M. 1984. In *Recent Developments in Nonequilibrium Thermodynamics* (J. Casa-Vazquez, ed.). Berlin: Springer.

Schenzle, A. and Tél, T. 1985. *Phys. Rev. A* **32**, 596–605.

Schimansky-Geier, L. 1988. *Phys. Lett. A* **126**, 455–8. Erratum *ibid.*, p. 126 (last issue).

Schimansky-Geier, L. and Ebeling, W. 1983. *Ann. der Phys.* **40**, 10–24.

Sejnowski, T. J., Kienker, P. K. and Hinton, G. E. 1986. *Physica* **22D**, 260–74.

Stratonovich, R. L. 1963. *Topics in the Theory of Random Noise*, vol. I. New York: Gordon and Breach.

Talkner, P. and Hänggi, P. 1984. *Phys. Rev. A* **29**, 768–73.

Talkner, P. and Ryter, D. 1983. In *Noise in Physical Systems and 1/f Noise* (M. Savelli, G. Lecoy and J-P. Nougier, eds.). New York: Elsevier.

Van den Broeck, C. and Hänggi, P. 1984. *Phys. Rev. A* **30**, 2730–6.

Van Kampen, N. G. 1981. *Stochastic Processes in Physics and Chemistry*. Amsterdam: North-Holland.

Vogel, K., Risken, H., Schleich, W., James, M., Moss, F. and McClintock, P. V. E. 1987. *Phys. Rev. A* **35**, 463–5.

Weiss, G. H. 1986. *J. Stat. Phys.* **42**, 3–36.

9   Colored noise in continuous dynamical systems: a functional calculus approach

PETER HÄNGGI

9.1   Introduction

Recent work on dye lasers (Fox and Roy, 1987; Jung and Risken, 1984; Lett, Short and Mandel, 1984; Roy, Yu and Zhu, 1985; Short, Mandel and Roy, 1982) and the optical ring laser gyroscope (Vogel *et al.*, 1987a, b) has emphasized the physically important role of colored noise sources. A well-known classical situation in which strongly colored noise has an impact on the physics is the phenomenon of motional narrowing in magnetic resonance (Kubo, 1962). Kubo has shown that a fluctuating magnetic field with very short noise correlation time (almost white noise) does typically not manifestly affect the motion of spins; on the contrary, if the fluctuations are correlated over a long time scale (colored noise) the motion of the spin becomes greatly modified.

Another area where there has been much recent activity addresses escape problems. These are currently in the limelight both from the theoretical viewpoint (Grote and Hynes,1980; Hänggi, 1986; Hänggi and Mojtabai, 1982; Hänggi and Riseborough, 1983; Hänggi, Mroczkowski, Moss and McClintock, 1985) as well as from an experimental point of view (Devoret, Martinis, Esteve and Clarke, 1984; Fleming, Courtney and Balk, 1986; Hänggi *et al.*, 1985; Maneke, Schroeder, Troe and Voss, 1985). In this latter case, a frequency-dependent friction, or noise of finite correlation time, can considerably modify the *classical barrier transmission*. Except for two-state noise (Hänggi and Talkner, 1985; Masoliver, Lindenberg and West, 1986; Rodriguez and Pesquera, 1986; Van den Broeck and Hänggi, 1984) there exist no exact analytic methods for truly nonlinear systems, being driven by correlated noise. Generally, the finite correlation time of the noise will affect not only dynamical aspects but also the form of the stationary probability. This fact has been exploited, for example, in recent studies of colored noise-induced transitions (see Chapter 8, Volume 2).

Our main interest in this chapter is the study of dynamical continuous-time systems of the form (with λ a set of control parameters)

$$\dot{x}_\alpha = f_\alpha(x, \lambda) + \sum_i g_{\alpha i}(x, \lambda)\xi_i(t) \tag{9.1.1}$$

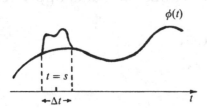

Figure 9.1. A pictorial view of a variation $\delta\phi_0(t)$ of a functional $\phi(t)$.

being driven by noise forces $\{\xi_i(t)\}$ of generally finite correlation time. For the sake of clarity only, we shall primarily restrict in the following the discussion to one-dimensional flows. Without loss of generality we shall confine ourselves to additive noise. Note that a multiplicative noise structure $\xi(t) \to g(x)\xi(t)$ can (in one dimension) be transformed into additive noise if we consider the transform $x \to \tilde{x} = \int^x g^{-1}(y)\mathrm{d}y$, $g(y) \neq 0$, observing that the probabilities transform according to $\tilde{p}_t(\tilde{x}) = p_t(x(\tilde{x}))|\mathrm{d}x/\mathrm{d}\tilde{x}|$.

9.2   Functional calculus

The stochastic dynamics in (9.1.1) is suitably described in terms of probability notions. If we now focus on one-dimensional flows of the form

$$\dot{x} = f(x) + \xi(t) \tag{9.2.1}$$

our interest is in evaluation of mean values $\langle x(t) \rangle$, variances, or the probability $p_t(x) = \langle \delta(x(t) - x) \rangle$ itself. The trajectory $x(t)$ depends, of course, on the whole previous history of the random force $\xi(s)$, $t \geqslant s \geqslant t_0 = 0$; i.e. $x(t)$ is a functional of the noise $\xi(s)$.

If $\rho_t(\xi)$ denotes the (normalized) probability function of the random variable $\xi_t \equiv \xi(t)$ at time t, we obtain for the mean value a functional integral over the realizations of the noise; i.e.

$$\langle x(t) \rangle = \int D\xi(t) x\{\xi(t)\} \rho_t(\xi). \tag{9.2.2}$$

For the probability $p_t(x)$ itself one obtains

$$p_t(x) = \langle \delta(x(t) - x) \rangle = \int D\xi(t) \rho_t(\xi) \delta(x(t) - x). \tag{9.2.3}$$

For the rate of change of $p_t(x)$ one finds

$$\dot{p}_t(x) = \int D\xi(t) \rho_t(\xi) \left\{ -\frac{\partial}{\partial x} \delta(x(t) - x) \right\} \dot{x}(t)$$

$$= -\frac{\partial}{\partial x} \langle \delta(x(t) - x)\dot{x}(t) \rangle$$

$$= -\frac{\partial}{\partial x}\{f(x)p_t(x)\} - \frac{\partial}{\partial x}\langle\delta(x(t)-x)\xi(t)\rangle. \tag{9.2.4}$$

This latter relation can be evaluated further if we make explicit use of the statistics of the noise $\xi(t)$. This task is suitably performed in terms of functional derivatives techniques (Hänggi, 1978b). An extended introduction into the techniques of functional derivatives together with its use in the description of noisy dynamical systems has been given by the author elsewhere (Hänggi, 1985). In the following we shall restrict ourselves to essentials only.

9.3   The functional derivative: a poor man's approach

Let us consider the values of a functional $F[\phi]$ for the functions $\phi(t)$ and $(\phi(t)+\delta\phi_0(t))$ with $\delta\phi_0(t)\neq 0$ within a small interval, $s - \Delta t/2 \leqslant t \leqslant s + \Delta t/2$. In other words, we vary the function $\phi(t)$ near the position s (see Figure 9.1).

The functional derivative is then defined as the limit

$$\frac{\delta F[\phi]}{\delta\phi(s)} \equiv \lim_{\Delta t\to 0} \frac{F[\phi+\delta\phi_0] - F[\phi]}{\int_{\Delta t}\delta\phi_0(t)\,dt}. \tag{9.3.1}$$

Before we proceed further, let us exercise this limiting procedure with an example of a quadratic functional $F[\phi]$

$$F[\phi] = \int\int g(t,s)\phi(t)\phi(s)\,dt\,ds.$$

Therefore we have

$$\Delta F[\phi] \equiv F[\phi+\delta\phi_0] - F[\phi] = \int\int g(t,s)[\phi(t)\delta\phi_0(s) + \phi(s)\delta\phi_0(t)]\,dt\,ds + O(\delta\phi^2).$$

Using for the second part of the integral a substitution of variables, i.e. $t \to s$, $s \to t$, we obtain

$$\Delta F[\phi] = \int\phi(t)\,dt\int ds\{g(t,s)+g(s,t)\}\,\delta\phi_0(s).$$

We now vary the function ϕ around the position $t = s \equiv \tau$. Next we apply the average theorem for integrals and get with $\tau+\Delta t/2 \geqslant \bar{s} \geqslant \tau - \Delta t/2$

$$\Delta F[\phi] = \int\phi(t)\{g(t,\bar{s})+g(\bar{s},t)\}\,dt\int_{\Delta t}\delta\phi_0(s)\,ds.$$

Observing that

$$A \equiv \int\int\{g(t,s)-g(s,t)\}\,dt\,ds$$

309

$$= \iint \{g(s,t) - g(t,s)\} \, dt \, ds = -A = 0,$$

only the symmetric part of $g(t,s)$, i.e. $g_s(t,s) = \frac{1}{2}(g(t,s) + g(s,t))$ enters the result

$$\frac{\delta F[\phi]}{\delta \phi(\tau)} = 2 \int g_s(t,\tau) \phi(t) \, dt.$$

Inspecting the above example one might wonder if the answer could not have been obtained more simply. We remember that a functional $F[\phi]$ can be looked upon as the continuum limit of a multivariable function. Thus we can perturb the functional at the position $t = \tau$ by varying the 'variable' $\phi_\tau = \phi(\tau)$ by an amount $\delta\phi(\tau) = \lambda\delta(t-\tau)$. We may then express the limit in (9.3.1) as the limit of

$$\frac{\delta F[\phi]}{\delta \phi(\tau)} = \frac{dF[\phi(t) + \lambda\delta(t-\tau)]}{d\lambda}\bigg|_{\lambda=0} \tag{9.3.2}$$

assuming that both sides (generally only in the sense of a distribution) exist. Clearly with this trick one readily recovers the result of the above example. Analogously one finds the useful relations:

(i) $\quad F[\phi] = f(\phi(t)) \rightarrow \dfrac{\delta F[\phi]}{\delta \phi(\tau)} = \dfrac{\partial f}{\partial \phi} \delta(t-\tau)$

(ii) $\quad F[\phi] = f(g(\phi)) \rightarrow \dfrac{\delta F[\phi]}{\delta \phi(\tau)} = \dfrac{\partial f}{\partial g} \dfrac{\delta g}{\delta \phi(\tau)}$

(iii) $\quad F[\phi] = \displaystyle\int f(\phi,\dot\phi) \, dt \rightarrow \dfrac{\delta F[\phi]}{\delta \phi(\tau)} = \dfrac{\partial f}{\partial \phi(\tau)} - \dfrac{d}{d\tau} \dfrac{\partial f}{\partial \dot\phi(\tau)}.$

Examples

The characteristic functional of a stochastic process $\xi(t)$ is defined by

$$\chi[\phi] = \left\langle \exp i\left(\int \phi(t)\xi(t) \, dt\right) \right\rangle, \quad \chi[\phi = 0] = 1.$$

The *moments* $m_n(t_1,\ldots,t_n) = \langle \xi(t_1)\xi(t_2)\ldots\xi(t_n) \rangle$ are then obtained via the nth order functional derivative of $\chi[\phi]$; i.e.

$$m_n(t_1,\ldots,t_n) = i^{-n} \frac{\delta^n \chi[\phi]}{\delta\phi(t_1)\ldots\delta\phi(t_n)}\bigg|_{\phi=0}. \tag{9.3.3}$$

The cumulant generating functional $\Psi[\phi]$ is defined by

$$\Psi[\phi] = \ln \chi[\phi],$$

yielding the *cumulants* $C_n(t_1,\ldots,t_n)$

$$C_n(t_1,\ldots,t_n) = i^{-n} \frac{\delta^n \Psi[\phi]}{\delta\phi(t_1)\ldots\delta\phi(t_n)}\bigg|_{\phi=0}. \tag{9.3.4}$$

9.4 A summary of important correlation formulae

The master equation in (9.2.4) explicitly introduces a correlation between the noise $\xi(t)$ and the functional $F[\xi] = \delta(x(t) - x)$ of the noise $\xi(s)$, $t \geqslant s \geqslant t_0$. The expression can only be disentangled further if we make explicit use of the statistical properties of the random force $\xi(s)$. In this context it should be emphasized that the *initial preparation of the system*, which in turn determines the statistical properties of the noise $\xi(s)$, is of *equal importance as the dynamical law* (Grabert, Hänggi and Talkner, 1980; Hänggi, Marchesoni and Grigolini, 1984; Marchesoni, 1984). It should be stressed that different initial correlations between the dynamical variable $x(t = 0)$ and the environment to which the dynamical variable is coupled imply different statistical properties for the noise $\xi(s)$. In particular, the statistical properties of $\xi(s)$ generally cannot be chosen to be independent of the form of initial probability $p_0(x)$ of the (reduced) dynamics $x(s)$. Throughout the rest of the chapter we shall assume that the coarse-grained dynamics $x(s)$ in (9.2.1) has been prepared at initial time $t_0 = 0$ without any memory of the past and without correlations between system and environment (correlation-free preparation); i.e. $p_0(x;$ environment$) = p_0(x) \cdot p_0$(environment).

Below we now list without proof some important relations. For the explicit derivation the interested reader is referred to the original papers (Hänggi, 1978b; Hänggi, 1985). The correlation between a functional $g[\xi]$ and the noise $\xi(t)$ is expressed in terms of the cumulants C_n of the noise $\xi(t)$ as

(i)
$$\langle \xi(t)g[\xi]\rangle = \sum_{n=0}^{\infty} \frac{1}{n!} \int_0^t \ldots \int_0^t dt_1 \ldots dt_n$$

$$\times C_{n+1}(t, t_1, t_2, \ldots, t_n)\left\langle \frac{\delta^n g[\xi]}{\delta\xi(t_1) \ldots \delta\xi(t_n)}\right\rangle. \qquad (9.4.1)$$

This relation can be generalized to a correlation between two functionals $f[\xi]$ and $g[\xi]$ (Hänggi, 1985) to give

(ii)
$$\langle f[\xi]g[\xi]\rangle = \langle f[\xi]\rangle\langle g[\xi]\rangle$$

$$+ \sum_{n=1}^{\infty} \frac{1}{n!} \prod_{i=1}^{n} \sum_{m_i=1}^{\infty} \sum_{k_i=1}^{\infty} \frac{1}{m_i!k_i!}$$

$$\times \int_0^t \underset{m_i+k_i}{\ldots} \int_0^t dt_1^{(i)} \ldots dt_{m_i}^{(i)} \ldots ds_1^{(i)} \ldots ds_{k_i}^{(i)}$$

$$\times C_{m_i+k_i}(t_1^{(i)}, \ldots, t_{m_i}^{(i)}; s_1^{(i)}, \ldots, s_{k_i}^{(i)})\left\langle \frac{\delta^{m_1+\cdots+m_n}f[\xi]}{\delta\xi(t_1^{(1)}) \ldots \delta\xi(t_{m_n}^{(n)})}\right\rangle$$

$$\times \left\langle \frac{\delta^{k_1+\cdots+k_n}g[\xi]}{\delta\xi(s_1^{(1)}) \ldots \delta\xi(s_{k_n}^{(n)})}\right\rangle. \qquad (9.4.2)$$

For a *Gaussian* process there exist only two comulants $C_1(t) = \langle \xi(t)\rangle$ and

311

$C_2(t,s) = \langle \xi(t)\xi(s) \rangle - \langle \xi(t) \rangle \langle \xi(s) \rangle$. Then the above relations simplify considerably to give (Dubkov and Malakhov, 1975; Furutsu, 1963; Hänggi, 1985; Novikov, 1965)

(iii) $\quad \langle \xi(t)g[\xi] \rangle = \langle \xi(t) \rangle \langle g[\xi] \rangle + \int_0^t C_2(t,s) \left\langle \dfrac{\delta g[\xi]}{\delta \xi(s)} \right\rangle ds \qquad (9.4.3)$

and (Dubkov and Malakhov, 1975; Hänggi, 1985)

(iv) $\quad \langle f[\xi]g[\xi] \rangle = \langle f[\xi] \rangle \langle g[\xi] \rangle$

$$+ \sum_{n=1}^{\infty} \frac{1}{n!} \int_0^t \overset{..}{\underset{2n}{}} \int_0^t \left\langle \frac{\delta^n f[\xi]}{\delta \xi(t_1)\dots\delta \xi(t_n)} \right\rangle$$

$$\times \left\langle \frac{\delta^n g[\xi]}{\delta \xi(s_1)\dots\delta \xi(s_n)} \right\rangle \prod_{i=1}^{n} C_2(t_i, s_i)\, dt_i\, ds_i. \qquad (9.4.4)$$

The set of above formulae remains true also for white noise (i.e. δ-correlated random forces) if the latest time, s (e.g. $\xi(t) \to \xi(s)$ with $s < t$ in (i)) of the random force occurring in the functionals is less than the final observation time t. Some care is needed for white random forces if functionals of the noise up to the final observation time $s \leqslant t$ are considered; in this latter case it is advantageous to consider curtailed generating functionals (for details see Hänggi, 1978b).

9.5 Functional derivatives for dynamical flows

On inspecting (9.2.4), we find that the master equation in (9.2.4) can be evaluated further by use of relations such as (9.4.1) or (9.4.3). This procedure thus involves the functional derivative

$$\frac{\delta}{\delta \xi(s)} \delta(x(t) - x) = -\frac{\partial}{\partial x} \delta(x(t) - x) \frac{\delta x(t)}{\delta \xi(s)}. \qquad (9.5.1)$$

A dynamical flow with *multiplicative* noise, i.e.

$$\dot{x} = f(x) + g(x)\xi(t) \qquad (9.5.2)$$

then yields via its integral

$$x(t) = \int_0^t \{f(x(s)) + g(x(s))\xi(s)\}\, ds + x(0) \qquad (9.5.3)$$

the following integral equation (Hänggi, 1978b):

$$\frac{\delta x(t)}{\delta \xi(s)} = \theta(t - s)\left\{ g(x(s)) + \int_s^t du\, \frac{\partial \dot{x}(u)}{\partial x(u)} \frac{\delta x(u)}{\delta \xi(s)} \right\}, \qquad (9.5.4)$$

where $\theta(t)$ is the step function expressing causality. This integral equation is

312

readily solved to yield the formal expression

$$\frac{\delta x(t)}{\delta \xi(s)} = \theta(t - s)g(x(s))\exp \int_s^t \frac{\partial \dot{x}(u)}{\partial x(u)}\,du \tag{9.5.5a}$$

$$= \theta(t - s)g(x(s))$$

$$\times \exp \int_s^t \left\{ \frac{\partial f(x(u))}{\partial x(u)} + \frac{\partial g(x(u))}{\partial x(u)}\xi(u) \right\} du. \tag{9.5.5b}$$

Equation (9.5.5b) can be recast in alternative form if we make use of a trick due to Fox (1986b). Observing

$$\dot{g} = g'\dot{x} = \frac{g'}{g}(f + g\xi)g$$

one finds the formal solution

$$g(x(s)) = g(x(t))\exp \int_t^s du \frac{g'}{g}(f + g\xi).$$

Thus, the result in (9.5.5b) is rewritten as

$$\frac{\delta x(t)}{\delta \xi(s)} = \theta(t - s)g(x(t))$$

$$\times \exp \int_s^t \left\{ f'(x(u)) - f(x(u))\frac{g'(x(u))}{g(x(u))} \right\} du, \tag{9.5.5c}$$

where the prime indicates differentiation with respect to $x(u)$. This latter form is preferred to (9.5.5b) because the multiplicative coupling enters at the final time t. Before we proceed further we now present a few useful examples.

Examples

(1) Consider (9.1.1) and show that in this multidimensional case one obtains for (9.5.4)

$$\frac{\delta x_\alpha(t)}{\delta \xi_i(s)} = \theta(t - s)\left\{ \int_s^t du \left[\frac{\partial f_\alpha}{\partial x_n} + \frac{\partial g_{\alpha m}}{\partial x_n}\xi_m \right] \frac{\delta x_n(u)}{\delta \xi_i(s)} + g_{\alpha i}(x(s)) \right\}, \tag{9.5.6}$$

where a summation convention over equal indices is implied.

(2) Show that for linear flow $\dot{x} = a(t)x(t) + \xi(t)$ one finds the explicit answer

$$\frac{\delta x(t)}{\delta \xi(s)} = \theta(t - s)\exp \int_s^t a(u)\,du. \tag{9.5.7}$$

(3) For a drift-free flow $\dot{x} = a(t)g(x(t))\xi(t)$ one finds (Hänggi, 1981)

$$\frac{\delta x(t)}{\delta \xi(s)} = \theta(t - s)\,a(s)\,g(x(t)). \tag{9.5.8}$$

313

(4) For a Mori-type flow

$$\dot{x}(t) = - \int_0^t \gamma(t - s) x(s) \, ds + \xi(t)$$

we obtain in terms of the Green's function χ, obeying

$$\dot{\chi}(t) = - \int_0^t \gamma(t - s) \chi(s) \, ds, \quad \chi(0) = 1$$

for the functional derivative the result

$$\frac{\delta x(t)}{\delta \xi(s)} = \theta(t - s) \chi(t - s). \tag{9.5.9}$$

Generally, it will not be possible to obtain a closed answer for the functional derivative $\delta x(t)/\delta \xi(s)$ taken at different times $s < t$. At this stage we have now collected enough material to evaluate the colored noise master equation in (9.2.4) in terms of the statistics of the random force.

9.6 The colored noise master equation (Hänggi, 1978b)

The result in (9.2.4) can be evaluated further if we invoke the correlation formula in (9.4.1) with $g[\xi] = \delta(x(t) - x)$. For non-Gaussian noise with vanishing mean, $\langle \xi(t) \rangle = C_1(t) = 0$, we find from (9.2.4) and (9.5.2)

$$\dot{p}_t(x) = -\frac{\partial}{\partial x} \{ f(x) p_t(x) \} - \frac{\partial}{\partial x} g(x) \sum_{n=1}^{\infty} \frac{1}{n!} \int_0^t \dots \int_0^t dt_1 \dots dt_n$$

$$\times C_{n+1}(t, t_1, \dots, t_n) \left\langle \frac{\delta^n}{\delta \xi(t_1) \dots \delta \xi(t_n)} \delta(x(t) - x) \right\rangle. \tag{9.6.1}$$

Note here the typical non-Markovian character, i.e. the dependence on the initial time $t_0 = 0$ of preparation, yielding for the initial rate of change of probability:

$$\dot{p}_{t_0=0}(x) = -\frac{\partial}{\partial x} \{ f(x) p_{t_0=0}(x) \}.$$

The last term in (9.6.1) can now be written out more explicitly by observing the chain rule:

$$\frac{\delta}{\delta \xi(s)} \delta(x(t) - x) = -\frac{\partial}{\partial x} \delta(x(t) - x) \frac{\delta x(t)}{\delta \xi(s)}.$$

In the remainder of this chapter we restrict ourselves to *Gaussian noise* only. Then we obtain from (9.4.3) the colored noise master equation

$$\dot{p}_t(x) = -\frac{\partial}{\partial x} \{ f(x) p_t(x) \}$$

$$+ \frac{\partial}{\partial x} g(x) \frac{\partial}{\partial x} \int_0^t ds\, C_2(t,s) \left\langle \delta(x(t) - x) \frac{\delta x(t)}{\delta \xi(s)} \right\rangle. \qquad (9.6.2a)$$

For *additive* noise, (9.2.1), we find with $g(x) \equiv 1$ from (9.5.5b)

$$\dot{p}_t(x) = - \frac{\partial}{\partial x} \{ f(x) p_t(x) \}$$

$$+ \frac{\partial^2}{\partial x^2} \int_0^t ds\, C_2(t,s) \left\langle \delta(x(t) - x) \exp \int_s^t \frac{\partial f(u)}{\partial x(u)} du \right\rangle. \quad (9.6.2b)$$

Note that for the *multiplicative* stochastic flow in (9.5.2) the corresponding colored noise master equation reads with (9.5.5c)

$$\dot{p}_t(x) = - \frac{\partial}{\partial x} \{ f(x) p_t(x) \} + \frac{\partial}{\partial x} g(x) \frac{\partial}{\partial x} g(x) \int_0^t ds\, C_2(t,s)$$

$$\times \left\langle \delta(x(t) - x) \exp \int_s^t [f' - (fg'/g)]\, du \right\rangle. \qquad (9.6.2c)$$

At this stage we generally cannot simplify the master equation any further. In view of the δ-function $\delta(x(t) - x)$ a closed expression is possible only if the functional derivative $\delta x(t)/\delta \xi(s)$ either does not involve the process $x(s)$ itself, or depends on the process $x(s)$ solely on the 'Markovian' end-point $s = t$. In the examples discussed in (9.5.7)–(9.5.9) we obtain thus a *closed colored noise master equation*, which for Gaussian noise $\xi(t)$ is of Fokker–Planck type. We leave it as an exercise to the reader to verify the following results.

(i) A linear flow $\dot{x} = a(t)x(t) + \xi(t)$ yields with (9.5.7) the exact result (Hänggi, 1978b)

$$\dot{p}_t(x) = - a(t) \frac{\partial}{\partial x} (x p_t(x))$$

$$+ \left(\int_0^t ds\, C_2(t,s) \exp \left[\int_s^t a(u)\, du \right] \right) \frac{\partial^2}{\partial x^2} p_t(x). \qquad (9.6.3)$$

More generally, (9.6.3) holds for a nonlinear process y, which up to a nonlinear transformation, $y \to x = f(y; t)$, coincides with a Gaussian process (linear flow) (see Hänggi, 1978a).

(ii) For a drift-free flow one finds in view of (9.5.8)

$$\dot{p}_t(x) = \left(a(t) \int_0^t C_2(t,s) a(s)\, ds \right) \frac{\partial}{\partial x} g(x) \frac{\partial}{\partial x} g(x) p_t(x). \qquad (9.6.4a)$$

Note the typical non-Markovian character in (9.6.3) and (9.6.4a). The Fokker–Planck type equation depends – in contrast to the Markovian case – on the initial time point $t_0 = 0$ of preparation. By use of the time scale

$$\hat{t} = \int^t ds\, a(s) \int_0^s C_2(s,r) a(r)\, dr$$

315

PETER HÄNGGI

we can recast (9.6.4a) into the 'Markovian form'

$$\dot{p}_i(x) = \frac{\partial}{\partial x} g(x) \frac{\partial}{\partial x} g(x) p_i(x), \qquad (9.6.4b)$$

with all non-Markovian features transformed away. Nevertheless the process $x(\hat{t})$ is, of course, still non-Markovian. Its single event evolution just happens to coincide on the time scale, $t \to \hat{t}$, with the single-event evolution of a corresponding time-homogeneous Markov process with Fokker–Planck operator (9.6.4b).

(iii) For a Mori-type flow

$$\dot{x} = -\Omega^2 x - \int_0^t ds\, \hat{\gamma}(t-s) x(s) + \xi(t)$$

$$= -\int_0^t \gamma(t-s) x(s)\, ds + \xi(t),$$

$\gamma(t-s) = \hat{\gamma}(t-s) + 2\Omega^2 \delta(t-s)$ one finds, in view of (9.5.9) and the explicit solution

$$x(t) = \chi(t) x_0 + \int_0^t \chi(t-u) \xi(u)\, du$$

for the colored noise master equation, the explicit result (Hänggi, 1985)

$$\dot{p}_t(x) = \left(\int_0^t ds\, C_2(t,s) \chi(t-s) \right) \frac{\partial^2}{\partial x^2} p_t(x)$$

$$+ \left(\langle x(0) \rangle \int_0^t \gamma(t-s) \chi(s)\, ds \right) \frac{\partial}{\partial x} p_t(x)$$

$$- \left\{ \int_0^t ds\, \gamma(t-s) \int_0^s \int_0^t dt_1\, dt_2\, \chi(s-t_1) \right.$$

$$\left. \times \chi(t-t_2) C_2(t_1,t_2) \right\} \frac{\partial^2}{\partial x^2} p_t(x). \qquad (9.6.5)$$

Except for times $\{\bar{t}\}$ satisfying $\chi(\bar{t}) = 0$, the Mori-flow can be recast into time-convolutionless form

$$\dot{x} = \left[\frac{\dot{\chi}(t)}{\chi(t)} \right] x(t) + \chi(t) \frac{d}{dt} \int_0^t \frac{\chi(t-s)}{\chi(t)} \xi(s)\, ds$$

yielding (Hänggi, 1978b)

$$\dot{p}_t(x) = \frac{-\dot{\chi}(t)}{\chi(t)} \frac{\partial}{\partial x} (x p_t(x)) - \frac{\dot{\chi}(t)}{\chi(t)} \left(\int_0^t ds\, \chi(t-s) \right.$$

$$\left. \times \int_0^t du\, \chi(t-u) C_2(s,u) \right) \frac{\partial^2}{\partial x^2} p_t(x)$$

316

$$+ \left(\int_0^t ds\, \dot{\chi}(t-s) \int_0^t du\, \chi(t-u)\, C_2(s,u) \right) \frac{\partial^2}{\partial x^2} p_t(x)$$

$$+ \left(\int_0^t du\, \chi(t-u)\, C_2(t,u) \right) \frac{\partial^2}{\partial x^2} p_t(x). \qquad (9.6.6)$$

Example

Consider the colored noise Brownian motion of a particle with mass m

$$\dot{x} = u$$

$$\dot{u} = -\omega^2 x - \int_0^t \gamma(t-s) u(s)\, ds + \xi(t)$$

obeying the fluctuation–dissipation relation

$$\langle \xi(t)\xi(s) \rangle = \frac{kT}{m} \gamma(t-s). \qquad (9.6.7)$$

With

$$\chi(t) = L^{-1} \left[\frac{1}{z + \tilde{\gamma}(z)} \right],$$

where L^{-1} is the inverse Laplace transform being denoted by $f(t) \to \tilde{f}(z)$, we obtain

$$u(t) = \chi(t) u(0) + \int_0^t \chi(t-s)\xi(s)\, ds$$

$$-\omega^2 \int_0^t \chi(t-s) x(s)\, ds. \qquad (9.6.8a)$$

For $x(t)$ we find with

$$\eta(t) = L^{-1} \left[\frac{1}{z + \omega^2 \tilde{\chi}(z)} \right]$$

$$x(t) = \eta(t) x(0) + \int_0^t ds\, \eta(t-s) \int_0^s \chi(t-u)\xi(u) du + u(0)$$

$$\times \int_0^t \eta(t-s)\chi(s)\, ds. \qquad (9.6.8b)$$

Using the time-convolutionless form for $x(t)$ and observing (9.6.7) one arrives from (9.6.6) at the result

$$\dot{p}_t(x) = \frac{\partial}{\partial x} \left[-\frac{\dot{\eta}(t)}{\eta(t)} x p_t(x) + \frac{u(0)}{\omega^2} \eta(t) \left(\frac{d}{dt} \frac{\dot{\eta}(t)}{\eta(t)} \right) p_t(x) \right]$$

$$+ \frac{\partial^2}{\partial x^2} \left(\left\{ \frac{-kT}{m\omega^2} \frac{\dot{\eta}(t)}{\eta(t)} - \frac{kT}{2m\omega^4} \eta^2(t) \left[\frac{d}{dt} \left(\frac{\dot{\eta}(t)}{\eta(t)} \right)^2 \right] \right\} p_t(x) \right). \qquad (9.6.9)$$

317

For an equilibrium Maxwell distribution of the initial probability $u(0)$ we obtain from (9.6.8a)

$$\dot{x}(t) = u(t) = -\omega^2 \int_0^t \mathrm{d}s\, \chi(t-s)x(s) + \hat{\xi}(t),$$

where

$$\langle \hat{\xi}(t)\hat{\xi}(s) \rangle = \frac{kT}{m}\chi(t-s). \qquad (9.6.10a)$$

This then yields the simple answer (see Hänggi, 1978b; San Miguel and Sancho, 1980)

$$\dot{p}_t(x) = -\frac{\dot{\eta}(t)}{\eta(t)}\left(\frac{\partial}{\partial x}(xp_t(x)) + \frac{kT}{m\omega^2}\frac{\partial^2}{\partial x^2}p_t(x) \right). \qquad (9.6.10b)$$

For the non-Markovian master equation for the joint probability $p_t(x, u)$ of the colored noise Brownian motion

$$\ddot{x} = \pm \omega^2 x - \int_0^t \gamma(t-s)u(s)\,\mathrm{d}s + \xi(t), \quad \dot{x}(s) = u(s)$$

we refer the reader to the original papers (Adelman, 1976; Hänggi and Mojtabai, 1982).

These selected examples of exactly solvable cases make it clear that in general the functional derivative cannot be evaluated explicitly. Thus, in most cases of physical interest we are stuck with (9.6.2b) or (9.6.2c). Therefore one needs approximative schemes which are tailored to the specific noise parameters under consideration, such as the noise correlation time τ and/or the noise strength D, i.e.

$$D = \int_0^\infty \langle \xi(t)\xi(0) \rangle\,\mathrm{d}t.$$

Generally, these approximative schemes become useful in practice only if they reduce to a Fokker–Planck form.

9.7   Approximation schemes

In the following we shall report, extend and interpret various approximative schemes for colored noise driven nonlinear stochastic flows. First we start with the widely used *small correlation time expansion*.

9.7.1   Small correlation time expansion

If the noise $\xi(t)$ is close to the white noise limit (i.e. colored noise with very small correlation time) it seems appropriate to expand the functional derivative around its Markovian value which it attains for δ-correlated noise. Thus, we expand $\delta x(t)/\delta \xi(s)$ into a Taylor series around the Markovian end point t, i.e.

$$\frac{\delta x(t)}{\delta \xi(s)} = \frac{\delta x(t)}{\delta \xi(t)} + \sum_{n=1}^{\infty} \frac{(-1)^n}{n!} \left[\frac{d^n}{ds^n} \frac{\delta x(t)}{\delta \xi(s)} \right]_{st} (t-s)^n. \qquad (9.7.1)$$

For the multiplicative stochastic flow in (9.5.2) we obtain from (9.5.5a–c)

$$\frac{\delta x(t)}{\delta \xi(s)} = \theta(t-s) \left[g(x(t)) + \{ g'(x(t)) f(x(t)) - g(x(t)) f'(x(t)) \} (t-s) \right.$$

$$\left. + \left\{ f^2 \left[f \left(\frac{g}{f} \right)' \right]' - g^2 \left[g \left(\frac{f}{g} \right)' \right]' \right\} \xi(t) \right\} (t-s)^2 + \cdots \right] \qquad (9.7.2)$$

Note that this expansion involves already in second order again the noise $\xi(t)$. This leads to new correlations, which in turn must be expanded again into functional derivatives, and so on, yielding never-ending sums of series of series. Most commonly, one truncates the series at the first order. Using exponentially correlated Gaussian noise of vanishing mean, i.e.

$$C_2(t-s) = \frac{D}{\tau} \exp - \frac{|t-s|}{\tau}, \qquad (9.7.3)$$

we find from (9.6.2a), by neglecting transients (i.e. we extend the time integration to infinity) the *small correlation time result*

$$\dot{p}_t(x; \tau) = - \frac{\partial}{\partial x} \{ f(x) p_t(x; \tau) \}$$

$$+ D \frac{\partial}{\partial x} g(x) \frac{\partial}{\partial x} g(x) \left[1 + \tau g(x) \left(\frac{f}{g} \right)' \right] p_t(x; \tau). \qquad (9.7.4)$$

This very result has been repeatedly derived in the literature by a variety of different, but equivalent methods (Dekker, 1982; Fox 1983; Horsthemke and Lefever, 1980; Kaneko, 1981; Lax, 1966; Lindenberg and West, 1983; Lugiato and Harowicz, 1985; Sancho and San Miguel, 1980; Sancho, San Miguel, Katz and Gunton, 1982; Schenzle and Tél, 1985; Stratonovich, 1963; Suzuki, 1980; Van Kampen, 1976; Zoller, Alber and Salvador, 1981). This same small correlation time approximation is also inherent within the large body of work performed on the problem of elimination of the velocity variable in nonlinear Brownian motion (for extended reviews on this body of work see Grigolini and Marchesoni, 1985; Marchesoni, 1985; Van Kampen, 1985). Unfortunately this approximation is of limited use (even for a linear flow) (Hänggi, Marchesoni and Grigolini, 1984); it does not converge uniformly in x, and the diffusion coefficient in (9.7.4) exhibits generally negative values for sufficiently large values of x (and/or τ), thereby introducing unphysical singularities (boundaries) into the problem. In other words (9.7.4) does not constitute a truly Markovian process with a well-defined corresponding Langevin equation driven by white Gaussian noise.

319

Some authors attempted to 'improve' the Fokker–Planck approximation in (9.7.4) by evaluating additional Fokker–Planck contributions proportional to $D\tau^n$ (Lindenberg and West, 1983; Sancho et al., 1982; Van Kampen, 1976) while at the same time neglecting noise-dependent contributions in (9.7.2) which yield additional Fokker–Planck terms together with non-Fokker–Planck terms. This procedure, however, does not cure the short-comings already present in (9.7.4). As pointed out previously (Hänggi, Marchesoni and Grigolini, 1984; see also Marchesoni, 1988) such a formal ordering of the τ-expansion according to powers of D and τ is fictitious: $D\tau^n$ is not a systematic expansion coefficient; it is the action of the operator with coefficient $D\tau^n$, $n = 1, 2, \ldots$, acting on $p_t(x; \tau)$ which must be compared with the terms neglected (note that $p_t(x; \tau)$ itself depends on D and τ). It then turns out that the regime of validity of (9.7.4) is given by: $\tau/D \ll 1$, $\tau \ll 1$ (in dimensionless units). An approximation which overcomes these difficulties (at least within a certain class of stochastic flows) will be presented in the following.

9.7.2 Unified colored noise approximation

There is certainly a need (see the introduction to this chapter) to consider not only very short correlation times τ but also moderate-to-large noise correlations τ. With this viewpoint in mind, there have been some recent advances in the theory. The result in (9.7.4) becomes of course exact in the limit $\tau \to 0$. In recent work (Jung and Hänggi, 1987) *we have been guided to seek an approximation which becomes exact both for $\tau \to 0$ and $\tau \to \infty$. For intermediate τ-values such a scheme will then hopefully give a useful approximation.* In the following we restrict ourselves to additive noise only; a multiplicative noise source can, as mentioned previously, be transformed into additive noise via $x \to \tilde{x} = \int^x g^{-1}(y)\, dy, g(y) \neq 0$. Starting with (9.2.1), i.e.

$$\dot{x} = f(x) + \xi(t),$$

with $\xi(t)$ Gaussian, exponentially correlated noise (see (9.7.3)) we can recast the one-dimensional flow as a two-dimensional Markovian flow of the form

$$\left.\begin{aligned}
\dot{x} &= f(x) + \xi \\
\dot{\xi} &= -\frac{1}{\tau}\xi + \frac{D^{1/2}}{\tau}Q(t),
\end{aligned}\right\} \tag{9.7.5}$$

$\langle Q(t)Q(s)\rangle = 2\delta(t - s)$. Upon elimination of ξ, i.e.

$$\ddot{x} + \dot{x}\left(\frac{1}{\tau} - f'(x)\right) - \frac{1}{\tau}f(x) = \frac{D^{1/2}}{\tau}Q(t)$$

and the new time scale, $\hat{t} = \tau^{-1/2}t$, we find the nonlinearly damped random oscillator motion

$$\ddot{x} + \dot{x}(\tau^{-1/2} + \tau^{1/2}[-f'(x)]) - f(x) = D^{1/2}Q(\tau^{1/2}\hat{t}). \tag{9.7.6}$$

The damping $\gamma(x, \tau) = (\tau^{-1/2} + \tau^{1/2}[-f'(x)])$ is positive in regions of local stability and approaches infinity both for $\tau \to 0$ and $\tau \to \infty$. If in addition $\gamma(x, \tau)$ attains a large positive value in some x-region it is justifiable to eliminate the fast velocity variable $\dot{x}$ by setting $\ddot{x} = 0$. Setting the change of velocity equal to zero implies in turn that the change of the force field over the characteristic length, $l = D^{1/2}/\gamma(x, \tau)$, is small. Therefore, the Smoluchowski dynamics for this very same x-region becomes a *truly Markovian Fokker–Planck process*

$$\dot{x} = \frac{f(x)}{\gamma(x, \tau)} + \frac{D^{1/2}\tau^{-1/4}}{\gamma(x, \tau)} Q(\hat{t}), \tag{9.7.7a}$$

with $\langle Q(\hat{t})Q(\hat{s}) \rangle = 2\delta(\hat{t} - \hat{s})$ being Gaussian δ-correlated (white) noise. Equation (9.7.7a) is valid on *time scales* $\hat{t} \gg \gamma^{-1}(x, \tau)$; i.e. $t \gg \tau/(1 - \tau f')$ and in *space regions* x obeying $l|\bar{f}'| \ll |\bar{f}|$, i.e. $\gamma(x, \tau) \gg D^{1/2}(|\bar{f}'(x)/\bar{f}(x)|)$ (the bar indicates the characteristic value of the quantity f'/f within the length l).

The reduced dynamics in (9.7.7a) implies the (Stratonovich) Fokker–Planck dynamics on the time scale $\hat{t}$ (Jung and Hänggi, 1987)

$$\dot{p}_{\hat{t}}(x; \tau) = -\frac{\partial}{\partial x}\left(\gamma^{-1}(x, \tau)\left\{ f(x) - D\tau^{-1/2}\frac{\partial}{\partial x}\gamma^{-1}(x, \tau) \right\} p_{\hat{t}}(x; \tau) \right), \tag{9.7.7b}$$

with stationary probability*

$$p_{st}(x; \tau) = N^{-1}|1 - \tau f'(x)| \exp\left[-\frac{1}{2D}\tau f^2(x) \right]$$

$$\times \exp\left[\frac{1}{D} \int^x f(y)\,dy \right]. \tag{9.7.7c}$$

In contrast to the previous approximation in (9.7.4) the diffusion coefficient in (9.7.7b) is strictly positive. Most importantly, the dynamics in (9.7.7b) does not restrict the τ-value to very small correlation times only. In regions of local stability; i.e. $-f'(x) > 0$, the theory is exact for $\tau = 0$ and again becomes exact with τ approaching infinity. The theory in (9.7.7a, b) clearly supersedes the conventional small correlation time approach in Section 9.7.1: For sufficiently (very) small τ-values, i.e. τ so small that it becomes justified to replace factors like $(1 - \tau f'(x)) \to (1 + \tau f'(x))^{-1}$ (which are formally divergent) the solution (9.7.7c) approaches the solution of (9.7.4).

We conclude this section on the unified colored noise approximation by

---

* For multiplicative noise, $\dot{x} = f(x) + g(x)\xi(t)$, one finds $\gamma(x, \tau) = [\tau^{-1/2} - \tau^{1/2}(f' - (g'/g)f)]$, and

$$p_{st}(x; \tau) = Z^{-1}|[1 - \tau(f' - (g'/g)f)]/g|$$

$$\times \exp\left\{ \int^x dy\, f(y)[1 - \tau(f' - (g'/g)f)]/(Dg^2) \right\}.$$

pointing out that the stationary probability in (9.7.7c) *precisely coincides* with the stationary probability of a recently improved small correlation time theory due to Fox (Fox, 1986a, b); the Fokker–Planck dynamics of the two theories, however, *differ*! In the small correlation time theory of Fox the diffusion is still plagued by possible negative values; nevertheless the two Fokker–Planck approximations possess identical stationary probabilities. Also it should be remarked that the approximative dynamics in (9.7.7b) is close to the equilibrium dynamics (more precisely, it corresponds approximately to the stationary preparation class; for details see Grabert, Hänggi and Talkner, 1980). Thus we also expect that (9.7.7b) provides *a reasonable* approximation for quantities such as equilibrium correlation functions.

9.7.3   Decoupling approximation

Although the unified colored noise approximation of Section 9.7.2 works generally for small-to-moderate-to-large noise correlation times τ; it is, however, still restricted to a positive damping $\gamma(x, \tau)$. Unfortunately, this novel approximation with $f'(x) > 0$ does not cover all situations such as the case of multistability at moderate-to-strong-noise color; nor can it describe exponentially large (or small) asymptotic statistical quantities such as escape times or rates at weak noise intensities D. On inspecting the colored noise master equation in (9.6.2c) a different approximation scheme would simply involve a repeatedly applied decoupling between the functional $\delta(x(t) - x)$ and the residual functional stemming from the functional derivative. Such a decoupling is particularly suitable if the probabilities possess small widths. In general, this latter condition implies small noise intensities D – a reasonable condition present in most applications – but otherwise does not restrict the value of the noise correlation time τ. For exponentially correlated Gaussian noise (9.7.3) this quasi-stationary decoupling approximation, put forward originally by Hänggi (Hänggi *et al.*, 1985) thus reads

$$\dot{p}_t(x; \tau) = -\frac{\partial}{\partial x}\{f(x)p_t(x; \tau)\} + \left\{\frac{D}{1 - \tau(\langle f' \rangle - \langle fg'/g \rangle)}\right\}$$
$$\times \frac{\partial}{\partial x}g(x)\frac{\partial}{\partial x}g(x)p_t(x; \tau), \qquad (9.7.8)$$

where the average $(\langle \cdots \rangle)$ is over the stationary probability. Comparing this equation to the corresponding Fokker–Planck dynamics for white noise (i. e. $\tau = 0$), we observe that the influence of colored noise is contained in a renormalized diffusion coefficient. Due to the neglect of transients (i.e. $t \to \infty$ in (9.6.2c)), the diffusion coefficient can be evaluated self-consistently via the stationary probability. It should also be noted that the averages themselves depend on τ and D. In addition, the dynamical law which determines the stationary probability guarantees that the diffusion stays positive. The approximation in (9.7.8) can actually be improved systematically if we invoke the correlation formula between two functionals in (9.4.4): The approximation in (9.7.8) simply refers to the zeroth order approximation of the correlation in (9.6.2c); higher orders involve the cumulant(s) of the noise and yield, via the functional derivatives of $\delta(x(t) - x)$, higher order non-Fokker–Planck terms.

In conclusion we emphasize that the approximative schemes in Sections 9.7.2 and 9.7.3 do *not* rely on an expansion in the correlation time τ.

9.7.4   *A case study: bistability driven by colored Gaussian noise*

The archetype of a colored bistable flow is given by the overdamped motion in a double-well potential driven by correlated Gaussian noise of zero mean (9.7.3), i.e.

$$\dot{x} = x - x^3 + \xi(t). \tag{9.7.9}$$

This example has been used as a basis for testing the quality of different colored noise approximation schemes (Hänggi, Marchesoni and Grigolini, 1984). This latter work has started a flurry of activity around this specific colored noise driven dynamics (Fox, 1986a, b, 1988; Fronzoni *et al.*, 1986; Hänggi *et al.*, 1985; Jung and Hänggi, 1988; Jung and Risken, 1985; Leiber, Marchesoni and Risken, 1987; Malchow and Schimansky–Geier, 1985; Marchesoni, 1987; Masoliver, West and Lindenberg, 1987; Moss and McClintock, 1985; Moss, Hänggi, Mannella and McClintock, 1986; Sancho, Sagués and San Miguel, 1986).

In Figure 9.2 we compare results of different approximation schemes (Jung and Hänggi, 1988) for the stationary probability $p_{st}(x;\tau)$ for a noise intensity $D = 0.1$ and various τ-values. The small correlation time approximation (Section 9.7.1) starts to break down at τ-values $\tau \geqslant 0.2$. Actually this small correlation time approximation *loses* its bistable character (i.e. two maxima and one minimum) for

$$p_{st}(x;\tau) \quad \text{at} \quad \tau \geqslant \{-1 + (1 + 18D)^{1/2}\}/(18D).$$

For $D = 0.1$ this occurs at $\tau \geqslant 0.3740\dots$. Notwithstanding claims to the contrary (Masoliver, West and Lindenberg, 1987), this theory can thus not describe mean sojourn times, or escape rates at weak-to-moderate noise color $0.3 \leqslant \tau \leqslant 1.5$.

The unified colored noise approximation with $\gamma(x = 0, \tau) \leqslant 0$ for $\tau \geqslant 1$ cannot describe the instable behavior around $x \approx 0$. This behavior is in contrast to the decoupling theory, which works qualitatively for all τ and increases in accuracy for probabilities of small widths, i.e. for very small D and/or large τ. Unfortunately, the continued fraction method used to evaluate $p_{st}(x;\tau)$ numerically (Jung and Risken, 1985) encounters convergence pro-

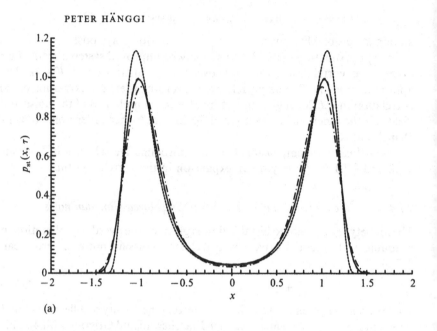

(a)

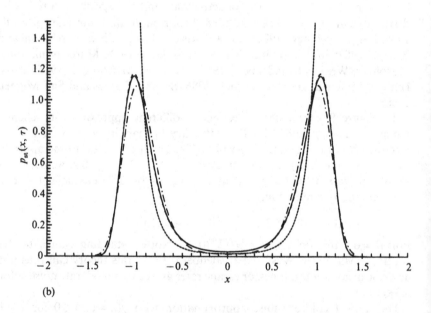

(b)

blems for small D-values, $D < 0.1$ (and also for very large τ-values). From this view point the numerical method and the decoupling scheme by Hänggi (Section 9.7.3) – which works best for $D \ll 1$ – complement each other. In the asymptotic regime of very small noise intensity $D < 0.05$, the ratio $R(\tau)$ of maximal over minimal stationary probability is from the decoupling scheme

324

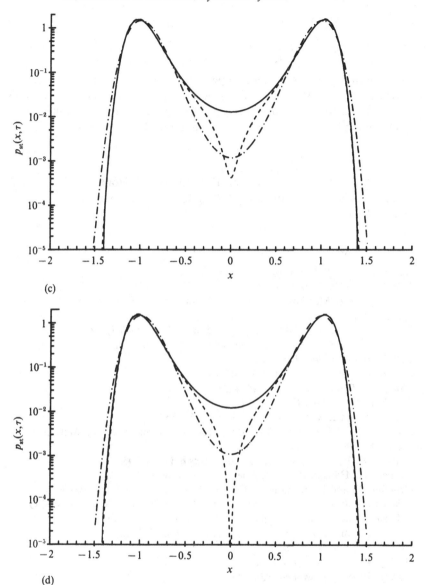

(c)

(d)

Figure 9.2. The stationary probability for the bistable colored noise dynamics in
(9.7.9). (a) $D = 0.1$, $\tau = 0.2$; (b) $D = 0.1$, $\tau = 0.4$; (c) $D = 0.1$, $\tau = 0.99$; (d) $D = 0.1$,
$\tau = 1$. In all (a)–(d): ———— numerical continued fraction evaluation of $p_{st}(x; \tau)$
(Jung and Hänggi, 1988; - - - - unified approximation in (9.7.7c); ·—·—· decoupling
approximation in (9.7.8); .... small correlation time approximation in (9.7.4);
there we depict $p_{st}(x; \tau)$ only for $\tau = 0.2$ where it has bistable character, and at
$\tau = 0.4$ where $p_{st}(x; \tau)$ is already no longer bistable; however it can be
normalized for $\tau < 0.5$. Note that the unified approximation (9.7.7c) in (a)
and (b) practically coincides with the numerical solution (solid line).

325

PETER HÄNGGI

(9.7.8) with $\langle x^2 \rangle \approx 1$ given by

$$R(\tau) = \exp\left(\frac{1}{4D}(1 + 2\tau)\right). \qquad (9.7.10)$$

This characteristic asymptotic dependence, i.e. $R(\tau) \propto \exp(\text{const} \cdot \tau/D)$, has independently been verified within a Kramers rate approach for weak noise (Marchesoni, 1987), and also via explicit numerical calculations in periodic, multistable potentials (Leiber, Marchesoni and Risken, 1987).

Acknowledgement

The author thanks his coworkers (P. Jung and W. Hontscha) for a careful reading of the manuscript, and P. Jung for providing numerical results for Section 9.7.4.

References

Adelman, S. A. 1976. *J. Chem. Phys.* **64**, 124.
Dekker, H. 1982. *Phys. Lett. A* **90**, 26.
Devoret, M. H., Martinis, J. M., Esteve, D. and Clarke, J. 1984. *Phys. Rev. Lett.* **53**, 1260.
Dubkov, A. A. and Malakhov, A. N. 1975. *Sov. Phys. Doklady* **20**, 401.
Fleming, G. R., Courtney, S. H. and Balk, M. W. 1986. *J. Stat. Phys.* **42**, 83–104.
Fox, R. F. 1983. *Phys. Lett.* **94A**, 281.
Fox, R. F. 1986a. *Phys. Rev. A* **33**, 467.
Fox, R. F. 1986b. *Phys. Rev. A* **34**, 4525.
Fox, R. F. 1988. *Phys. Rev. A* **37**, 911.
Fox, R. F. and Roy, R. 1987. *Phys. Rev. A* **35**, 1838.
Fronzoni, L., Grigolini, P., Hänggi, P., Moss, F., Mannella, R. and McClintock, P. V. E. 1986. *Phys. Rev. A* **33**, 3320.
Furutsu, K. 1963. *J. Res. Nat. Bur. Standards* **67D**, (3), 303.
Grabert, H., Hänggi, P. and Talkner, P. 1980. *J. Stat. Phys.* **22**, 537.
Grigolini, P. and Marchesoni, F. 1985. In *Memory Function Approaches to Stochastic Problems in Condensed Matter* (M. W. Evans, P. Grigolini and G. Pastori-Parravicini, eds.), pp. 29–79. New York: John Wiley.
Grote, R. F. and Hynes, J. T. 1980. *J. Chem. Phys.* **73**, 2715.
Hänggi, P. 1978a. *Z. Phys. B* **30**, 85.
Hänggi, P. 1978b. *Z. Phys. B* **31**, 407.
Hänggi, P. 1981. *Phys. Lett.* **83A**, 196.
Hänggi, P. 1985. In *Stochastic Processes Applied to Physics, the Functional Derivative and Its Use in the Description of Noisy Dynamical Systems* (L. Pesquera and M. A. Rodriguez eds.), pp. 69–95. Singapore: World Scientific.
Hänggi, P. 1986. *J. Stat. Phys.* **42**, 105–48; **42**, 1003–5.
Hänggi, P. and Mojtabai, F. 1982. *Phys. Rev. A* **26**, 1168.
Hänggi, P. and Riseborough, P. S. 1983. *Phys. Rev. A* **27**, 3379.
Hänggi, P. and Talkner, P. 1985. *Phys. Rev. A* **32**, 1934.
Hänggi, P., Marchesoni, F. and Grigolini, P. 1984. *Z. Phys. B* **56**, 333.

Colored noise in continuous dynamical systems

Hänggi, P., Mroczkowski, T. J., Moss, F. and McClintock, P. V. E. 1985. *Phys. Rev. A* **32**, 695.
Horsthemke, W. and Lefever, R. 1980. *Z. Phys. B* **40**, 241.
Jung, P. and Hänggi, P. 1987. *Phys. Rev. A* **35**, 4464.
Jung, P. and Hänggi, P. 1988. *J. Opt. Soc. Am. B* **5**, 1950.
Jung, P. and Risken, H. 1984. *Phys. Lett.* **103A**, 38.
Jung, P. and Risken, H. 1985. *Z. Phys. B* **61**, 367.
Kaneko, K. 1981. *Progr. Theor. Phys.* **66**, 129.
Kubo, R. 1962. In *Fluctuations, Relaxation and Resonance in Magnetic Systems* (D. ter Haar, ed.), p. 23. Edinburgh University Press.
Lax, M. 1966. *Rev. Mod. Phys.* **38**, 541.
Leiber, T., Marchesoni, F. and Risken, H. 1987. *Phys. Rev. Lett.* **59**, 1381.
Lett, P., Short, R. and Mandel, L. 1984. *Phys. Rev. Lett.* **52**, 34.
Lindenberg, K. and West, B. 1983. *Physica A* **119**, 485.
Lugiato, L. A. and Harowicz, R. J. 1985. *J. Opt. Soc.* **2**, 971.
Malchow, H. and Schimansky-Geier, L. 1985. In *Noise and Diffusion in Bistable Nonequilibrium Systems* (W. Ebeling, W. Meiling, A. Uhlmann and B. Wilhelmi, eds.), vol. 5, p. 83. Teubner-Texte.
Maneke, G., Schroeder, J., Troe, J. and Voss, F. 1985. *Ber. Bunsenges. Phys. Chemie* **89**, 896.
Marchesoni, F. 1984. *Phys. Lett.* **101A**, 11.
Marchesoni, F. 1985. In *Dynamical Processes in Condensed Matter* (M. W. Evans ed.), pp. 603–29. New York: John Wiley.
Marchesoni, F. 1987. *Phys. Rev. A* **36**, 4050.
Masoliver, J., Lindenberg, K. and West, B. 1986. *Phys. Rev. A* **34**, 2351.
Masoliver, J., West, B. J. and Lindenberg, K. 1987. *Phys. Rev. A* **35**, 3086.
Moss, F., Hänggi, P., Mannella, R. and McClintock, P. V. E. 1986. *Phys. Rev. A* **33**, 4459.
Moss, F. and McClintock, P. V. E. 1985. *Z. Phys. B* **61**, 381.
Novikov, E. A. 1965. *Sov. Phys. JETP* **20**, 1290.
Rodriguez, M. A. and Pesquera, L. 1986. *Phys. Rev. A* **34**, 4532.
Roy, R., Yu, A. W. and Zhu, S. 1985. *Phys. Rev. Lett.* **55**, 2794.
Sancho, J. M. and San Miguel, M. 1980. *Z. Phys. B* **36**, 357.
Sancho, J. M., Sagués, F. and San Miguel, M. 1986. *Phys. Rev. A* **33**, 3399.
Sancho, J. M., San Miguel, M., Katz, S. L. and Gunton, J. D. 1982. *Phys. Rev. A* **26**, 1589.
San Miguel, M. and Sancho, J. M. 1980. *J. Stat. Phys.* **22**, 605.
Schenzle, A. and Tél, T. 1985. *Phys. Rev. A* **32**, 596.
Short, R., Mandel, L. and Roy, R. 1982. *Phys. Rev. Lett.* **49**, 647.
Stratonovich, R. L. 1963. *Topics in the Theory of Random Noise*, vol. I. New York: Gordon and Breach.
Suzuki, M. 1980. *Suppl. Progr. Theor. Phys.* **69**, 160.
Van den Broeck, C. and Hänggi, P. 1984. *Phys. Rev. A* **30**, 2730.
Van Kampen, N. G. 1976. *Phys. Reports* **24C**, 171.
Van Kampen, N. G. 1985. *Phys. Reports* **124C**, 71.
Vogel, K., Leiber, T., Risken, H., Hänggi, P. and Schleich, W. 1987a. *Phys. Rev. A* **35**, 4882.

Vogel, K., Risken, H., Schleich, W., James, M., Moss, F. and McClintock, P. V. E. 1987b. *Phys. Rev. A* **35**, 463.
Zoller, P., Alber, G. and Salvador, R. 1981. *Phys. Rev. A* **24**, 398.

Note added in proof

Recent precise calculations on the problem of colored noise driven bistability (Jung and Hänggi, *Phys. Rev. Lett.* **61**, 11 (1988)), see section (9.7.4), show that the escape rate $\Gamma(\tau)$ undergoes two different crossover behaviors as τ increases from $\tau = 0$ to $\tau = \infty$. At very small $\tau \ll 1$, $\tau/D \ll 1$ and weak noise $D \ll 1$, one has $\Gamma(\tau) = (2\pi^2)^{-1/2}$ $(1 - 1.5\tau)\exp - (1/4D)$. This result is followed by a crossover to a behavior of the form $\Gamma(\tau) \propto \exp - (\alpha\tau/D)$, $\alpha \cong 0.1$ at small-to-moderate noise correlation time τ, being in qualitative agreement with the decoupling theory (see (9.7.10)). At even larger noise color $\tau \gg 1$, there occurs yet another very slow crossover to a limiting law $\Gamma(\tau) = (54\pi D(\tau + \frac{1}{2}))^{-1/2}\exp - \{(1/4D)[\frac{4}{9} + \frac{8}{27}\tau]\}$, as $\tau \to \infty$ (Hänggi, Jung and Marchesoni, 'Escape Driven by Strongly Correlated Noise', preprint, 1988).

Appendix. On the statistical treatment of dynamical systems*

L. PONTRYAGIN, A. ANDRONOV, and A. VITT

1   Formulation of the problem

Suppose we have a dynamical system determined by n differential equations of first order[†]

$$\frac{dx_i}{dt} = X^{(i)}(x_1, x_2, \ldots, x_n); \quad i = 1, 2, \ldots, n. \tag{I}$$

For given initial conditions, these equations uniquely determine the behavior of the point that 'represents' our system in the phase space.[‡] We assume that our system, which satisfies equation (I), is subject to 'impulses' or 'perturbations', which act in accordance with laws of chance (different probability hypotheses are possible here).

These 'random' impulses are considered for two reasons connected with the two problems that we pose in this paper.

First problem

There is no doubt that the processes in real dynamical systems are not completely represented by differential equations of the form (I); these equations determine the motion of the system only in the basic features, i.e. only approximately, and do not take into account random impulses and perturbations. Under favorable conditions, an experiment can reveal certain consequences of the existence of such random impulses. This leads to the following problem: *To establish the general behavior of a system in the presence of random impulses and, in particular, to give a theoretical construction that makes it possible to elucidate from experimental data the nature of the*

---

* Translated by Julian B. Barbour from Pontryagin, L., Andronov, A. and Vitt, A. (1933). *Zh. Eksp. Teor. Fiz.* **3**, pp. 165–80.

† We restrict the treatment to autonomous systems, i.e. systems for which the differential equations do not depend explicitly on time. An analogous treatment can also be given for nonautonomous systems.

‡ Translator's note: 'Phase space' has been translated literally; however, except in the final paragraph of the paper, it appears to mean configuration space.

329

'*random impulses*' *in real dynamical systems*. Such a problem was posed some years ago by L. I. Mandel'shtam as a regular problem in the theory of auto-oscillations.

Second problem

Hitherto in the general theory of motions (see, for example, Birkhoff), all treatments, including probability treatments, have been associated with a representation of the motion of the representative point along a definite phase trajectory. The random impulses that we have just discussed and whose possibility was always taken into account in dynamics when Lyapunov stability was being considered carry the representative point from one trajectory to another. This leads us naturally to the problem of *complementing Birkhoff's general theory of motions by considerations associated with allowance for random impulses and, in particular, identifying in the set of motions of a dynamical system the motions that are realized with greatest probability in the presence of such impulses.*

When the problem is posed in this way, the random impulses are merely an apparatus that serves to investigate the nature of the motions determined by equation (I).

Although neither the first nor the second problem is solved here in any generality, it nevertheless appears to us that the considerations that follow represent a certain step forward in these directions and, possibly, have a certain general interest besides the applications that they may have in the theory of auto-oscillations. In our view, it would also be interesting to trace the connection between the matters presented below and statistical mechanics, but we do not propose to do it in the present paper.

2 Equation for the probability density distribution

We consider first the simplest case when $n = 1$ and when the phase space is the straight line OX. Instead of the system (I), we obtain the single equation

$$\frac{dx}{dt} = X(x). \tag{1}$$

As we have already said, it is possible to specify different probability hypotheses for the random impulses. Suppose the impulses occur as follows: after every interval of time τ the phase point jumps instantaneously through distance a along a random direction (the directions to the right and to the left are equally probable), then moves for τ seconds in agreement with the equation of motion, then again makes a jump, etc.

Because the motion of the representative point is determined not only by equation (1) but also by the probability laws, it is impossible to regard x as a definite function of t, and one can speak only of the probability for our representative point to be in any particular region of the phase space.

330

On the statistical treatment of dynamical systems

It is easy to see an analogy between the problem posed here and the so-called 'problem of the motion of a completely drunk man', which was apparently first considered by Rayleigh* in connection with problems of the combination of vibrations. By analogy, our problem can be called the 'problem of the swimming of a completely drunk man in a channel in which regular flows exist'.

The necessary probabilistic treatment can be given comparatively easily in the limiting case in which we assume that a tends to zero together with τ in such a way that a^2/τ tends to a finite limit, which characterizes the intensity of the impulses.

Namely, going to the limit, we obtain the partial differential equation

$$\frac{\partial f}{\partial t} + \frac{\partial}{\partial x}\{X(x)f\} = \tfrac{1}{2}b\frac{\partial^2 f}{\partial x^2}, \quad b = \lim_{\tau \to 0} \frac{a^3}{\tau}, \tag{2}$$

which is satisfied by the probability density distribution $f(t, x)^\dagger$. In the example of Rayleigh just discussed, i.e., for $X(x) = 0$, equation (2) is transformed into the simple heat conduction equation. We have chosen a very special assumption about the nature of the random impulses. In the more general case, we must assume that we have a deterministic motion of the representative point in accordance with equation (1) and, superimposed on it, a random process that satisfies a certain statistical law that depends on the position of the representative point. If we assume that this statistical process does not have any intrinsic directionality and that the random perturbations are such that the probability of large displacements tends to zero sufficiently rapidly with decreasing time τ, then instead of equation (2) we obtain the somewhat more general equation

$$\frac{\partial f}{\partial t} + \frac{\partial}{\partial x}\{X(x)f\} = \frac{1}{2}\frac{\partial^2}{\partial x^2}\{b(x)f\}, \tag{3}$$

where $b(x)$, a coefficient that characterizes the disturbative strength of the statistical process, can be defined as $\lim_{\tau \to 0}(\overline{\xi^2}/\tau)$, where $\overline{\xi^2}$ is the mean square of the displacement during the time τ under the influence of the statistical process[‡]. In accordance with the meaning of the probability density, we are

---

* Rayleigh, *Theory of Sound*, Vol. 1, §42a.

[†] This partial differential equation and its generalizations that we shall consider later in the paper are well known through the work of Rayleigh, Fokker, Smoluchowski, Kolmogorov, and others (see, for example, the bibliography given by Zernicke in *Handbuch der Physik*, Vol. III, p. 457). Note that if $f(t, x)$ is the probability density distribution the probability of finding the representative point at the time t in the region G is $W(t, G) = \int_G f(t, x)\mathrm{d}x$.

[‡] Let $p(x, \tau, y)\mathrm{d}y$ be the probability of the representative point, situated at the position x, arriving *as a result of the random process* at a position from y to $y + \mathrm{d}y$ during the time interval τ. Then

$$b(x) = \lim_{\tau \to 0} \frac{1}{\tau} \int_{-\infty}^{+\infty} p(x, \tau, y)(y - x)^2\,\mathrm{d}y.$$

(*Continued on p. 332.*)

interested in only those solutions of equation (3) for which $f(t, x) \geqslant 0$ and which are normalized, i.e. for which*

$$\int_{-\infty}^{+\infty} f(t, x)\,\mathrm{d}x = 1. \tag{4}$$

To find a definite solution of equation (3), it is sufficient to know the function $f(t, x)$ at $t = 0$, i.e. to have the initial probability distribution. If one wishes to study the behavior of a representative point that at the initial time had the definite position ξ, then it is necessary to find a distribution function $f(t, x)$ that in the limit $t \to 0$ tends to zero at all points except for ξ and in addition satisfies the condition (4). Of course, the function determined in this manner depends on the point ξ; we shall denote it by $p(\xi, t, x)$; $p(\xi, t, x)\,\mathrm{d}x$ is the probability that a random point situated at time $t = 0$ at the position ξ goes over during the time t to a position between x and $x + \mathrm{d}x$.

It may happen (and we shall be primarily interested in systems for which it does) that every nonstationary distribution function $f(t, x)$ tends as $t \to \infty$ to a definite limit function $f(x)$. To find this stationary limit distribution, we must set $\partial f / \partial t = 0$ in equation (3) and consider the equation

$$\frac{\mathrm{d}}{\mathrm{d}x}\{X(x)f\} = \frac{1}{2}\frac{\mathrm{d}^2}{\mathrm{d}x^2}\{b(x)f\}, \tag{5}$$

which we shall call the stationary case of the Fokker equation. The solution of equation (5), which does not depend on the initial conditions, reflects the properties of the dynamical system (I) to the greatest degree. Therefore, in what follows we shall investigate this equation and its generalization to the case $n > 1$. With regard to equation (3), whose investigation, it is true, is of interest from the point of view of the problem of the nature of the impulses in real physical systems, we shall not consider it directly; such an investigation in general form is very difficult; a solution can be found only in a few special cases.[†] We note – and will require below – the fact that equation (3) can be

(Continued from p. 331.)

Since we have assumed that the statistical process does not have any directionality,

$$\lim_{\tau \to 0} \int_{-\infty}^{+\infty} p(x, \tau, y)(y - x)\,\mathrm{d}y = 0.$$

The rate of decrease of the probability of large deflections with decreasing τ is characterized by the relation

$$\lim_{\tau \to 0} \frac{1}{\tau} \int_{-\infty}^{+\infty} p(x, \tau, y)|y - x|^3\,\mathrm{d}y = 0.$$

* Under certain assumptions of a general nature regarding equation (3), it can be asserted that if $f(0, x)$ is everywhere positive and normalized at $t = 0$ then these conditions will also hold for all $t > 0$.

† See R. von Mises, *Wahrscheinlichkeitsrechnung*, §517 (1931); A. Kolmogorov, *Math. Annalen*, 104, 454 (1931).

given a different, purely statistical treatment and need not be regarded, as we just have, as the result of superposition of a statistical process on a dynamical process*.

We have hitherto assumed that we have one differential equation of the form (1) and that the corresponding phase space is a straight line. In the general case, we shall have instead of equation (1) the system (I) and instead of equation (3) the equation

$$\frac{\partial f}{\partial t} + \sum_{\alpha=1}^{n} \frac{\partial}{\partial x_\alpha}\{X^\alpha f\} = \frac{1}{2}\sum_{\alpha,\beta=1}^{n}\frac{\partial^2}{\partial x_\alpha \partial x_\beta}\{b^{\alpha\beta}f\}, \tag{II}$$

where $b^{ij}(x_1, x_2, \ldots, x_n)$ again characterizes a statistical process.[†] Note that we shall understand the coordinates $x_1, x_2, \ldots, x_n$ as Cartesian coordinates in Euclidean space.

Thus, if we know equation (I), which characterizes the dynamical system, and we know the functions b^{ij}, which characterize the random impulses, we can write down equation (II). We shall call equation (II) the Fokker equation corresponding to the system (I)[‡]. Obviously, we are interested in non-negative and normalization solutions of this equation. We obtain the stationary case

---

* Suppose the representative point is subject only to a statistical process, and suppose there exists a function $p(\xi, t, x)$ such that $p(\xi, t, x)\mathrm{d}x$ is the probability for the representative point to go over from the definite position ξ to a position between x and $x + \mathrm{d}x$ during time t. One can then show that under well-known assumptions for the function p and under the conditions

$$\lim_{\tau\to 0}\frac{1}{\tau}\int_{-\infty}^{+\infty} p(\xi,t,x)(x-\xi)\mathrm{d}x = X(\xi); \quad \lim_{\tau\to 0}\frac{1}{\tau}\int_{-\infty}^{+\infty} p(\xi,\tau,x)(x-\xi)^2\,\mathrm{d}x = b(\xi);$$

and

$$\lim_{\tau\to 0}\frac{1}{\tau}\int_{-\infty}^{+\infty} p(\xi,\tau,x)|x-\xi|^3\,\mathrm{d}x = 0$$

that $p(\xi, t, x)$ as a function of t and x satisfies equation (3). One can then also readily see that a distribution of general form $f(t,x)$ satisfies this equation, since

$$f(t,x) = \int_{-\infty}^{+\infty} f(\xi)p(\xi,t,x)\mathrm{d}\xi,$$

where $f(\xi)$ is the distribution at $t=0$.

[†] $b^{ij}(x_1, x_2, \ldots, x_n) = \lim_{\tau\to 0}\frac{1}{\tau}\int_{-\infty}^{+\infty}\ldots\int_{-\infty}^{+\infty} p(x_1, x_2, \ldots, x_n; \tau; y_1, y_2, \ldots, y_n)$
$$\times (y_i - x_i)(y_j - x_j)\mathrm{d}y_1\,\mathrm{d}y_2\ldots\mathrm{d}y_n,$$

where $p(x_1, x_2, \ldots, x_n; \tau; y_1, y_2, \ldots, y_n)\mathrm{d}y_1\,\mathrm{d}y_2\ldots\mathrm{d}y_n$ is the probability of the representative point passing from the position $x_1, x_2, \ldots, x_n$ *as a result of the random process* to a position from y_1 to $y_1 + \mathrm{d}y_1$, y_2 to $y_2 + \mathrm{d}y_2$, etc., during the time interval τ. The corresponding relations for the first and third moments can be written down by analogy with those for equation (3) (see footnote * on p. 332).

[‡] It was given in the form (II) by Kolmogorov, *l. c.*, p. 415.

of this equation by setting, as before, $\partial f/\partial t = 0$:

$$\sum_{\alpha=1}^{n} \frac{\partial}{\partial x_\alpha} \{X^\alpha f\} = \tfrac{1}{2} \sum_{\alpha,\beta=1}^{n} \frac{\partial^2}{\partial x_\alpha \partial x_\beta} \{b^{\alpha\beta} f\}. \tag{III}$$

Having now equation (II) and (III), we return to our problems posed in Section 1.

From the point of view of the first problem – that of studying random impulses in real dynamical systems* – it is necessary to find the b^{ij} that reflect in the best manner the experimental results. Here, the b^{ij} are determined by the investigated physical system.

From the point of view of the second problem – the study by means of equations (II) and (III) of the dynamical system determined by equation (I)– equation (III) must, in particular, reflect the properties of the system (I). Therefore, from the point of view of the second problem the coefficients b^{ij} are *auxiliary* quantities, which serve for the investigation of the system (I). In what follows, we shall, for example, investigate the behavior of the solutions of equation (III) as $b^{ij} \to 0$, and we shall consider at the same time how the behavior of these solutions depends on the manner in which the b^{ij} tend to zero.

3 Equation for the mathematical expectation of the transition time

Besides the distribution function $f(t, x)$, there are other functions important for characterizing the behavior of the random point.

We begin the study of these new functions by again considering the simplest one-dimensional case. Suppose the phase space is the straight line OX. Suppose that along this line there moves a random point for which the corresponding probability density distribution satisfies the Fokker equation[†].

We calculate the probability that the random point, situated at the initial time $t = 0$ at some point x within the interval ab, leaves this interval during the time t, passing (at least once) either the point a or the point b[‡].

We denote the required probability by $\varphi(t, x)$ and investigate $\varphi(t + \tau, x)$. Since at the initial time $t = 0$ the representative point had a fixed position x, at the time τ it has the probability density distribution $p(x, \tau, \xi)$. Since the probability of the random point leaving the interval ab during the short time τ

---

* We do not at all assert that all random perturbations in real dynamical systems can be investigated through the Fokker equation.
[†] In the following considerations, we will adopt the purely statistical scheme (see footnote † on p. 333).
[‡] One can also pose the problem of the probability of the random point passing only through the left or only through the right end of the interval or, finally, the probability of not leaving the interval. It is readily seen that the equation remains the same, only the boundary conditions being changed.

On the statistical treatment of dynamical systems

is very small we can, having in mind a subsequent passage to the limit, ignore it and write

$$\varphi(t+\tau,x) = \int_a^b p(x,\tau,\xi)\varphi(t,\xi)\,d\xi. \tag{6}$$

Expanding $\varphi(t,\xi)$ in a Taylor series, we find

$$\varphi(t,\xi) = \varphi(t,x) + \varphi'_x(t,x)(\xi-x) + \tfrac{1}{1\cdot 2}\varphi''_{xx}(t,x)(\xi-x)^2$$
$$+ \tfrac{1}{1\cdot 2\cdot 3}\varphi'''_{xxx}[t,x+\theta(\xi-x)](\xi-x)^3,$$

whence in accordance with (4)

$$\varphi(t+\tau,x) = \varphi(t,x)\int_a^b p(x,\tau,\xi)\,d\xi + \varphi'_x(t,x)\int_a^b p(x,\tau,x)(\xi-x)\,d\xi$$

$$+\tfrac{1}{2}\varphi''_{xx}(t,x)\int_a^b p(x,\tau,\xi)(\xi-x)^2\,d\xi$$

$$+\tfrac{1}{6}\varphi'''_{xxx}[t,x+\theta(\xi-x)]\int_a^b p(x,\tau,\xi)(\xi-x)^3\,d\xi.$$

Dividing by τ and going to the limit $\tau\to 0$, we can extend the integrals that appear here to the complete line*. After the passage to the limit, we obtain for the function $\varphi(t,x)$ the partial differential equation[†]

$$\frac{\partial\varphi}{\partial t} = X(x)\frac{\partial\varphi}{\partial x} + \tfrac{1}{2}b(x)\frac{\partial^2\varphi}{\partial x^2}. \tag{7}$$

We now find the initial and boundary conditions for $\varphi(t,x)$. It is obvious that at the initial time $\varphi(0,x)=0$ if x, as was assumed, lies within the interval (a,b). We also assume that $\varphi(t,a)=\varphi(t,b)=1$ for any t, since if the random point approaches a or b then it is natural to assume that the probability of its leaving the interval (a,b) approaches unity at the same time.

We now pose the question of the mathematical expectation $M(x)$ of the *exiting time*, i.e. the time needed for a representative point initially situated at some point x of the interval (a,b) to leave this interval either through the point a or through the point b. Since the probability of the random point doing this during the interval of time from t to $t+dt$ is $(\partial\varphi/\partial t)\,dt$, the required mathematical expectation is

$$M(x) = \int_0^\infty t\frac{\partial\varphi}{\partial t}\,dt. \tag{8}$$

* The justification for doing this follows from the condition imposed on the third moment (see footnote † on p. 333).
† Equations (7) and (9), like their generalizations to the case $n>1$, were obtained by L. S. Pontryagin.

335

To obtain a differential equation for M, we differentiate equation (5) with respect to t, then multiply both sides of it by t, and integrate from 0 to ∞. We obtain

$$\int_0^\infty t \frac{\partial^2 \varphi}{\partial t^2} dt = X(x) \frac{\partial}{\partial x} \int_0^\infty t \frac{\partial \varphi}{\partial t} dt + \tfrac{1}{2} b(x) \frac{\partial^2}{\partial x^2} \int_0^\infty t \frac{\partial \varphi}{\partial t} dt$$

or, since*

$$\int_0^\infty t \frac{\partial^2 \varphi}{\partial t^2} dt = \left[t \frac{\partial \varphi}{\partial t} \right]_0^\infty - \int_0^\infty \frac{\partial \varphi}{\partial t} dt = -[\varphi(t,x)]_{t=0}^{t=\infty} = -1,$$

the required differential equation will have the form

$$\tfrac{1}{2} b(x) \frac{d^2 M}{dx^2} + X(x) \frac{dM}{dx} = -1. \tag{9}$$

It is readily seen that the corresponding boundary conditions are

$$M(a) = 0; \quad M(b) = 0.$$

In addition, the formulation of the problem requires that $M(x) > 0$. Note that if we are interested in the mathematical expectation of the transition time of a random point of position $a(a < x)$ then we must find the solution of the problem just posed and then go to the limit $b \to \infty$.

The foregoing considerations can be readily generalized to the case $n > 1$. Suppose we have a random point whose motion satisfies the Fokker equation corresponding to the general case. Let G be a certain region of the corresponding phase space, A be the boundary of this region, and α be a part of this boundary.

Let $\varphi(t, x_1, x_2, \ldots, x_n)$ be the probability that the random point situated at time $t = 0$ at the position $x_1, x_2, \ldots, x_n$ in the region G leaves the region G during the time t, crossing the part α of the boundary as it leaves the region. Proceeding as before, we can readily find for the function $\varphi(t, x_1, x_2, \ldots, x_n)$ the partial differential equation

$$\frac{\partial \varphi}{\partial t} = \sum_{\alpha=1}^n X^\alpha(x_1, x_2, \ldots, x_n) \frac{\partial \varphi}{\partial x_\alpha} + \frac{1}{2} \sum_{\alpha,\beta=1}^n b^{\alpha\beta} \frac{\partial^2 \varphi}{\partial x_\alpha \partial x_\beta} \tag{IV}$$

with the following initial and boundary conditions: $\varphi(0, x_1, x_2, \ldots, x_n) = 0$ for all points lying within the region, G; $\varphi(t, x_1, x_2, \ldots, x_n) = 1$ for points belonging to the part α of the boundary, and $\varphi(t, x_1, x_2, \ldots, x_n) = 0$ for points belonging to the remaining part A.

If we wish to investigate the question of the random point leaving the region G through the part α of the boundary, not during the definite time t, but during all the time that elapses after the initial time, then we must go to the limit

---

* $\varphi(0, x) = 0$, since by assumption x lies within the interval (a, b); $\varphi(\infty, x) = 1$, since this is the probability that the random point *at some time* leaves the interval (a, b), and this is certain.

$t \rightarrow \infty$. In this case, as we did in Section 2, we must set $\lim_{t \rightarrow \infty}(\partial \varphi / \partial t) = 0$. Therefore, the question reduces to finding the solution $\varphi(x_1, x_2, \ldots, x_n)$ of the equation

$$\sum_{\alpha=1}^{n} X^{\alpha} \frac{\partial \varphi}{\partial x_{\alpha}} + \frac{1}{2} \sum_{\alpha,\beta=1}^{n} b^{\alpha\beta} \frac{\partial^2 \varphi}{\partial x_{\alpha} \partial x_{\beta}} = 0 \qquad \text{(IV')}$$

subject to the same boundary conditions as we have just considered. If α coincides with A, the functions $\varphi(t, x_1, x_2, \ldots, x_n)$ and $\varphi(x_1, x_2, \ldots, x_n)$ become equal to unity along the complete boundary A. In this case, unity satisfies equation (IV'); this means that the probability for the random point to leave the region G at some time and at some place is equal to unity.

The equation for the mathematical expectation of the exiting time in this case has the form

$$\frac{1}{2} \sum_{\alpha,\beta=1}^{n} b^{\alpha\beta}(x_1, x_2, \ldots, x_n) \frac{\partial^2 M}{\partial x_{\alpha} \partial x_{\beta}} + \sum_{\alpha=1}^{n} X^{\alpha}(x_1, x_2, \ldots, x_n) \frac{\partial M}{\partial x_{\alpha}}$$

$$= -\varphi(x_1, x_2, \ldots, x_n), \qquad \text{(V)}$$

where $\varphi(x_1, x_2, \ldots, x_n)$ is the corresponding solution of equation (IV), for the boundary conditions of $M(x_1, x_2, \ldots, x_n)$ on the complete boundary A^*. If α coincides with A, then, as we know $\varphi(x_1, x_2, \ldots, x_n) = 1$, and, therefore, equation (V) takes the form

$$\frac{1}{2} \sum_{\alpha,\beta=1}^{n} b^{\alpha\beta} \frac{\partial^2 M}{\partial x_{\alpha} \partial x_{\beta}} + \sum_{\alpha=1}^{n} X^{\alpha} \frac{\partial M}{\partial x_{\alpha}} = -1$$

with the same boundary conditions.

4   The case of one first-order equation

A   Stationary probability density distribution

As we already know, a stationary distribution is a non-negative normalized solution of the equation

$$\frac{\mathrm{d}}{\mathrm{d}x} \left\{ X(x) f - \frac{1}{2} \frac{\mathrm{d}}{\mathrm{d}x} [b(x) f] \right\} = 0. \qquad (5)$$

In the general case, one cannot rely on the existence of a stationary solution. The most natural conditions under which one can expect the existence of a

---

* Since $\varphi(t, x_1, x_2, \ldots, x_n)$ is a constant on the boundary A, it follows that

$$M(x_1, x_2, \ldots, x_n) = \int_{0}^{\infty} t \frac{\partial \varphi}{\partial t} \mathrm{d}t$$

vanishes on the boundary A.

L. PONTRYAGIN, A. ANDRONOV and A. VITT

stationary solution, and which we shall in what follows assume are satisfied, are these: $(a) c'' > b(x) > c' > 0$; $(b) X(x)$ for sufficiently large positive x is negative, for sufficiently large negative x is positive, and in both cases is greater in modulus that some constant $g > 0$. The first integration of equation (5) gives

$$X(x)f - \frac{1}{2}\frac{\mathrm{d}}{\mathrm{d}x}\{b(x)f\} = C_1.$$ (10)

If conditions (a) and (b) are satisfied and if $f(x)$ is non-negative and normalizable, then $C_1 = 0^*$.

Integrating a second time, we find

$$f(x) = \frac{C}{b(x)} e^{\varphi(x)},$$ (11)

where

$$\varphi(x) = 2 \int_0^x \frac{X(\xi)}{b(\xi)} \mathrm{d}\xi,$$

and C is a new constant of integration or a normalization coefficient. We note first of all the circumstance that for constant $b(x)$ the maxima (and, respectively, minima) of our solution coincide with points of stable (and, respectively, unstable) equilibrium for the original dynamical system determined by equation (1). Indeed, since the function e^z is monotonic, the maxima and minima of our solution coincide with those of the function $\varphi(x)$; with regard to this last function, for it the considered fact can be very readily established.

We now investigate the solution under the assumption that $b(x)$ decreases uniformly. For this, we represent $b(x)$ in the form $b(x) = \lambda q(x)$, where $q(x)$ $(q'' > q(x) > q' > 0)$ is an unchanged function, while λ is a parameter which we shall allow to tend to zero.

Then the solution can be written in the form

$$f(x) = \frac{C(\lambda)}{q(x)} e^{\psi(x)/\lambda},$$ (12)

where

$$\psi(x) = \int_0^x \frac{X(\xi)}{q(\xi)} \mathrm{d}\xi, \quad \frac{1}{C(\lambda)} = \int_{-\infty}^{+\infty} \frac{e^{\psi(\xi)/\lambda}}{q(\xi)} \mathrm{d}\xi.$$ (13)

Suppose there exists a unique point (which we shall take to be the origin) at which $\psi(x)$ attains an absolute maximum. One can show that in the limit $\lambda \to 0$ the function $f(x)$ tends everywhere except at the origin to zero, while at the origin it tends to infinity. Indeed, in this case one can find an estimate of $C(\lambda)$

---

* Namely, one can show that under these conditions f and its derivative decrease so rapidly as $x \to \infty$ that the left-hand side of the equation tends to zero and, therefore, $C_1 = 0$.

338

for *sufficiently small* λ:

$$\frac{1}{C(\lambda)} = s(\lambda)\sqrt[n]{\lambda},$$

where n is an even positive integer and $s(\lambda)$ lies between positive limits.* Therefore,

$$f(x) = \frac{e^{\psi(x)/\lambda}}{q(x)[s(\lambda)\sqrt[n]{\lambda}]}. \tag{14}$$

If $x \neq 0$, then the numerator of this expression decreases much more rapidly as $\lambda \to 0$ than the denominator, and the entire expression tends to zero.

But if $x = 0$, then the numerator does not depend on λ and, therefore, the entire expression tends to infinity.

Having the given function $X(x)$, we can choose the function $q(x)$ in such a way that the absolute maximum of $\psi(x)$ occurs at any of the points of stable equilibrium of equation (1). The stationary distribution will tend everywhere to zero as $\lambda \to 0$ except at the point of stable equilibrium, where $\psi(x)$ attains an absolute maximum.

It should be noted that this property is not a local property of the given stable equilibrium point, i.e. it may happen that at the point that is locally most stable, i.e. where $|X'(x)|$ has the greatest value, the function $\varphi(x)$ does not attain an absolute maximum even if $b(x)$ is constant.

$$* \quad \frac{1}{C(\lambda)} = \int_{-\infty}^{+\infty} \frac{1}{q(\xi)} \cdot e^{\psi(\xi)/\lambda}\,\mathrm{d}\xi = \int_{-h}^{+h} \frac{1}{q(\xi)} \cdot e^{\psi(\xi)/\lambda}\,\mathrm{d}\xi + \int_{-\infty}^{-h} \frac{1}{q(\xi)} \cdot e^{\psi(\xi)/\lambda}\,\mathrm{d}\xi + \int_{+h}^{+\infty} \frac{1}{q(\xi)} \cdot e^{\psi(\xi)/\lambda}\,\mathrm{d}\xi.$$

Let h be a very small positive number. It is easy to show, by virtue of the properties of $\psi(\xi)$, that in the limit $\lambda \to 0$ the last two integrals are infinitesimally small compared with the first integral.

In a sufficiently small interval $(-h, +h)$, the function $\psi(x)$ can be approximately represented in the form $-kx^n$, where n is a positive even integer, i.e.

$$-k'x^n > \psi(x) > k''x^n,$$

where $k' > k'' > 0$.

Thus, in the same interval we have

$$\frac{1}{q''} \cdot e^{-k''x^n/\lambda} < \frac{1}{q(x)} \cdot e^{\psi(x)/\lambda} < \frac{1}{q'} \cdot e^{-k'x^n/\lambda}. \tag{A}$$

Since in the equation

$$\int_{-h}^{+h} e^{-k\xi^n/\lambda}\,\mathrm{d}\xi = \int_{-\infty}^{+\infty} e^{-k\xi^n/\lambda}\,\mathrm{d}\xi - 2\int_{h}^{\infty} e^{-k\xi^n/\lambda}\,\mathrm{d}\xi$$

we can again (as we have just done) ignore the second term on the right-hand side, and since

$$\int_{-\infty}^{+\infty} e^{-k\xi^n/\lambda}\,\mathrm{d}\xi = \sqrt[n]{(\lambda)} \int_{-\infty}^{+\infty} e^{-kz^n}\,\mathrm{d}z = p(k)\sqrt[n]{(\lambda)},$$

where $p(k)$ does not depend on λ, our assertion now follows from this in conjunction with (A).

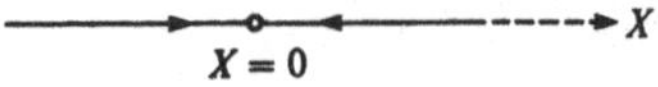

Figure 1.

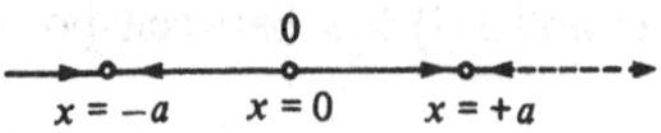

Figure 2.

We consider a pair of simple examples for a stationary distribution.

Example 1

$$\frac{dx}{dt} = -kx; \quad k > 0; \quad b(x) = 2D = \text{const}, \tag{15}$$

where D is a constant.

The phase space in this case is an infinite straight line with unique equilibrium state $x = 0$ (see Figure 1). This equilibrium state is stable. Conditions (*a*) and (*b*) from near the beginning of Section 4A are satisfied. The dynamical equation is solved by the function $x = Ce^{-kt}$. If at $t = 0$ we have $x = x_0$, then the solution $x = x_0 e^{-kt}$

$$\varphi(x) = 2 \int \frac{X(\xi)}{b(\xi)} d\xi = -2 \int \frac{k\xi}{2D} d\xi = -\frac{k}{2D} \xi^2 \tag{16}$$

with

$$f(x) = Ce^{-k\xi^2/2D},$$

$$\frac{1}{C} = \int_{-\infty}^{+\infty} e^{-k\xi^2/2D} d\xi = \sqrt{\left(\frac{2\pi D}{k}\right)}. \tag{17}$$

We have obtained an ordinary Gaussian distribution (Figure 3), which has a greater spread the stronger the impulses and the larger D, and is more compressed the larger k and the greater the stability of the equilibrium state*.

Example 2

$$\frac{dx}{dt} = kx(a^2 - x^2); \quad k > 0; \quad b(x) = 2D, \tag{18}$$

where D is a constant.

In this case the phase space is an infinite straight line with three states of equilibrium: $x = 0$, $x = a$, $x = -a$ (Figure 2). Among them, the state $x = 0$ is unstable, while $x = a$ and $x = -a$ are stable. Conditions (*a*) and (*b*) are

---

* This example is well known. See, for example, R. von Mises, *Wahrscheinlichkeitsrechnung*, §517, (1931).

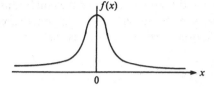

Figure 3.

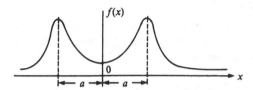

Figure 4.

satisfied. The dynamical equation is solved by the function

$$x^3 = \frac{a^2}{1 + C'e^{-2a^2kt}}.$$

If at $t = 0$ we have $x = x_0$, then

$$C' = \frac{a^2 - x_0^2}{x_0^2}$$

$$\varphi(x) = 2\int_0^x \frac{ka^2\xi - k\xi^3}{2D}\,d\xi = \frac{k}{D}\left\{\frac{a^2x^2}{2} - \frac{x^4}{4}\right\} \tag{19}$$

$$f(x) = Ce^{(k/4D)\{2a^2x^2 - x^4\}},$$

with

$$\frac{1}{C} = \int_{-\infty}^{+\infty} e^{(k/4D)\{2a^2\xi^2 - \xi^4\}}\,d\xi. \tag{20}$$

We have obtained a stationary distribution with two maxima ($x = +a$ and $x = -a$) and a minimum $x = 0$ (Figure 4).

B   *Mathematical expectation of the transition time*

We calculate for the considered one-dimensional case the mathematical expectation $M_q(x)$ of the time of transition of the representative point from the position x to the position $q(q < x)$. As we have shown, $M(x)$ satisfies the equation

$$\tfrac{1}{2}b(x)\frac{d^2M}{dx^2} + X(x)\frac{dM}{dx} + 1 = 0. \tag{9'}$$

We must find the solution of this equation under the conditions $M(q) = 0$ and that our solution increases in the slowest possible manner as $x \to \infty$. The derivative of the general solution of equation (9′) has the form

$$\frac{\mathrm{d}M}{\mathrm{d}x} = \left\{ C + \int_x^\infty \frac{2}{b(\xi)} e^{\varphi(\xi)} \, \mathrm{d}\xi \right\} e^{-\varphi(x)},$$

where

$$\varphi(x) = 2 \int_0^x \frac{X(\xi)}{b(\xi)} \, \mathrm{d}\xi.$$

It is obvious that when x increases so does $M(x, q)$ and, therefore, $C \geqslant 0$. But since the solution in which we are interested must increase in the slowest possible manner, $C = 0$. Therefore*,

$$\frac{\mathrm{d}M}{\mathrm{d}x} = \left\{ \int_x^\infty \frac{2}{b(\xi)} e^{\varphi(\xi)} \, \mathrm{d}\xi \right\} e^{-\varphi(x)}, \tag{21}$$

whence, integrating and bearing in mind that $M(q) = 0$ for $x = q$, we find

$$M_q(x) = \int_q^x \left\{ \int_\xi^\infty \frac{2}{b(y)} e^{\varphi(y)} \, \mathrm{d}y \right\} e^{-\varphi(\xi)} \, \mathrm{d}\xi. \tag{22}$$

It is readily seen that the function

$$\psi(x) = \left\{ \int_x^\infty \frac{2}{b(y)} e^{+\varphi(y)} \, \mathrm{d}y \right\} e^{-\varphi(x)} \tag{23}$$

is the reciprocal of the mean velocity with which the point moves from the position x to the position q from the right to the left. Similarly, we find that the mathematical expectation $M_p(x)$ of the time of transition of the representative point from the position x to the position $p(p > x)$ is represented by the integral

$$M_p(x) = \int_x^p \left\{ \int_{-\infty}^\xi \frac{2}{b(y)} e^{+\varphi(y)} \, \mathrm{d}y \right\} e^{-\varphi(\xi)} \, \mathrm{d}\xi, \tag{24}$$

where

$$\bar{\psi}(x) = \left\{ \int_{-\infty}^x \frac{2}{b(y)} e^{+\varphi(y)} \, \mathrm{d}y \right\} e^{-\varphi(x)} \tag{25}$$

is the reciprocal of the mean velocity with which the point moves from the position x to the position p from the left to the right.

One can show that in the limit of impulses that tend to zero this mean velocity tends to the velocity obtained from the dynamical equation (1) if the

---

* $X(x)$ and $b(x)$ must be such that

$$\int_x^\infty \frac{2}{b(\xi)} e^{+\varphi(\xi)}$$

is defined.

direction we have chosen (from x to q or from x to p) coincides with the dynamical direction of motion and tends to zero if our chosen direction and the dynamical direction are opposite.

We again consider a pair of examples that illustrate our theory, considering in fact the same cases for which we have just investigated the stationary probability distribution.

Example 3

$$\varphi(x) = -\frac{kx^3}{4D}$$

$$\psi(x) = \frac{1}{D}\left\{\int_x^\infty e^{-ky^2/2D}\,dy\right\}e^{+kx^2/2D},$$

$$\bar{\psi}(x) = \frac{1}{D}\left\{\int_{-\infty}^x e^{-ky^2/2D}\,dy\right\}e^{+kx^2/2D}. \tag{26}$$

The mathematical expectation of the time of transition from the position $x = 0$ to the position $x = p(0 \leqslant p)$ is

$$M_p(0) = \frac{1}{D}\int_0^p\left\{\int_{-\infty}^\xi e^{-ky^2/2D}\,dy\right\}e^{+k\xi^2/2D}\,d\xi. \tag{27}$$

Example 4

$$\varphi(x) = \frac{k}{4D}\{2a^2x^2 - x^4\}$$

$$\psi(x) = \frac{1}{D}\left\{\int_x^\infty e^{(k/4D)(2a^2y^2 - y^4)}\,dy\right\}e^{-(k/4D)(2a^2x^2 - x^4)}$$

$$\bar{\psi}(x) = \frac{1}{D}\left\{\int_{-\infty}^x e^{(k/4D)(2a^2y^2 = y^4)}\,dy\right\}e^{-(k/4D)(2a^2x^2 - x^4)}. \tag{28}$$

The mathematical expectation of the time of transition from the point $x = -a$ to the point $x = p(-a \leqslant p \leqslant +a)$ is

$$M_p(-a) = \frac{1}{D}\int_{-a}^p\left\{\int_{-\infty}^\xi e^{(k/4D)(2a^2y^2 - y^4)}\,dy\right\}e^{-(k/4D)(2a^2\xi^2 - \xi^4)}\,d\xi. \tag{29}$$

5   The case of two first-order equations

We restrict ourselves here to some remarks relating to the stationary problem and to a single heuristic example.

If the dynamical system is characterized by the equations

$$\frac{dx_1}{dt} = X^{(1)}(x_1, x_2); \quad \frac{dx_2}{dt} = X^{(2)}(x_1, x_2), \tag{30}$$

then the probability density distribution in the stationary problem satisfies the equation

$$\frac{\partial}{\partial x_1}\{X^{(1)}f\} + \frac{\partial}{\partial x_2}\{X^{(2)}f\}$$

$$= \frac{1}{2}\left[\frac{\partial^2\{b^{11}f\}}{\partial x_1^2} + 2\frac{\partial^2\{b^{12}f\}}{\partial x_1\,\partial x_2} + \frac{\partial^2\{b^{22}f\}}{\partial x_2^2}\right]. \tag{31}$$

We are interested in a non-negative solution of this equation whose integral taken over the complete plane is equal to unity, i.e.

$$\int_{-\infty}^{+\infty}\int_{-\infty}^{+\infty} f(x_1 x_2)\,dx_1\,dx_2 = 1.$$

The problem of when equation (31) has such a normalized solution has not yet been clarified. If $b^{ij}(x_1, x_2)$ are bounded above and below, i.e. if $0 < c' < b^{ij} < c''$, then one must expect such a solution to exist in the case when an infinitely distant point of the plane is completely stable for the system (30).

The qualitative nature of the solutions of the system (30) is well known through the work of Poincaré et al.* The singular points, limit cycles, and separatrices are the distinctive motions that are the determining elements of such a system. What is the part played by these motions from the point of view of equation (31), from the point of view of the statistical treatment of the dynamical system? One must expect that near stable singular points maxima of the function $f(x_1, x_2)$ will be formed, near the points of completely unstable singular points minima, near stable limit cycles elevations in the form of ridges, etc. What will happen when the impulses tend to zero? What solutions will then be distinguished as the most probable? To all these questions we have not been able to give such exhaustive answers as for $n = 1$.

Making the assumption that a solution of the kind that we seek does exist, one can prove the following proposition, the proof of which we here omit.

Let a be a point of the phase plane. If: (1) through a it is possible to describe a cycle without contact†, or (2) a lies on an unstable limit cycle, or (3) a is an unstable focus or an unstable node, then there exists a small neighborhood g of the point a such that

$$\lim_{\lambda \to 0} \int_g f(x_1, x_2; \lambda)\,dx_1\,dx_2 \to 0, \tag{32}$$

where $f(x_1, x_2; \lambda)$ is a stationary distribution, and λ is a parameter that characterizes the magnitude of the impulses $\{b^{ij}(x_1, x_2) = \lambda q^{ij}(x_1, x_2)\}$. It

---

* H. Poincaré, Oeuvres, Vol. 1, Paris, (1928).
† Translator's note: or 'without tangency'.

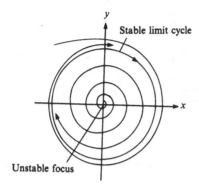

Figure 5.

follows from this that in the limit $\lambda \to 0$ the random point will be situated with probability arbitrarily close to unity in an arbitrarily small neighborhood of the stable focuses and nodes, stable limit cycles, separatrices, and saddles*. It is evident that here too, as for the case $n = 1$, one can, choosing appropriate $q^{ij}(x_1, x_2)$, create (in the limit $\lambda \to 0$) an absolute maximum of the probability in the neighborhood of one or other of these motions.

In conclusion, we consider a simple example that clearly illustrates the probability density distribution in the two-dimensional case. Suppose that our impulses are constant and isotropic ($b^{12} = 0$, $b^{11} = b^{22} = b_0$)†. Then equation (31) takes the simple form

$$\frac{\partial X f}{\partial x} + \frac{\partial Y f}{\partial y} = \frac{b_0}{2}\left(\frac{\partial^2 f}{\partial x^2} + \frac{\partial^2 f}{\partial y^2}\right). \tag{33}$$

where

$$X(x, y) = \frac{\mathrm{d}x}{\mathrm{d}t}, \quad Y(x, y) = \frac{\mathrm{d}y}{\mathrm{d}t}. \tag{34}$$

As a characteristic example, we now choose a system of equations (34) having an unstable infinity, a stable limit cycle, and an unstable singular point at the origin (Figure 5). We consider the special case

$$\frac{\mathrm{d}x}{\mathrm{d}t} = X(x, y) = \{1 - (x^2 + y^2)\}x - y;$$

$$\frac{\mathrm{d}y}{\mathrm{d}t} = Y(x, y) = \{1 - (x^2 + y^2)\}y + x, \tag{34'}$$

---

* One can give examples in which the random point arrives with probability unity in the limit $\lambda \to 0$ in the neighborhood of a stable node, stable focus, or stable limit cycle; however, it has not yet proved possible to clarify completely the role played by a saddle.
† Constant impulses can be made isotropic by a linear change of the variables.

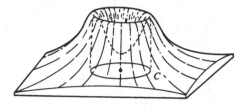

Figure 6.

which can be very readily solved by going over to polar coordinates r and φ ($x = r\cos\varphi$; $y = r\sin\varphi$).

Indeed, going over to the polar coordinates, we obtain

$$\frac{dr}{dt} = R(r, \varphi) = r(1 - r^2); \quad \frac{d\varphi}{dt} = \Phi(r, \varphi) = 1. \tag{35}$$

On the other hand, it is easy to express equation (33) in polar coordinates with functions R and Φ instead of $X(x, y)$ and $Y(x, y)$:

$$\frac{1}{r}\frac{\partial}{\partial r}\{rRf\} + \frac{1}{r}\frac{\partial}{\partial \varphi}\{r\Phi f\} = \frac{b_0}{2}\left[\frac{1}{r}\frac{\partial}{\partial r}\left\{r\frac{\partial f}{\partial r}\right\} + \frac{1}{r^2}\frac{\partial^2 f}{\partial \varphi^2}\right]. \tag{36}$$

In agreement with the fact that in our example R and Φ do not depend on φ, we shall seek a symmetric φ-independent solution of equation (36). Such a solution satisfies the equation

$$\frac{d}{dr}\{rRf\} = \frac{b_0}{2}\frac{d}{dr}\left\{r\frac{df}{dr}\right\}, \tag{37}$$

and integrating this equation we find

$$Rf = \tfrac{1}{2}b_0\frac{df}{dr} + \frac{C}{r}.$$

Assuming that in the limit $r \to \infty$ the function f and df/dr tend to zero sufficiently rapidly, we can set $C = 0$, and we then obtain

$$f = Ce^{(2/b_0)\int_0^r R\,dr} = Ce^{(1/b_0)\{r^2 - (1/2)r^4\}}$$

with

$$\frac{1}{C} = 2\pi\int_0^\infty e^{(1/b_0)\{r^2 - (1/2)r^4\}}r\,dr.$$

It is easy to picture the probability density distribution that we obtain. We have a 'crater-shaped' surface with a minimum at the point $r = 0$ and maxima that form a circle lying on the limit cycle $r = 1$ (Figure 6).

346

6   Some general comments

Comments relating to the first problem

(1) The scheme with 'impulses' or, more precisely, with 'jumps' on the phase plane is an abstract scheme. It must be adapted in accordance with the particular conditions of the problem*. For example, the ordinary 'impulses' of classical mechanics are impulses or 'jumps' with respect to the velocities but there is no jump of the representative point with respect to the coordinates.

(2) In every derivation of the Fokker equation it is assumed that during an arbitrarily short interval of time there can be arbitrarily large 'jumps', but with very small probability. This corresponds to allowing arbitrarily fast motions of the phase point. Naturally, this assumption is reflected in the result, namely if we start with an initial Dirac distribution (at $t = 0$ zero everywhere except at one point), we obtain a solution that for arbitrarily small values of t is everywhere nonzero. However, this solution tends to zero extremely rapidly as we move away from the initial value. Even if it is assumed that in real cases arbitrarily rapid motions of the phase point are impossible, the solutions we have obtained correspond to reality sufficiently well, since they allow arbitrarily large impulses only with a small probability. It is well known that there are similar things in the ordinary theory of heat conduction.

(3) We also note the following: that it is not always obvious in a particular system what influence is to be regarded as statistical and what dynamical. For example, in the case of the motion of a Brownian particle in a field of conservative forces it is necessary to assume that the particle is subject to not only the conservative forces but also a regular frictional force proportional to the velocity generated by the random impulses of the molecules. It is sensible to take into account this frictional force when writing down the dynamical equations.

Comments relating to the second problem

We have considered the behavior of the stationary solution for impulses that tend to zero for the cases $n = 1$ and $n = 2$. In more general cases with $n > 2$ it is natural to assume that by an appropriate limiting process it is possible to distinguish particular asymptotic stable (in the sense of approach to orbits) sets of recurrent trajectories; however, there is no doubt that such a limiting process may also distinguish other classes of motions. All this applies to essentially nonconservative cases, which do not possess integral invariants. If the system admits an integral invariant and if the phase space of the system is closed, then the picture of the behavior of the distribution function for impulses that tend to zero is evidently quite different. Here an interesting

* We have already emphasized that by no means all random perturbations in dynamical systems can be investigated in accordance with the scheme of the Fokker equation.

question arises: for impulses that tend to zero, does the stationary distribution tend to one of the integral invariants of the system? This is suggested by the fact that for impulses that tend to zero the Fokker equation goes over into the equation of an integral invariant.

A positive answer to this question could be used to justify* some conclusions of classical statistical mechanics based on the quasiergodic hypothesis. Indeed, let R be the phase space of some mechanical Hamiltonian problem, and M be the manifold of constant energy in it. Suppose M is closed. Since every trajectory that intersects M lies entirely on M, a system of differential equations is thereby established on M. If this system of differential equations has a unique integral invariant, then it is known – it is the phase area on M. Thus, in the case of sufficiently small impulses the probability of finding the random point in a particular region is approximately proportional to the corresponding phase area.

---

* We do not here consider the question of the physical worth of such a justification based on the assumption of statistical perturbations.

Index

activated reaction, 11
adiabatic elimination, 73, 205
amplitude equation, 234
analogue computer simulations, 131, 132
Andronov, xv, 5, 7, 113, 329
annealing, simulated, 4
attraction, the local rate of, 271
attractors, 232
  coexistence of two, 263
  Lorenz, 267
  strange, 267
autocorrelation function, of the velocity, 177

backward Kolmogorov equation, 113, 157
barrier crossing, 11
Bénard instability, 5, 235
bifurcation
  codimension-two, xvi, 243
  diagram, 246
  homoclinic, 250
  Hopf, xvi, 233, 240
  pitchfork, xvi, 233, 235
Birkhoff's theory, 330
bistability
  dispersive optical, 194
  driven by coloured noise, 323
  optical, 235
bistable system, 340
  interaction with oscillator, 289
bound process, 154
boundaries, artificial, 79–81
boundary conditions, 145
  absorbing, 114
  natural, 194
  of Fokker–Planck equations, 194
boundary value, 114
Brinkman's hierarchy, 206
Brownian motion, 92, 192, 347
  driven by coloured noise, 317
  in a double well potential, 214
  in a tilted periodic potential, 208

Cantor structure, 271
Cantor substructure, 270, 275
centre manifold theory, 234, 244
Chaikin, 7
charge density waves, 192
circuits
  electrical, 1
  memory, 9
  nonlinear, 3
coefficient functions, of the Fokker–Planck
  equation, 21
coherence lifetime, 164
competing states, 4
contracted systems, xiii
contraction, over set of ignorable variables,
  163
coordinate transformation, curvilinear, 17
correlation
  between two functionals, 311
  formulae, 311
  functions, 83
  time, 119
cumulant
  function, 118
  generating functional, 309
  resummation, 118
  summation, xv
  techniques, 79
cumulants, 120, 309

decay, nonstationary, xvi
decay rate, xvi
  effective, 91
decoupling ansatz, 87
  Stratonovich, 88, 89, 99
decoupling approximation, 322
degrees of freedom
  irrelevant, 170
  many, 294
density, see probability density
detailed balance, 230, 235
  lack of, 283

Index

phase (*contd.*)
space, reduced, 126
transition, 4
transitions, Landau theory of, 235
phase-locked loops, 192
physical chemistry, 12
Phythian, 33
Poincaré, 344
cross-section, 253–4
Poisson process, 65
Pontryagin, xv, 5, 113, 283, 329, 335
Pontryagin equation, 283
potential
bimodal, 137
conditions, 227
conditions of the Fokker–Planck
equation, 199
double-well, *see* also potential, bimodal,
192
effective, 270
macroscopic, xv, xvi, 227
macroscopic, bifurcations of, 225
nonequilibrium, 227, 230
periodic, 12
sinusoidal, 12
tilted periodic, 192
unimodal, 137
potentiality condition, 58, 60
prelimit sum, 29
probability current, 193
probability density, *see* also probability
distribution, reduced
bimodal, 91
conditional, 17
crater-shaped, 346
equation for, 330
functional, 27, 60
initial, 332
joint, 74, 83–4
scalar, 59
stationary, 17, 57, 60, 337, 344, 348
steady state, 128
time dependent, 16
two-dimensional, xvi, 345–6
probability distribution, reduced, 168
projection method, 163
for approximate solutions of Fokker–
Planck equations, 161
projection operator, xv, xvi, 116, 126, 168
pump parameter, 99

quasi-harmonic approximation, 181–2
quasi-Markovian, *see* also small
correlation time expansion *and* small
tau theory
approximation, 79, 85–6, 90–1
Fokker–Planck theories, 84

Radon–Nicodim theorem, 31
reaction, activated, 11
reaction-diffusion, 296
recursion relation, 119
reduced space, 115
reflecting barrier, 66
relaxation time, xvi, 83
renewal process, 141
resistors, 3
response function, 76–8, 92
retarding points, 62, 68
reversibility, lack of, 226
Roekaert, 40, 54

S-equation, *see* also Stratonovich equation,
16, 20
S-form, *see* also Stratonovich form, of the
Fokker–Planck equation, 23
saddle point, 68, 345
method, 10
method, multicomponent, 17
sample paths
non-smooth, 17
set of, 29
Schrödinger eigenvalue problem, 295, 300
self-organization, 5
separatrix, 252, 259, 344–5
smooth, 253
wildly oscillating, 254, 259
series resistance method, 12
sigma algebra, 31
simulations
analogue, 131–2
digital, 137
sine-Gordon, 10
singular point
stable, 344
unstable, 345
singularities, movable, 256
skewing effect, xvi
small correlation time expansion, 318
small noise strength approximation, 61,
78–9, 81
small parameter, 25, 59
small tau theory, xv, xvi, 33, 75, 139, 184,
318
Smoluchowski, 2
Smoluchowski equation, 195, 205, 210
derivation of, 207
used to calculate the mean drift velocity,
211
solitons, 10
stability
condition, 60
relative, 9, 237
stable point, 65, 68
statistical processes, superposition of, 333